# Manual of Structural Kinesiology

## R. T. Floyd EdD, ATC, CSCS

Director of Athletic Training and Sports Medicine
Distinguished Professor of Physical Education
  and Athletic Training
Chair, Department of Physical Education and
  Athletic Training
  The University of West Alabama
  Livingston, Alabama

TWENTIETH EDITION

Mc
Graw
Hill
Education

MANUAL OF STRUCTURAL KINESIOLOGY, TWENTIETH EDITION

Published by McGraw-Hill Education, 2 Penn Plaza, New York, NY 10121. Copyright © 2018 by McGraw-Hill Education. All rights reserved. Printed in the United States of America. Previous editions © 2015, 2012, and 2009. No part of this publication may be reproduced or distributed in any form or by any means, or stored in a database or retrieval system, without the prior written consent of McGraw-Hill Education, including, but not limited to, in any network or other electronic storage or transmission, or broadcast for distance learning.

Some ancillaries, including electronic and print components, may not be available to customers outside the United States.

This book is printed on acid-free paper.

1 2 3 4 5 6 7 8 9 LWI 21 20 19 18 17

ISBN 978-1-259-87043-9
MHID 1-259-87043-X

Chief Product Officer, SVP Products & Markets: *G. Scott Virkler*
Vice President, General Manager, Products & Markets: *Michael Ryan*
Vice President, Content Design & Delivery: *Betsy Whalen*
Managing Director: *Gina Boekeker*
Brand Manager: *Jamie Laferrera*
Director, Product Development: *Meghan Campbell*
Marketing Manager: *Meredith Leo*
Director, Content Design & Delivery: *Terri Schiesl*
Program Manager: *Jennifer Shekelton*
Content Project Managers: *Kelly Hart, George Theofanopoulos, Sandra Schnee*
Buyer: *Jennifer Pickel*
Design: *Studio Montage, St. Louis, MO*
Content Licensing Specialist: *Shannon Mandersheid*
Cover Image: *©Стас Толстнев/123RF*
Compositor: *Lumina Datamatics, Inc.*
Printer: *LSC Communications - Willard*

All credits appearing on page or at the end of the book are considered to be an extension of the copyright page.

**Library of Congress Cataloging-in-Publication Data**

Names: Floyd, R. T., author.
Title: Manual of structural kinesiology / R.T. Floyd, EdD, ATC, CSCS,
   Director of Athletic Training and Sports Medicine, Distinguished Professor of
   Physical Education and Athletic Training, Chair, Department of Physical
   Education and Athletic Training, The University of West Alabama (formerly
   Livingston University), Livingston, Alabama.
Description: 20th edition. | New York, NY : McGraw-Hill Education, [2017] |
   Includes bibliographical references and index.
Identifiers: LCCN 2016041135 | ISBN 9781259870439 (paper) | ISBN 9781259956027
   (ebook)
Subjects: LCSH: Kinesiology. | Human locomotion. | Muscles. | BISAC:
   EDUCATION / Physical Education.
Classification: LCC QP303 .T58 2017 | DDC 612.7/6–dc23 LC record available at https://lccn.loc.
gov/2016041135

The Internet addresses listed in the text were accurate at the time of publication. The inclusion of a website does not indicate an endorsement by the authors or McGraw-Hill Education, and McGraw-Hill Education does not guarantee the accuracy of the information presented at these sites.

mheducation.com/highered

# Contents

# Preface

I greatly appreciate the loyalty that the numerous faculty members, students, and other professionals have shown this work over the years. Conversations with some of you have generated ideas to improve and advance the presentation of the content which I continue to incorporate while attempting to maintain the successful presentation approach the late Dr. Clem Thompson established from 1961 through 1989. I first used this book as an undergraduate and later in my teachings over the years. Having developed great respect for this text and Dr. Thompson's style, it is my intention to continue to preserve the effectiveness of this time-honored text, while adding material pertinent to the professions working with today's ever-growing physically active population. I have attempted to maintain and improve clarity while continuing with a concise, simple, and straightforward presentation method. I have drawn upon my career experiences, both as a clinician and an educator, to assist in this process and sincerely hope that this approach is beneficial to those who use this text.

This text, now in its 70th year, has undergone many revisions over the years. My goal continues to be making the material as applicable as possible to physical activity and to make it more understandable and easier to use for the student and professional. While reading this text, I challenge kinesiology students and professionals to immediately apply the content to physical activities with which they are individually familiar. I hope that the reader will simultaneously palpate his or her own moving joints and contracting muscles to gain application. Concurrently, I encourage readers to palpate the joints and muscles of colleagues to gain a better appreciation of the wide range of normal anatomy and, when possible, appreciate the variation from normal found in injured and pathological musculoskeletal anatomy. Additionally, with the tremendous growth of information and media available via the Internet and other technological means, I encourage careful and continuous exploration of these resources. These resources should be helpful, but must be reviewed with a critical eye, as all information should be.

## Audience

This text is designed for students in an undergraduate structural kinesiology course after completing courses in human anatomy and physiology. While primarily utilized in physical education, human performance, movement science, exercise science, athletic training, physical therapy, and massage therapy curriculums, it is often used as a continuing reference by other clinicians and educators in addressing musculoskeletal concerns of the physically active. Applied kinesiologists, athletic trainers, athletic coaches, physical educators, physical therapists, occupational therapists, health club instructors, strength and conditioning specialists, personal trainers, massage therapists, physicians, and others who are responsible for evaluating, improving, and maintaining the muscular strength, endurance, flexibility, and overall health of individuals will benefit from this text.

With the ever-continuing growth in the number of participants of all ages in a spectrum of physical activity, it is imperative that medical, health, fitness, and education professionals involved in providing instruction and information to the physically active be correct and accountable for the teachings that they advance. The variety of exercise machines, techniques, strengthening and flexibility programs, and training programs is continuously expanding and changing, but the musculoskeletal system is constant in its design and architecture. Regardless of the goals sought or the approaches used in exercise activity, the human body is the basic factor and must be thoroughly understood and considered to maximize performance capabilities and minimize undesirable results. Most advances in kinesiology and exercise science continue to result from a better understanding of the body and how it functions. I believe that an individual in this field can never learn enough about the structure and function of the human body and that this is typically best learned through practical application.

Those who are charged with the responsibility of providing examination, instruction and consultation to the physically active will find this text a helpful and valuable resource in their never-ending quest for knowledge and understanding of human movement.

## New to this edition

The most obvious change in this edition is the combination of what was previously Chapter 8 and 13 into a new Chapter 12, Muscular Analysis of Selected Exercises and Related Concepts. This change was designed to assemble the content primarily relevant to movement analysis in one location for improved focus. Chapters 4 through 11 now have a reasonably extensive table in each chapter that provides details regarding the names, origin, insertion, and function of most ligaments that provide static stability to each joint. Another addition to each of these chapters is an illustration detailing the innervation of the muscles and their related movements. A number of figures have been tweaked for more detail. Some tables have been refined, along with an extensive review and editing of the content throughout. Additional references have been added along with some revisions and additions to the review and laboratory exercises, and end-of-chapter worksheets. Additional questions and exercises will continue to be added to the Online Learning Center. Finally, a few new terms have been added to the Glossary.

**connect**

The 20th edition of *Manual of Structural Kinesiology* is now available online with Connect, McGraw-Hill Education's integrated assignment and assessment platform. Connect also offers SmartBook for the new edition, which is the first adaptive reading experience proven to improve grades and help students study more effectively. All of the title's website and ancillary content are also available through Connect, including:

- Downloadable PowerPoint presentations
- Image bank
- Test bank questions
- End-of-chapter exercise and worksheet answers
- Student Success Strategies

## Acknowledgments

I am very appreciative of the numerous comments, ideas, and suggestions provided by the eight reviewers. These reviews have been a most helpful guide in this revision and the suggestions have been incorporated to the extent possible when appropriate. These reviewers are:

William R. Barfield, *College of Charleston*

Jerry LeBron Bell, *Tennessee Tech University*

Phillip E. Morgan, *Washington State University*

Wendy L. Pelto, *Rhode Island College*

Erin VanderBunt, *Southern Illinois University Edwardsville*

Suzanne Walsh, *Concordia University Chicago*

Carl Wilson, *Lincoln University*

Jennifer Zuberbier, *University of Wisconsin Oshkosh*

I would like to especially thank the kinesiology, exercise science and athletic training students and faculty of the University of West Alabama for their suggestions, advice, and input throughout this revision. Their assistance and suggestions have been very helpful. I am particularly grateful to Britt Jones of Livingston, Alabama, for his outstanding photography. I also acknowledge John Hood and Lisa Floyd of Birmingham and Livingston, Alabama, respectively, for the fine photographs. Special thanks to Linda Kimbrough of Birmingham, Alabama, for her superb illustrations and insight. I appreciate the models for the photographs, Audrey Crawford, Fred Knighten, Darrell Locket, Amy Menzies, Matthew Phillips, Emma Powell, Jay Sears, Marcus Shapiro, and David Whitaker. My thanks also go to Poornima H and Erin Guendelsberger, Jamie Laferrera, Jasmine Staton, and the McGraw-Hill staff who have been most helpful in their assistance and suggestions in preparing the manuscript for publication.

R. T. Floyd

## McGraw-Hill Connect®
## Learn Without Limits

Connect is a teaching and learning platform that is proven to deliver better results for students and instructors.

Connect empowers students by continually adapting to deliver precisely what they need, when they need it, and how they need it, so your class time is more engaging and effective.

> 73% of instructors who use **Connect** require it; instructor satisfaction **increases** by 28% when **Connect** is required.

# Analytics

## Connect Insight®

Connect Insight is Connect's new one-of-a-kind visual analytics dashboard—now available for both instructors and students—that provides at-a-glance and immediately actionable information regarding student performance. By presenting assignment, assessment, and topical performance results together with a time metric that is easily visible for aggregate or individual results, Connect Insight gives the user the ability to take a just-in-time approach to teaching and learning, which was never before available. Connect Insight presents data that empowers students and helps instructors improve class performance in a way that is efficient and effective.

# Mobile

Connect's new intuitive mobile interface gives students and instructors flexible and convenient anytime–anywhere access to all components of the Connect platform.

### Connect's Impact on Retention Rates, Pass Rates, and Average Exam Scores

Retention Rates: without Connect 71.3%, with Connect 89.9%
Course Pass Rates: without Connect 72.5%, with Connect 85.2%
Average Exam Scores: without Connect 71.0%, with Connect 80.1%

without Connect · with Connect

> Using **Connect** improves passing rates by **12.7%** and retention by **19.8%**.

### Impact on Final Course Grade Distribution

| | without Connect | with Connect |
|---|---|---|
| A | 22.9% | 31.0% |
| B | 27.4% | 34.3% |
| C | 22.9% | 18.7% |
| D | 11.5% | 6.1% |
| F | 15.4% | 9.9% |

> Students can view their results for any **Connect** course.

# Adaptive

©Getty Images/iStockphoto

## THE **ADAPTIVE** **READING EXPERIENCE** DESIGNED TO TRANSFORM THE WAY STUDENTS READ

More students earn **A**s and **B**s when they use McGraw-Hill Education **Adaptive** products.

## SmartBook®

Proven to help students improve grades and study more efficiently, SmartBook contains the same content within the print book, but actively tailors that content to the needs of the individual. SmartBook's adaptive technology provides precise, personalized instruction on what students should do next, guiding students to master and remember key concepts while both targeting gaps in knowledge and offering customized feedback, and driving students toward comprehension and retention of subject matter. Available on smartphones and tablets, SmartBook puts learning at the student's fingertips—anywhere, anytime.

Over **5.7 billion questions** have been answered, making McGraw-Hill Education products more intelligent, reliable, and precise.

**STUDENTS WANT**

Mc Graw Hill Education **SMARTBOOK**®

**95%** of students reported **SmartBook** to be a more effective way of reading material

**100%** of students want to use the Practice Quiz feature available within **SmartBook** to help them study

**100%** of students reported having reliable access to off-campus wifi

**90%** of students say they would purchase **SmartBook** over print alone

**95%** reported that **SmartBook** would impact their study skills in a positive way

Mc Graw Hill Education

*Findings based on a 2015 focus group survey at Pellissippi State Community College administered by McGraw-Hill Education

**www.mheducation.com**

# About the Author

R. T. Floyd is in his forty-third year of providing athletic training services for the University of West Alabama. Currently, he serves as the Director of Athletic Training and Sports Medicine for the UWA Athletic Training and Sports Medicine Center, Program Director for UWA's CAATE accredited curriculum, and as a distinguished professor in the Department of Physical Education and Athletic Training, which he chairs. He has taught numerous courses in physical education, exercise science and athletic training, including kinesiology, at both the undergraduate and graduate levels since 1980.

Floyd has maintained an active professional life throughout his career. He is currently serving as President of the National Athletic Trainers' Association (NATA) Research & Education Foundation after serving in multiple roles on the Board of Directors since 2002. He previously finished eight years of service on the NATA Board of Directors representing District IX, the Southeast Athletic Trainers' Association (SEATA). Previously, he served as the District IX representative to the NATA Educational Multimedia Committee from 1988 to 2002. He has served as the Convention Site Selection Chair for District IX from 1986 to 2004 and has directed the annual SEATA Competencies in Athletic Training Student Workshop since 1997. He has also served as a NATA BOC examiner for well over a decade and has served as a Joint Review Committee on Educational Programs in Athletic Training site visitor several times. He has provided over a hundred professional presentations at the local, state, regional, and national levels and has published several articles and videos related to the practical aspects of athletic training. He began authoring the *Manual of Structural Kinesiology* in 1992 with the twelfth edition after the passing of Dr. Clem W. Thompson, who authored the fourth through the eleventh editions. In 2010, much of the content of this text was incorporated into *Kinesiology for Manual Therapies,* which he co-authored with Nancy Dail and Tim Agnew.

Floyd is a certified member of the National Athletic Trainers' Association, a Certified Strength & Conditioning Specialist, and a Certified Personal Trainer in the National Strength and Conditioning Association. He is also a Certified Athletic Equipment Manager in the Athletic Equipment Managers' Association, a member of the American College of Sports Medicine, the American Orthopaedic Society for Sports Medicine, the American Osteopathic Academy of Sports Medicine, the American Sports Medicine Fellowship Society, and SHAPE America. Additionally, he is licensed in Alabama as an Athletic Trainer and an Emergency Medical Technician.

Floyd was presented the NATA Athletic Trainer Service Award in 1996, the Most Distinguished Athletic Trainer Award by the NATA in 2003, and received the NATA Sayers "Bud" Miller Distinguished Educator Award in 2007. In 2013 he was inducted into the NATA Hall of Fame. He received the District IX Award for Outstanding Contribution to the field of Athletic Training by SEATA in 1990 and the Award of Merit in 2001 before being inducted into the organization's Hall of Fame in 2008. In 1997, the UWA Faculty and Board of Trustees recognized Floyd for outstanding achievement in scholarship, teaching, and leadership by presenting him with a Loraine McIlwain Bell Trustee Professorship. In 2001, he was inducted into the Honor Society of Phi Kappa Phi and the University of West Alabama Athletic Hall of Fame. He was inducted into the Alabama Athletic Trainers' Association Hall of Fame in May 2004. In 2012, he was presented the Great Minds Award by the UWA Phi Kappa Phi Honor Society and the Nellie Rose McCrory Service Excellence Award by UWA. In 2014, he was inducted into the Society of the Golden Key at UWA and was selected as the first recipient of the UWA Distinguished Professorship. Upon completion of its expansion in 2014, the athletic training facility was named the R.T. Floyd Athletic Training and Sports Medicine Center by the UWA Board of Trustees.

To

my family,
Lisa, Robert Thomas, Jeanna, Rebecca, and Kate
who understand, support, and allow me to
pursue my profession

and to my parents,
Ruby and George Franklin,
who taught me the importance of a strong work ethic
with quality results

R.T.F.

# CHAPTER 1

# FOUNDATIONS OF STRUCTURAL KINESIOLOGY

## Objectives

■ To review the anatomy of the skeletal system

■ To review and understand the terminology used to describe body part locations, reference positions, and anatomical directions

■ To review the planes of motion and their respective axes of rotation in relation to human movement

■ To describe and understand the various types of bones and joints in the human body and their functions, features, and characteristics

■ To describe and demonstrate the joint movements

Kinesiology may be defined as the study of the principles of anatomy (active and passive structures), physiology, and mechanics in relation to human movement. The emphasis of this text is **structural kinesiology**—the study of muscles, bones, and joints as they are involved in the science of movement. To a much lesser degree, certain physiological and mechanical principles are addressed to enhance the understanding of the structures discussed.

Bones vary in size and shape, which factors into the amount and type of movement that occurs between them at the joints. The types of joint vary in both structure and function. Muscles also vary greatly in size, shape, and structure from one part of the body to another.

Anatomists, athletic trainers, physical therapists, occupational therapists, physicians, nurses, massage therapists, coaches, strength and conditioning specialists, performance enhancement specialists, personal trainers, physical educators, and others in health-related fields should have an adequate knowledge and understanding of all the large muscle groups so they can teach others how to strengthen, improve, and maintain optimal function of the human body. This knowledge forms the basis of exercise programs followed to strengthen and maintain all the muscles. In most cases, exercises that involve the larger primary movers also involve the smaller muscles, however, in certain instances more detailed programs are needed to address certain muscles.

More than 600 muscles are found in the human body. In this book, an emphasis is placed on the larger muscles that are primarily involved in movement of the joints. Details related to many of the small muscles located in the hands, feet, and spinal column are provided to a lesser degree.

Fewer than 100 of the largest and most important muscles, primary movers, are considered in this text. Some small muscles in the human body, such as the multifidus, plantaris, scalenus, and serratus posterior, are omitted because they are exercised with other larger primary movers. In addition, most small muscles of the hands and feet are not given the full attention provided to the larger muscles. Many small muscles of the spinal column, and the facial muscles, are beyond the scope of this text, and are not considered in full detail.

Kinesiology students frequently become so engrossed in learning individual muscles that they lose sight of the total muscular system. They miss the "big picture"—that muscle groups move joints in given movements necessary for bodily action and skilled performance. Although it is vital to learn the small details of muscle attachments, it is even more critical to be able to apply the information to real-life situations. Once the information can be applied in a useful manner, the specific details are usually much easier to understand and appreciate.

## Reference positions

It is crucial for kinesiology students to begin with a reference point in order to better understand the musculoskeletal system, its planes of motion, joint classification, and joint movement terminology. Two reference positions can be used as a basis from which to describe joint movements. The **anatomical position** is the most widely used and is accurate for all aspects of the body. Fig. 1.1 demonstrates

this reference position, with the subject standing in an upright posture, facing straight ahead, with feet parallel and close and palms facing forward. The **fundamental position** is essentially the same as the anatomical position, except that the arms are at the sides with the palms facing the body.

## Reference lines

To further assist in understanding the location of one body part in relation to another, certain imaginary reference lines may be used. Some examples follow in Fig 1.2.

**Mid-axillary line:** A line running vertically down the surface of the body passing through the apex of the axilla (armpit)

**Mid-sternal line:** A line running vertically down the surface of the body passing through the middle of the sternum

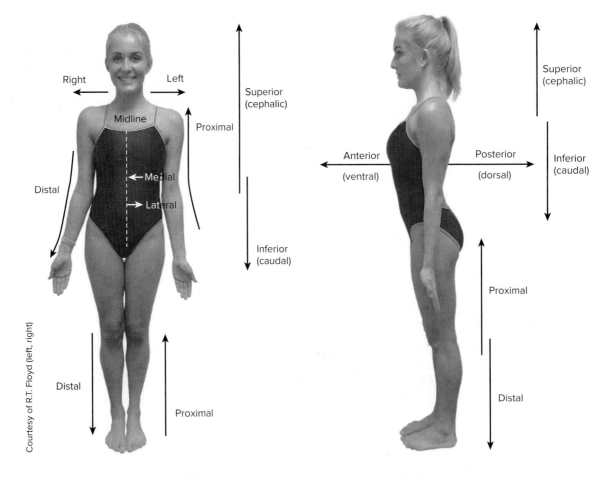

FIG. 1.1 • Anatomical position and anatomical directions. Anatomical directions refer to the position of one body part in relation to another.

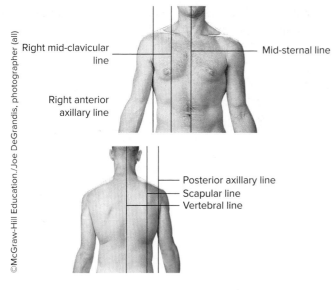

FIG. 1.2 • Reference lines.

**Anterior axillary line:** A line that is parallel to the mid-axillary line and passes through the anterior axillary skinfold

**Posterior axillary line:** A line that is parallel to the mid-axillary line and passes through the posterior axillary skinfold

**Mid-clavicular line:** A line running vertically down the surface of the body passing through the mid-point of the clavicle

**Mid-inguinal point:** A point midway between the anterior superior iliac spine and the pubic symphysis

**Scapula line:** A line running vertically down the posterior surface of the body passing through the inferior angle of the scapula

**Vertebral line:** A line running vertically down through the spinous processes of the spine

## Anatomical directional terminology   FIGS. 1.1, 1.3, 1.4

It is important that we all be able to find our way around the human body. To an extent, we can think of this as similar to giving or receiving directions about how to get from one geographic location to another. Just as we use the terms *left, right, south, west, northeast,* etc. to describe geographic directions, we have terms such as *lateral, medial, inferior, anterior, inferomedial,* etc. to use for anatomical directions. With geographic directions we may use *west* to indicate the west end of a street or the western United States. The same is true when we use anatomical directions. We may use *superior* to indicate the end of a bone in our lower leg closest to the knee, or we may be speaking about the top of the skull. It all depends on the context at the time. Just as we combine *south* and *east* to get *southeast* for the purpose of indicating somewhere

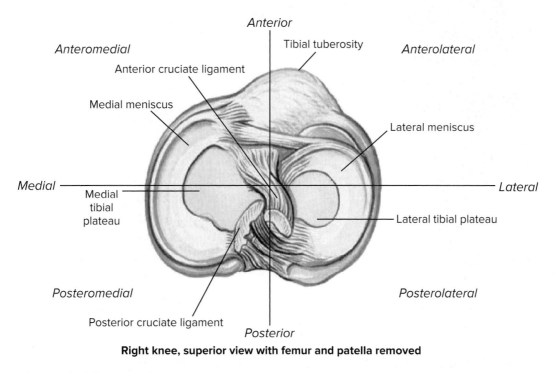

**Right knee, superior view with femur and patella removed**

FIG. 1.3 • Anatomical directional terminology.

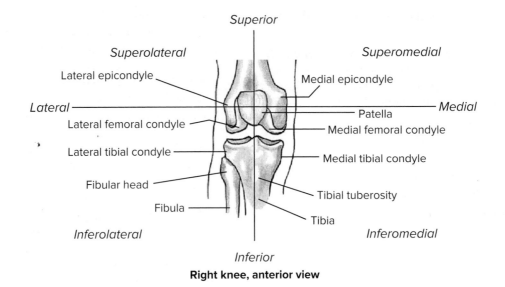

**Right knee, anterior view**

FIG. 1.4 • Anatomical directional terminology.

in between these directions, we may combine *anterior* and *lateral* to get *anterolateral* for the purpose of describing the general direction or location "in the front and to the outside." Figs. 1.3 and 1.4 provide further examples.

**Anterior:** In front or in the front part

**Anteroinferior:** In front and below

**Anterolateral:** In front and to the outside

**Anteromedial:** In front and toward the inner side or midline

**Anteroposterior:** Relating to both front and rear

**Anterosuperior:** In front and above

**Bilateral:** Relating to the right and left sides of the body or of a body structure such as the right and left extremities

**Caudal:** Below in relation to another structure; inferior

**Caudocephalad:** Directionally from tail to head in the long axis of the body

**Cephalic:** Above in relation to another structure; higher, superior

**Cephalocaudal:** Directionally from head to tail in the long axis of the body

**Contralateral:** Pertaining or relating to the opposite side

**Deep:** Beneath or below the surface; used to describe relative depth or location of muscles or tissue

**Dexter:** Relating to, or situated to the right or on the right side of, something

**Distal:** Situated away from the center or midline of the body, or away from the point of origin

**Dorsal (dorsum):** Relating to the back, being or located near, on, or toward the back, posterior part, or upper surface of; also relating to the top of the foot

**Fibular:** Relating to the fibular (lateral) side of the knee, leg, ankle, or foot

**Inferior (infra):** Below in relation to another structure; caudal

**Inferolateral:** Below and to the outside

**Inferomedial:** Below and toward the midline or inside

**Ipsilateral:** On the same side

**Lateral:** On or to the side; outside, farther from the median or midsagittal plane

**Medial:** Relating to the middle or center; nearer to the median or midsagittal plane

**Median:** Relating to, located in, or extending toward the middle; situated in the middle, medial

**Palmar:** Relating to the palm or volar aspect of the hand

**Plantar:** Relating to the sole or undersurface of the foot

**Posterior:** Behind, in back, or in the rear

**Posteroinferior:** Behind or in back and below

**Posterolateral:** Behind and to one side, specifically to the outside

**Posteromedial:** Behind and to the inner side

**Posterosuperior:** Behind or in back and above

**Prone:** Face-downward position of the body; lying on the stomach

**Proximal:** Nearest the trunk or the point of origin

**Proximodistal:** From the center of the body out toward the distal ends of appendages

**Radial:** Relating to the radial (lateral) side of the forearm or hand

**Scapular plane:** In line with the normal resting position of the scapula as it lies on the posterior rib cage; movements in the scapular plane are in line with the scapular, which is at an angle of 30 to 45 degrees from the frontal plane

**Sinister:** Relating to, or situated to the left or on the left side of, something

**Superficial:** Near the surface; used to describe relative depth or location of muscles or tissue

**Superior (supra):** Above in relation to another structure; higher, cephalic

**Superolateral:** Above and to the outside

**Superomedial:** Above and toward the midline or inside

**Supine:** Face-upward position of the body; lying on the back

**Tibial:** Relating to the tibial (medial) side of the knee, leg, ankle, or foot

**Ulnar:** Relating to the ulnar (medial) side of the forearm or hand

**Ventral:** Relating to the belly or abdomen, on or toward the front, anterior part of

**Volar:** Relating to palm of the hand or sole of the foot

## Alignment variation terminology

**Anteversion:** Abnormal or excessive rotation forward of a structure, such as femoral anteversion

**Kyphosis:** Increased curving of the spine outward or backward in the sagittal plane

**Lordosis:** Increased curving of the spine inward or forward in the sagittal plane

**Recurvatum:** Bending backward, as in knee hyperextension

**Retroversion:** Abnormal or excessive rotation backward of a structure, such as femoral retroversion

**Scoliosis:** Lateral curving of the spine

**Valgus:** Outward angulation of the distal segment of a bone or joint, as in knock-knees

**Varus:** Inward angulation of the distal segment of a bone or joint, as in bowlegs

## Planes of motion

When we study the various joints of the body and analyze their movements, it is helpful to characterize them according to specific planes of motion (Fig. 1.5). A plane of motion may be defined as an imaginary two-dimensional surface through which a limb or body segment is moved.

There are three specific, or **cardinal**, planes of motion in which the various joint movements can be classified. The specific planes that divide the body exactly into two halves are often referred to as cardinal planes. The cardinal planes are the sagittal, frontal, and transverse planes. There are an infinite number of planes within each half that are parallel to the cardinal planes. This is best understood in the following examples of movements in the sagittal plane. Sit-ups involve the spine and, as a result, are performed in the cardinal sagittal plane, which is also known as the **midsagittal** or **median** plane. Biceps curls and knee extensions are performed in **parasagittal** planes, which are parallel to the midsagittal plane. Even though these latter examples are not in the cardinal plane, they are thought of as movements in the sagittal plane.

Although each specific joint movement can be classified as being in one of the three planes of motion, our movements are usually not totally in one specific plane but occur as a combination of motions in more than one plane. These movements in the combined planes may be described as occurring in diagonal, or oblique, planes of motion and involve joints that are capable of movement in two or more planes.

### Sagittal, anteroposterior, or AP plane

The sagittal, anteroposterior, or AP plane bisects the body from front to back, dividing it into right and left symmetrical halves. Generally, flexion and extension movements such as biceps curls, knee extensions, and sit-ups occur in this plane.

### Frontal, coronal, or lateral plane

The frontal plane, also known as the coronal or lateral plane, bisects the body laterally from side to side, dividing it into front (ventral) and back (dorsal) halves. Abduction and adduction movements such as jumping jacks (shoulder and hip) and spinal lateral flexion occur in this plane.

### Transverse, axial, or horizontal plane

The transverse plane, also known as the axial or horizontal plane, divides the body into superior (cephalic) and inferior (caudal) halves. Generally, rotational movements such as forearm pronation and supination and spinal rotation occur in this plane.

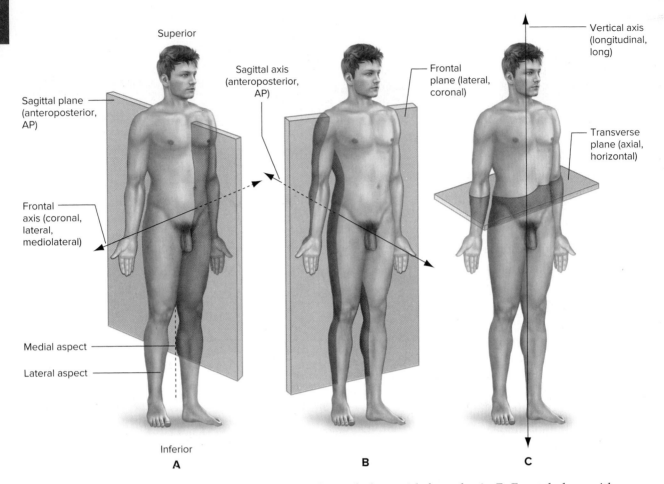

FIG. 1.5 • Planes of motion and axes of rotation. **A**, Sagittal plane with frontal axis; **B**, Frontal plane with sagittal axis; **C**, Transverse plane with vertical axis.

**Diagonal or oblique plane**   FIG. 1.6

The diagonal or oblique plane is a combination of more than one plane of motion. In reality, most of our movements in sporting activities fall somewhere between parallel and perpendicular to the previously described planes and occur in a diagonal plane. To further delineate, all movements in diagonal planes occur in a high diagonal plane or one of two low diagonal planes. The high diagonal plane is utilized for overhand movements in the upper extremity, whereas the two low diagonal planes are used to differentiate upper-extremity underhand movements from lower-extremity diagonal movements.

It should be noted that for a joint to move in a diagonal plane, the joint must be capable of movements in at least two planes. In other words, diagonal plane movements involve combining motions in one plane with motions of one or more other planes. Less commonly mentioned are diagonal movements that involve biaxial joints. For example,

combining wrist flexion and ulnar deviation or metacarpophalangeal extension with radial deviation when pointing with the index finger.

## Axes of rotation

As movement occurs in a given plane, the joint moves or turns about an axis that has a 90-degree relationship to that plane. The axes are named in relation to their orientation (Fig. 1.5). Table 1.1 lists the planes of motion with their axes of rotation.

### Frontal, coronal, lateral, or mediolateral axis

If the sagittal plane runs from anterior to posterior, then its axis must run from side to side. Since this axis has the same directional orientation as the frontal plane of motion, it is named similarly. As the elbow flexes and extends in the sagittal plane during a biceps curl, the forearm is actually rotating about a frontal axis that runs laterally through the elbow joint. The frontal axis may also be referred to as the bilateral axis.

FIG. 1.6 • Diagonal planes and axes of rotation. **A**, Upper-extremity high diagonal plane movement and axis; **B**, Upper-extremity low diagonal plane movement and axis; **C**, Lower-extremity low diagonal plane movement and axis.

TABLE 1.1 • **Planes of motion and their axes of rotation**

| Plane | Description of plane | Axis of rotation | Description of axis | Common movements |
|---|---|---|---|---|
| Sagittal (anteroposterior or AP) | Divides the body into right and left halves | Frontal (coronal, lateral, or mediolateral) | Runs medial/lateral | Flexion, extension |
| Frontal (coronal or lateral) | Divides the body into anterior and posterior halves | Sagittal (anteroposterior or AP) | Runs anterior/posterior | Abduction, adduction |
| Transverse (axial, horizontal) | Divides the body into superior and inferior halves | Vertical (longitudinal or long) | Runs superior/inferior | Internal rotation, external rotation |

### Sagittal or anteroposterior axis

Movement occurring in the frontal plane rotates about a sagittal axis. This sagittal axis has the same directional orientation as the sagittal plane of motion and runs from front to back at a right angle to the frontal plane of motion. As the hip abducts and adducts during jumping jacks, the femur rotates about an axis that runs front to back through the hip joint.

### Vertical or longitudinal axis

The vertical axis, also known as the longitudinal or long axis, runs straight down through the top of the head and is at a right angle to the transverse plane of motion. As the head rotates or turns from left to right when indicating disapproval, the skull and cervical vertebrae are rotating around an axis that runs down through the spinal column.

### Diagonal or oblique axis  FIG. 1.6

The diagonal axis, also known as the oblique axis, runs at a right angle to the diagonal plane. As the glenohumeral joint moves from diagonal abduction to diagonal adduction in overhand throwing, its axis runs perpendicular to the plane through the humeral head.

## Body regions

As mentioned later under the skeletal system, the body can be divided into axial and appendicular regions. Each of these regions may be further divided into different subregions, such as the cephalic, cervical, trunk, upper limbs, and lower limbs. Within each of these regions are many more subregions and specific regions. Table 1.2 details a breakdown of these regions and their common names, illustrated in Fig. 1.7.

TABLE 1.2 • **Body parts and regions**

| | Region name | Common name | Subregion | Specific region name | Common name for specific region |
|---|---|---|---|---|---|
| Axial | Cephalic | Head | Cranial (skull) | Frontal | Forehead |
| | | | | Occipital | Base of skull |
| | | | Facial (face) | Orbital | Eye |
| | | | | Otic | Ear |
| | | | | Nasal | Nose |
| | | | | Buccal | Cheek |
| | | | | Oral | Mouth |
| | | | | Mental | Chin |
| | Cervical | Neck | | Nuchal | Posterior neck |
| | | | | Throat | Anterior neck |
| | Trunk | Thoracic | Thorax | Clavicular | Collar bone |
| | | | | Pectoral | Chest |
| | | | | Sternal | Breastbone |
| | | | | Costal | Ribs |
| | | | | Mammary | Breast |
| | | Dorsal | Back | Scapula | Shoulder blade |
| | | | | Vertebral | Spinal column |
| | | | | Lumbar | Lower back or loin |
| | | Abdominal | Abdomen | Celiac | Abdomen |
| | | | | Umbilical | Navel |
| | | Pelvic | Pelvis | Inguinal | Groin |
| | | | | Pubic | Genital |
| | | | | Coxal | Hip |
| | | | | Sacral | Between hips |
| | | | | Gluteal | Buttock |
| | | | | Perineal | Perineum |
| Appendicular | Upper limbs | Shoulder | | Acromial | Point of shoulder |
| | | | | Omus | Deltoid |
| | | | | Axillary | Armpit |
| | | | | Brachial | Arm |
| | | | | Olecranon | Point of elbow |
| | | | | Cubital | Elbow |
| | | | | Antecubital | Front of elbow |
| | | | | Antebrachial | Forearm |
| | | | | Carpal | Wrist |
| | | Manual | | Palmar | Palm |
| | | | | Dorsal | Back of hand |
| | | | | Digital | Finger |
| | Lower limbs | | | Femoral | Thigh |
| | | | | Patella | Kneecap |
| | | | | Popliteal | Back of knee |
| | | | | Sural | Calf |
| | | | | Crural | Leg |
| | | Pedal | Foot | Talus | Ankle |
| | | | | Calcaneal | Heel |
| | | | | Dorsum | Top of foot |
| | | | | Tarsal | Instep |
| | | | | Plantar | Sole |
| | | | | Digital | Toe |

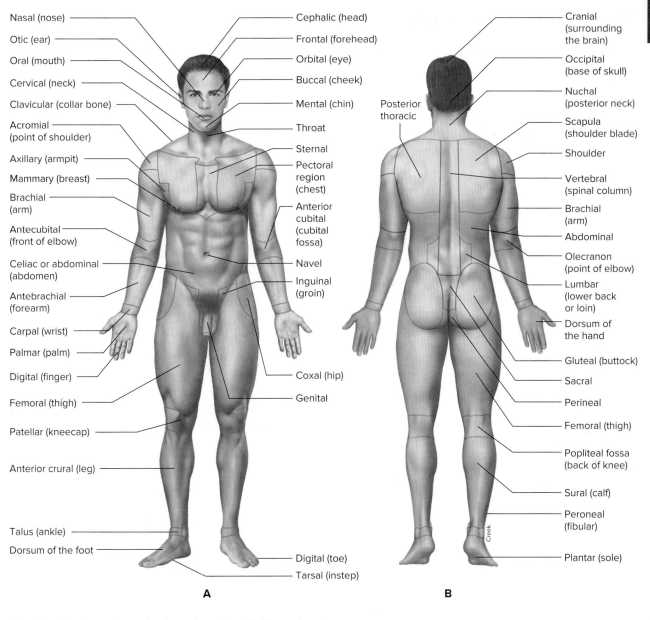

Nasal (nose)
Otic (ear)
Oral (mouth)
Cervical (neck)
Clavicular (collar bone)
Acromial (point of shoulder)
Axillary (armpit)
Mammary (breast)
Brachial (arm)
Antecubital (front of elbow)
Celiac or abdominal (abdomen)
Antebrachial (forearm)
Carpal (wrist)
Palmar (palm)
Digital (finger)
Femoral (thigh)
Patellar (kneecap)
Anterior crural (leg)
Talus (ankle)
Dorsum of the foot

Cephalic (head)
Frontal (forehead)
Orbital (eye)
Buccal (cheek)
Mental (chin)
Throat
Sternal
Pectoral region (chest)
Anterior cubital (cubital fossa)
Navel
Inguinal (groin)
Coxal (hip)
Genital
Digital (toe)
Tarsal (instep)

Posterior thoracic

Cranial (surrounding the brain)
Occipital (base of skull)
Nuchal (posterior neck)
Scapula (shoulder blade)
Shoulder
Vertebral (spinal column)
Brachial (arm)
Abdominal
Olecranon (point of elbow)
Lumbar (lower back or loin)
Dorsum of the hand
Gluteal (buttock)
Sacral
Perineal
Femoral (thigh)
Popliteal fossa (back of knee)
Sural (calf)
Peroneal (fibular)
Plantar (sole)

Creek

**A**     **B**

FIG. 1.7 • Body regions. **A**, Anterior view; **B**, Posterior view.

## Skeletal systems

Fig. 1.8 shows anterior and posterior views of the skeletal system. Some 206 bones make up the skeletal system, which provides support and protection for other systems of the body and provides for attachments of the muscles to the bones, by which movement is produced. Additional skeletal functions are mineral storage and hemopoiesis, which involves blood cell formation in the red bone marrow. The skeleton may be divided into the appendicular and the axial skeletons. The appendicular skeleton is composed of the appendages, or the upper and lower extremities, and the shoulder and pelvic girdles. The axial skeleton

consists of the skull, vertebral column, ribs, and sternum. Most students who take this course will have had a course in human anatomy, but a brief review is desirable before beginning the study of kinesiology. Later chapters provide additional information and more detailed illustrations of specific bones.

## Osteology

The adult skeleton, consisting of approximately 206 bones, may be divided into the axial skeleton and the appendicular skeleton. The axial skeleton contains 80 bones, which include the skull, spinal column, sternum, and ribs. The appendicular skeleton contains

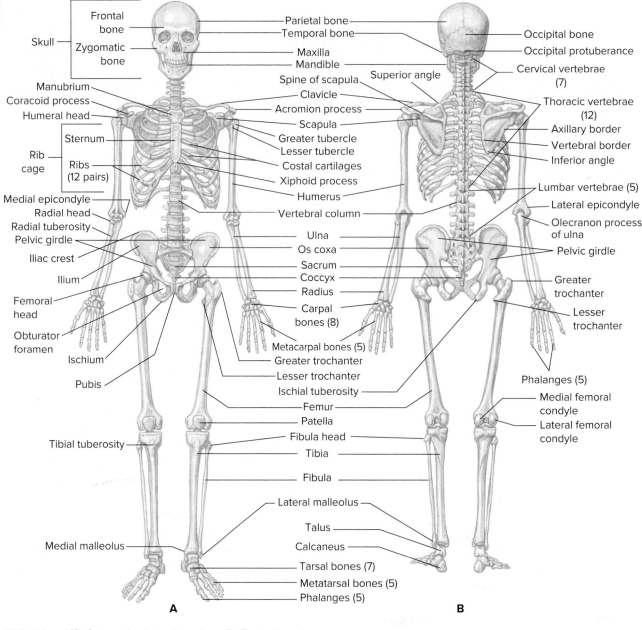

FIG. 1.8 • Skeleton. **A**, Anterior view; **B**, Posterior view.

126 bones, which include all the bones of the upper and lower extremities. The pelvis is sometimes classified as being part of the axial skeleton due to its importance in linking the axial skeleton with the lower extremities of the appendicular skeleton. The exact number of bones as well as their specific features occasionally varies from person to person.

## Skeletal functions

The skeleton has five major functions:

1. Protection of vital soft tissues such as the heart, lungs, and brain

2. Support to maintain posture
3. Movement by serving as points of attachment for muscles and acting as levers
4. Storage for minerals such as calcium and phosphorus
5. Hemopoiesis, which is the process of blood formation that occurs in the red bone marrow located in the vertebral bodies, femur, humerus, ribs, and sternum

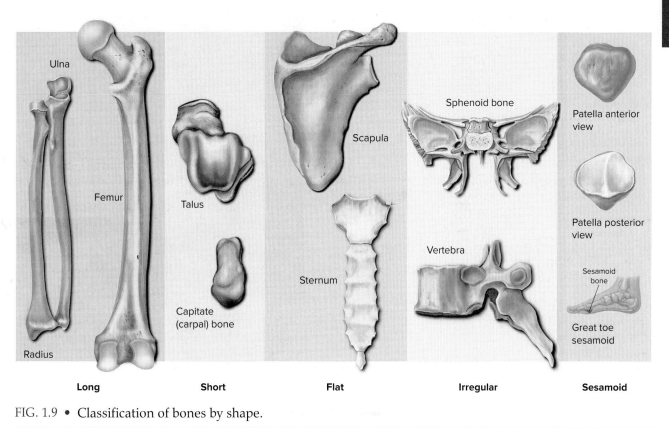

FIG. 1.9 • Classification of bones by shape.

## Types of bones

Bones vary greatly in shape and size but can be classified in five major categories (Fig. 1.9).

**Long bones:** Composed of a long cylindrical shaft with relatively wide, protruding ends; serve as levers. The shaft contains the medullary cavity. Examples include phalanges, metatarsals, metacarpals, tibia, fibula, femur, radius, ulna, and humerus.

**Short bones:** Small cube-shaped, solid bones that usually have a proportionally large articular surface in order to articulate with more than one bone. Short bones provide some shock absorption and include the carpals and tarsals.

**Flat bones:** Usually having a curved surface and varying from thick (where tendons attach) to very thin. Flat bones generally provide protection and include the ilium, ribs, sternum, clavicle, and scapula.

**Irregular bones:** Irregular-shaped bones serve a variety of purposes and include the bones throughout the entire spine and the ischium, pubis, and maxilla.

**Sesamoid bones:** Small bones embedded within the tendon of a musculotendinous unit that provide protection as well as improve the mechanical advantage of musculotendinous units. In addition to the patella, there are small sesamoid bones within the flexor tendons of the great toe and the thumb. Sesamoid bones are sometimes referred to as accessory bones however, accessory ossicles have no known function and often originate from unfused ossification centers. Both sesamoid and accessory bones occur in varying numbers from one individual to the next. They are most commonly found in smaller joints in the distal extremities of the foot, ankle, and hand but are not always symmetrical in an individual.

## Typical bony features

Long bones possess features that are typical of bones in general, as illustrated in Fig. 1.10. Long bones have a shaft or **diaphysis**, which is the long cylindrical portion of the bone. The diaphysis wall, formed from hard, dense, compact bone, is the **cortex**. The outer surface of the diaphysis is covered by a dense, fibrous membrane known as the **periosteum**. A similar fibrous membrane known as the **endosteum** covers the inside of the cortex. Between the walls of the diaphysis lies the **medullary** or marrow cavity, which contains yellow or fatty marrow. At each end of a long bone is the **epiphysis**, which is usually enlarged and shaped specifically to join with the epiphysis of an adjacent bone at a joint. The epiphysis is formed from spongy or **cancellous** or **trabecular** bone. During bony growth the diaphysis and the epiphysis are

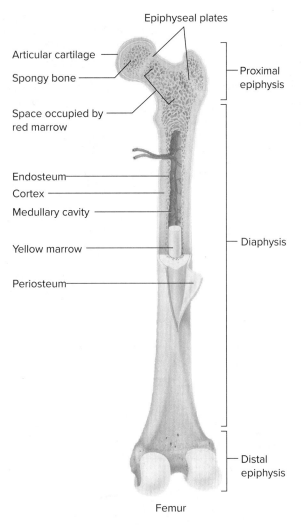

FIG. 1.10 • Major parts of a long bone.

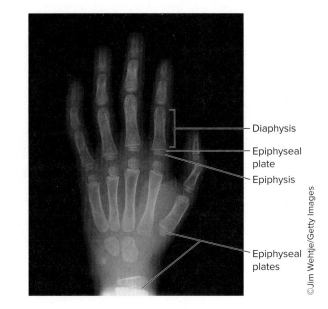

FIG. 1.11 • The presence of epiphyseal plates, as seen in a radiograph of a child's hand, indicates that the bones are still growing in length.

TABLE 1.3 · **Epiphyseal closure timetables**

| Approximate age | Bones |
|---|---|
| 7–8 | Inferior rami of pubis and ischium (almost complete) |
| 15–17 | Scapula, lateral epicondyle of humerus, olecranon process of ulna |
| 18–19 | Medial epicondyle of humerus, head and shaft of radius |
| About 20 | Humeral head, distal ends of radius and ulna, distal ends of femur and fibula, proximal end of tibia |
| 20–25 | Acetabulum in pelvis |
| 25 | Vertebrae and sacrum, clavicle, proximal end of fibula, sternum and ribs |

Adapted from Goss CM: *Gray's anatomy of the human body*, ed 29, Philadelphia, 1973, Lea & Febiger.

separated by a thin plate of cartilage known as the **epiphyseal plate**, commonly referred to as a growth plate (Fig. 1.11). As skeletal maturity is reached, on a timetable that varies from bone to bone as detailed in Table 1.3, the plates are replaced by bone and are closed. To facilitate smooth, easy movement at joints, the epiphysis is covered by **articular** or **hyaline** cartilage, which provides a cushioning effect and reduces friction.

## Bone development and growth

Most of the skeletal bones of concern to us in structural kinesiology are **endochondral bones**, which develop from hyaline cartilage. As we develop from an embryo, these hyaline cartilage masses grow rapidly into structures shaped similarly to the bones they will eventually become. This growth continues, and the cartilage gradually undergoes

significant change to develop into long bone, as detailed in Fig. 1.12.

Bones continue to grow longitudinally as long as the epiphyseal plates are open. These plates begin closing around adolescence and disappear. Most close by age 18, but some may be open until age 25. Growth in diameter continues throughout life. This is done by an internal layer of periosteum building new concentric layers on old layers. Simultaneously, bone around the sides of the medullary

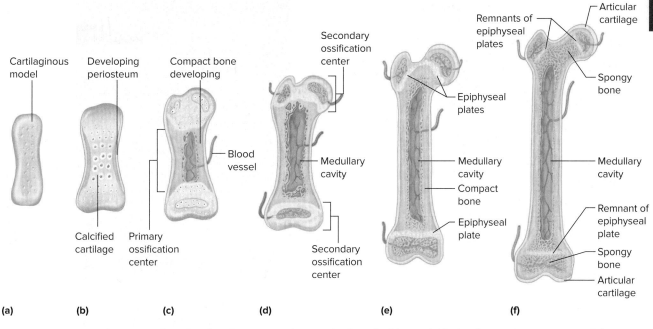

FIG. 1.12 • Major stages **a–f** in the development of an endochondral bone (relative bone sizes not to scale).

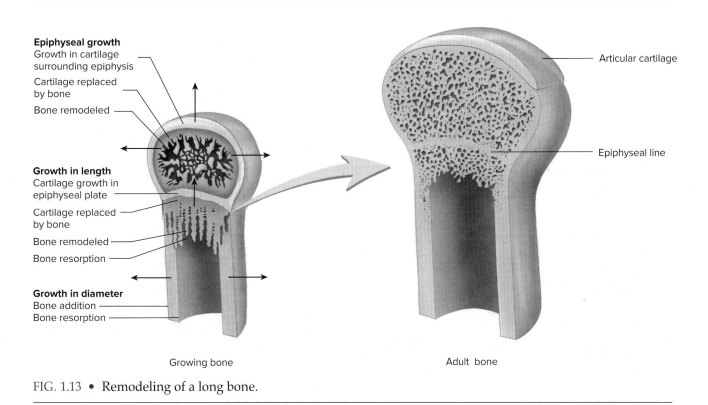

FIG. 1.13 • Remodeling of a long bone.

cavity is resorbed so that the diameter is continually increased. New bone is formed by specialized cells known as **osteoblasts**, whereas the cells that resorb old bone are **osteoclasts**. This bone remodeling, as depicted in Fig. 1.13, is necessary for continued bone growth, changes in bone shape, adjustment of bone to stress, and bone repair.

# Bone properties

Calcium carbonate, calcium phosphate, collagen, and water are the basis of bone composition. About 60% to 70% of bone weight is made up of calcium carbonate and calcium phosphate, with water making up approximately 25% to 30% of bone weight.

Collagen provides some flexibility and strength in resisting tension. Aging causes progressive loss of collagen and increases bone brittleness, resulting in increased likelihood of fractures.

Most outer bone is cortical; cancellous bone is underneath. Cortical bone is harder and more compact, with only about 5% to 30% of its volume being porous, with nonmineralized tissue. In contrast, cancellous bone is spongy, with around 30% to 90% of its volume being porous. Cortical bone is stiffer; it can withstand greater stress, but less strain, than cancellous bone. Due to its sponginess, cancellous bone can undergo greater strain before fracturing.

Bone size and shape are influenced by the direction and magnitude of forces that are habitually applied to them. Bones reshape themselves based on the stresses placed upon them, and their mass increases over time with increased stress.

This concept of bone adaptation to stress is known as **Wolff's law**, which essentially states that bone in a healthy individual will adapt to the loads it is placed under. When a particular bone is subjected to increased loading, the bone will remodel itself over time to become stronger to resist that particular type of loading. As a result, the external cortical portion of the bone becomes thicker. The opposite is also true: when the loading on a bone decreases, the bone will become weaker.

## Bone markings

Bones have specific markings that exist to enhance their functional relationship with joints, muscles, tendons, nerves, and blood vessels. Many of these markings serve as important bony landmarks in determining muscle location and attachment and joint function. Essentially, all bone markings may be divided into

1. Processes (including elevations and projections), which either form joints or serve as a point of attachment for muscles, tendons, or ligaments, and
2. Cavities (depressions), which include openings and grooves that contain tendons, vessels, nerves, and spaces for other structures.

Detailed descriptions and examples of many bony markings are provided in Table 1.4.

## Types of joints

The articulation of two or more bones allows various types of movement. The extent and type of movement determine the name applied to the joint. Bone structure limits the kind and amount of movement in each joint. Some joints or **arthroses** have no movement, others are only very slightly movable, and others are freely movable with a variety of movement ranges. The type and range of movements are similar in all humans; but the freedom, range, and vigor of movements are limited by the configuration of the bones where they fit together, and by ligaments and muscles.

Articulations may be classified according to the structure or function. Classification by structure places joints into one of three categories: fibrous, cartilaginous, or synovial. Functional classification also results in three categories: synarthrosis (synarthrodial), amphiarthrosis (amphiarthrodial), and diarthrosis (diarthrodial). There are subcategories in each classification. Due to the strong relationship between structure and function, there is significant overlap between the classification systems. That is, there is more similarity than difference between the two members in each of the following pairs: fibrous and synarthrodial joints, cartilaginous and amphiarthrodial joints, and synovial and diarthrodial joints. However, not all joints fit neatly into both systems. Table 1.5 provides a detailed listing of all joint types according to both classification systems. Since this text is concerned primarily with movement, the more functional system (synarthrodial, amphiarthrodial, and diarthrodial joints) will be used throughout, following a brief explanation of structural classification.

Fibrous joints are joined together by connective tissue fibers and are generally immovable. Subcategories are suture and gomphosis, which are immovable, and syndesmosis, which allows a slight amount of movement. Cartilaginous joints are joined together by hyaline cartilage or fibrocartilage, which allows very slight movement. Subcategories include synchondrosis and symphysis. Synovial joints are freely movable and generally are diarthrodial. Their structure and subcategories are discussed in detail under diarthrodial joints.

The articulations are grouped into three classes based primarily on the amount of movement possible, with consideration given to their structure.

### Synarthrodial (immovable) joints    FIG. 1.14

Structurally, these articulations are divided into two types:

#### Suture

Found in the sutures of the cranial bones. The sutures of the skull are truly immovable beyond infancy.

TABLE 1.4 · **Bone markings**

| | Marking | Description | Examples | Page |
|---|---|---|---|---|
| **Processes that form joints** | Condyle | Large, rounded projection that usually articulates with another bone | Medial or lateral condyle of femur | 266 |
| | Facet | Small, flat or nearly flat surface | Articular facet of vertebra | 331 |
| | Head | Prominent, rounded projection of the proximal end of a bone, usually articulating | Head of femur, head of humerus | 220, 222, 224, 117 |
| **Processes to which muscles, tendons, or ligaments attach** | Angle | Bend or protruding angular projection | Superior and inferior angle of scapula | 92, 94 |
| | Border or margin | Edge or boundary line of a bone | Lateral and medial border of scapula | 92, 94 |
| | Crest | Prominent, narrow, ridgelike projection | Iliac crest of pelvis | 222 |
| | Epicondyle | Projection located above a condyle | Medial or lateral epicondyle of humerus | 150 |
| | Line | Ridge of bone less prominent than a crest | Linea aspera of femur | 222 |
| | Process | Any prominent projection | Acromion process of scapula, olecranon process of humerus | 92, 94, 117, 118 150 |
| | Ramus | Part of an irregularly shaped bone that is thicker than a process and forms an angle with the main body | Superior and inferior ramus of pubis | 221 |
| | Spine (spinous process) | Sharp, slender projection | Spinous process of vertebra, spine of scapula | 330, 331, 94 |
| | Suture | Line of union between bones | Sagittal suture between parietal bones of skull | 16 |
| | Trochanter | Very large projection | Greater or lesser trochanter of femur | 220, 222 |
| | Tubercle | Small, rounded projection | Greater and lesser tubercles of humerus | 117 |
| | Tuberosity | Large, rounded or roughened projection | Radial tuberosity, tibial tuberosity | 150, 266 |
| **Cavities (depressions)** | Facet | Flattened or shallow articulating surface | Intervertebral facets in cervical, thoracic, and lumbar spine | 331 |
| | Foramen | Rounded hole or opening in bone | Obturator foramen in pelvis | 220, 221 |
| | Fossa | Hollow, depressed, or flattened surface | Supraspinatus fossa, iliac fossa | 94, 220 |
| | Fovea | Very small pit or depression | Fovea capitis of femur | 224 |
| | Meatus | Tubelike passage within a bone | External auditory meatus of temporal bone | 346 |
| | Notch | Depression in the margin of a bone | Trochlear and radial notch of the ulna | 150 |
| | Sinus | Cavity or hollow space within a bone | Frontal sinus | |
| | Sulcus (groove) | Furrow or groovelike depression on a bone | Intertubercular (bicipital) groove of humerus | 117 |

TABLE 1.5 • **Joint classification by structure and function**

| | | Structural classification | | |
| --- | --- | --- | --- | --- |
| | | Fibrous | Cartilaginous | Synovial |
| Functional classification | Synarthrodial | Gomphosis Suture | —— | —— |
| | Amphiarthrodial | Syndesmosis | Symphysis Synchondrosis | —— |
| | Diarthrodial | —— | —— | Arthrodial Condyloidal Enarthrodial Ginglymus Sellar Trochoidal |

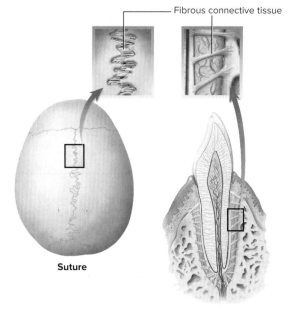

FIG. 1.14 • Synarthrodial joints.

### Gomphosis

Found in the sockets of the teeth. The socket of a tooth is often referred to as a gomphosis (type of joint in which a conical peg fits into a socket). Normally, there should be essentially no movement of the teeth in the mandible or maxilla.

**Amphiarthrodial (slightly movable) joints**   FIG. 1.15

Structurally, these articulations are divided into three types:

### Syndesmosis

Type of joint held together by strong ligamentous structures that allow minimal movement between the bones. Examples are the coracoclavicular joint and the inferior tibiofibular joint.

### Symphysis

Type of joint separated by a fibrocartilage pad that allows very slight movement between the bones. Examples are the symphysis pubis and the intervertebral disks.

FIG. 1.15 • Amphiarthrodial joints. **A**, Syndesmosis joint; **B**, Symphysis joint; **C**, Synchondrosis joint.

### Synchondrosis

Type of joint separated by hyaline cartilage that allows very slight movement between the bones. Examples are the costochondral joints of the ribs with the sternum.

### Diarthrodial (freely movable) joints  FIG. 1.16

Diarthrodial joints, also known as synovial joints, are freely movable. A sleevelike covering of ligamentous tissue known as the **joint capsule** surrounds the bony ends forming the joints. This ligamentous capsule is lined with a thin vascular synovial capsule that secretes synovial fluid to lubricate the area inside the joint capsule, known as the **joint cavity**. In certain areas the capsule is thickened to form tough, nonelastic **ligaments** that provide additional support against abnormal movement or joint opening. These ligaments vary in location, size, and strength depending upon the particular joint. Ligaments, in connecting bones to bones, provide static stability to joints.

In many cases, additional ligaments, not continuous with the joint capsule, provide further support. In some cases, these additional ligaments may be contained entirely within the joint capsule; or intraarticularly, such as the anterior cruciate ligament in the knee; or extraarticularly, such as the fibular collateral ligament of the knee, which is outside the joint capsule.

The articular surfaces on the ends of the bones inside the joint cavity are covered with layers of articular or **hyaline cartilage** that helps protect the ends of the bones from wear and damage. This cartilage is quite resilient because it is slightly compressible and elastic, which enables it to absorb compressive and shear forces. The articular surface, thanks in part to lubrication from synovial fluid, has a very low amount of friction and is very durable. When the joint surfaces are unloaded or distracted, this articular cartilage slowly absorbs a slight amount of the joint synovial fluid, only to slowly secrete it during subsequent weight bearing and compression. Articular cartilage has a very limited blood supply and as a result depends on joint movement to provide its nutrition through this synovial flow. Therefore, maintaining and utilizing

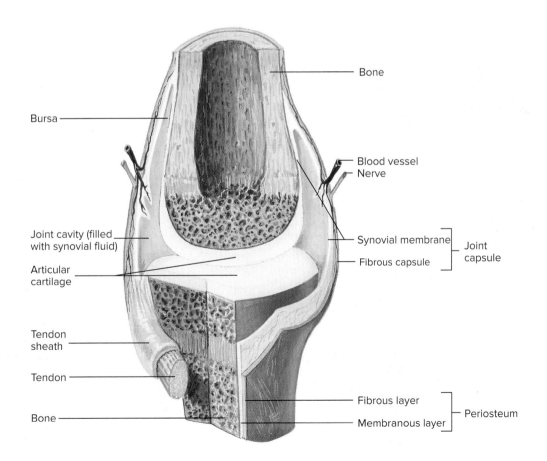

FIG. 1.16 • Structure of a diarthrodial synovial joint.

a joint through its normal range of motion are important to sustaining joint health and function.

Additionally, some diarthrodial joints have a fibro-cartilage disk between their articular surfaces to provide additional shock absorption and load distribution, and to further enhance joint stability. Examples are the knee's medial and lateral menisci and the acetabular and glenoid labrum of the hip and shoulder joints, respectively. Structurally, this type of articulation can be divided into six groups, as shown in Fig. 1.17.

Diarthrodial joints have motion possible in one or more planes. Those joints having possible motion in one plane are said to have one degree of freedom of motion, whereas joints having motion in two and three planes are described as having two and three degrees of freedom of motion, respectively. Refer to Table 1.6 for a comparison of diarthrodial joint features by subcategory.

### Arthrodial (gliding, plane) joint

This joint type is characterized by two flat, or plane, bony surfaces that butt against each other. This type of joint permits limited gliding movement.

Examples are the carpal bones of the wrist and the tarsometatarsal joints of the foot.

### Condyloidal (ellipsoid, ovoid, biaxial ball-and-socket) joint

This is a type of joint in which the bones permit movement in two planes without rotation. Examples are the wrist (radiocarpal joint) between the radius and the proximal row of the carpal bones or the second, third, fourth, and fifth metacarpophalangeal joints.

### Enarthrodial (spheroidal, multiaxial ball-and-socket) joint

This type of joint is most like a true ball-and-socket in that it permits movement in all planes. Examples are the shoulder (glenohumeral) and hip (acetabularfemoral) joints.

### Ginglymus (hinge) joint

This is a type of joint that permits a wide range of movement in only one plane. Examples are the elbow (humeroulnar), ankle (talocrural), and knee (tibiofemoral) joints.

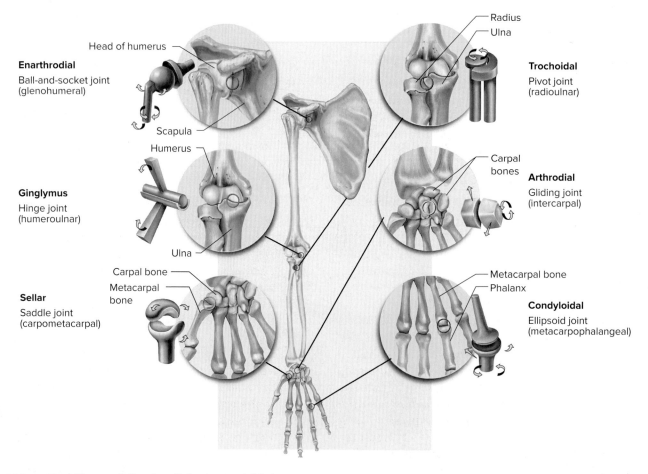

FIG. 1.17 • Types of diarthrodial or synovial joints.

## TABLE 1.6 · Diarthrodial joint classification

| Classification name | Number of axes | Degrees of freedom | Typical movements | Joint examples | Plane for examples | Axis for examples |
|---|---|---|---|---|---|---|
| Ginglymus (hinge) | Uniaxial | One | Flexion, extension | Elbow joint (humeroulnar) Ankle joint (talocrural) | Sagittal | Frontal |
| Trochoidal (pivot, screw) | | | Internal rotation, external rotation | Proximal and distal radioulnar joint Atlantoaxial joint | Transverse | Vertical |
| Condyloidal (ellipsoid, ball-and-socket, ovoid) | Biaxial | Two | Flexion, extension, abduction, adduction | Wrist (radiocarpal) 2nd–5th metacarpophalangeal joints | Sagittal Frontal | Frontal Sagittal |
| Arthrodial (gliding, plane) | Multiaxial | Three | Flexion, extension, abduction, adduction, internal rotation, external rotation | Transverse tarsal joint Vertebral facets in spine Intercarpal joints in wrist | Variable Frontal Variable | Variable Sagittal Variable |
| Enarthrodial (ball-and-socket, spheroidal) | | | | Glenohumeral joint Hip joint (acetabularfemoral) | Sagittal Frontal Transverse | Frontal Sagittal Vertical |
| Sellar (saddle) | | | | 1st carpometacarpal joint | Sagittal Frontal Transverse | Frontal Sagittal Vertical |

### Sellar (saddle) joint

This type of reciprocal reception is found in the thumb at the carpometacarpal joint and permits ball-and-socket movement, with the exception of slight rotation. Some anatomists also classify the sternoclavicular joint as a sellar joint.

### Trochoidal (pivot, screw) joint

This is a type of joint with a rotational movement around a long axis. An example is the rotation of the radius on the ulna at the proximal and distal radioulnar joints.

### Stability and mobility of diarthrodial joints

Generally, the more mobile a joint is, the less stable it is, and vice versa. This is true when comparing the same joints between individuals, but also when comparing one joint versus another in the same individual. Both heredity and developmental factors (Wolff's law for bone and Davis's law for soft tissue) contribute to these variances. In a manner similar to the adaption of bone to loading, as previously discussed in Wolff's law, soft tissue also adapts to stress or the lack thereof. This corollary to Wolff's law is known as **Davis's law**, which essentially states that ligaments, muscle,

and other soft tissue when placed under appropriate tension will adapt over time by lengthening, and conversely, when maintained in a loose or shortened state over a period of time will gradually shorten.

Five major factors affect the total stability, and consequently the mobility, of a joint (see Fig. 1.18).

- Bones—Although bones are usually very similar in bilateral comparisons within an individual, the actual anatomical configuration at the joint surfaces in terms of depth and shallowness may vary significantly between individuals.
- Cartilage—The structures of both hyaline cartilage and specialized cartilaginous structures, such as the knee menisci, glenoid labrum, and acetabular labrum, further assist in joint congruency and stability. As with bones, these structures normally are the same in bilateral comparisons within an individual, but may vary between individuals in size and configuration.
- Ligaments and connective tissue—Ligaments and connective tissue provide static stability to joints. As with bones and cartilage, variances exist between individuals in the degree of restrictiveness of ligamentous tissue. An

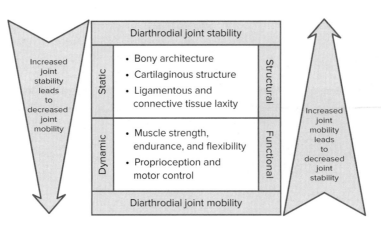

FIG. 1.18 • Factors affecting diarthrodial joint stability.

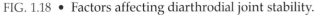

individual's amount of hypo- or hyperlaxity is primarily due to the proportional amount of elastin versus collagen within the joint structures. Simply put, individuals with proportionally higher elastin-to-collagen ratios are hyperlax, or "loose-jointed," whereas individuals with proportionally lower ratios are tighter.

- Muscles—Muscles provide dynamic stability to joints when actively contracting. Without active tension via contraction, muscles provide minimal static stability. Consequently, strength and endurance are significant factors in stabilizing joints, whereas muscle flexibility may affect the total range of joint motion possible.
- Proprioception and motor control—Proprioception is the subconscious mechanism by which the body is able to regulate posture and movements by responding to stimuli originating in the proprioceptors embedded in joints, tendons, muscles, and the inner ear. Motor control is the process by which bodily actions and movements are organized and executed. To determine the appropriate amount of muscular forces and joint activations needed, sensory information from the environment and the body must be integrated and then coordinated in a cooperative manner between the central nervous system and the musculoskeletal system. Muscle strength and endurance are not very useful in providing joint stability unless they can be activated precisely when needed.

The integrity of any of these structures may be affected by acute or chronic injury. These structures adapt over time both positively and negatively to the specific biomechanical demands placed upon them. When any of the above factors are compromised, additional demands are placed on the remaining structures to provide stability, which in turn may compromise their integrity, resulting in abnormal mobility. This abnormal mobility, whether hypermobility or hypomobility, may lead to further pathological conditions such as tendinitis, bursitis, arthritis, internal derangement, and joint subluxations.

## Movements in joints

In many joints, several different movements are possible. Some joints permit only flexion and extension; others permit a wide range of movements, depending largely on the joint structure. We refer to the area through which a joint may normally be freely and painlessly moved as the **range of motion (ROM)**. The specific amount of movement possible in a joint or range of motion may be measured by using an instrument known as a **goniometer** to compare the change in joint angles. The goniometer has a moving arm, a stationary arm, and an axis or fulcrum. Measuring the available range of motion in a joint or the angles created by the bones of a joint is known as **goniometry**.

The goniometer axis, or hinge point, is placed even with the axis of rotation at the joint line. The stationary arm is held in place either along or parallel to the long axis of the more stationary bone (usually the more proximal bone), and the moving arm is placed either along or parallel to the long axis of the bone that moves the most (usually the more distal bone). The joint angle can then be read from the goniometer, as shown in Fig. 1.19. As an example, we could measure the angle between the femur and the trunk in the anatomical position (which would usually be zero), and

FIG. 1.19 • Goniometric measurement of knee joint flexion.

then ask the person to flex the hip as far as possible. If we measured the angle again at full hip flexion, we would find a goniometer reading of around 130 degrees.

Depending on the size of the joint and its movement potential, different goniometers may be more or less appropriate. Fig. 1.20 depicts a variety of goniometers that may be utilized to determine the range of motion for a particular joint. Inclinometers may also be used to measure range of motion, particularly in the spine.

Please note that the normal range of motion for a particular joint varies to some degree from person to person. Appendix 1 provides the average normal ranges of motion for all joints.

When using movement terminology, it is important to understand that the terms are used to describe the actual change in position of the bones

FIG. 1.20 • Various goniometers used for measuring joint range of motion.

relative to each other. That is, the angles between the bones change, whereas the movement occurs between the articular surfaces of the joint. We may say, in describing knee movement, "flex the leg at the knee"; this movement results in the leg moving closer to the thigh. Some describe this as leg flexion occurring at the knee joint and may say "flex the leg," meaning flex the knee. Additionally, movement terms are utilized to describe movement occurring throughout the full range of motion or through a very small range. Using the knee flexion example again, we may flex the knee through the full range by beginning in full knee extension (zero degrees of knee flexion) and flexing it fully, so that the heel comes in contact with the buttocks; this would be approximately 140 degrees of flexion. We may also begin with the knee in 90 degrees of flexion and then flex it 30 degrees more; this movement results in a knee flexion angle of 120 degrees, even though the knee flexed only 30 degrees. In both examples, the knee is in different degrees of flexion. We may also begin with the knee in 90 degrees of flexion and extend it 40 degrees, which would result in a flexion angle of 50 degrees. Even though we extended the knee, it is still flexed, only less so than before.

In this example, we more commonly move the distal extremity in relation to the proximal extremity, which is usually more stationary. However, there are examples in every joint where the distal segment may be more stationary and we move the proximal segment in relation to it. An example is the knee in doing a squat from the standing position. As the squat occurs, the thigh moves toward the stabler leg, still resulting in knee flexion that could be stated as flexing the thigh at the knee.

Some movement terms may be used to describe motion at several joints throughout the body, whereas other terms are relatively specific to a joint or group of joints (Fig. 1.21). Rather than list the terms alphabetically, we have chosen to group them according to the body area and pair them with opposite terms where applicable. Additionally, the prefixes **hyper-** and **hypo-** may be combined with these terms to emphasize motion beyond and below normal, respectively. Of these combined terms, **hyperextension** is the most commonly used.

## Terms describing general movements

**Abduction:** Lateral movement away from the midline of the trunk in the frontal plane. An example is

A          B          C

Courtesy of R.T. Floyd (all)

FIG. 1.21 • Joint movements. **A**, Examples of sagittal plane movements: extension of left toes, ankle (plantar flexion), knee, hip, shoulder, elbow, wrist, fingers, lumbar and cervical spine; flexion of right toes, ankle (dorsiflexion), knee, hip, shoulder, elbow, wrist, and fingers. **B**, Examples of frontal plane movements: abduction of left transverse tarsal/subtalar joints (eversion), shoulder, wrist, fingers, and shoulder girdle (upward rotation), lumbar (lateral flexion to right) and cervical spine (lateral flexion to left), and right hip; adduction of right transverse tarsal/subtalar joints (inversion), shoulder, wrist, fingers, and shoulder girdle (downward rotation). **C**, Examples of transverse plane movements: internal rotation of right shoulder, right radioulnar joints (pronation); external rotation of right knee, right hip, left shoulder, left radioulnar joints (supination); and lumbar (left rotation) and cervical spine (left rotation).

raising the arms or thighs to the side away from the anatomical position.

**Adduction:** Movement medially toward and/or across the midline of the trunk in the frontal plane. An example is lowering the arm to the side or the thigh back to the anatomical position.

**Flexion:** Bending movement that results in a decrease of the angle in a joint by bringing bones together, usually in the sagittal plane. An example is the elbow joint when the hand is drawn to the shoulder.

**Extension:** Straightening movement that results in an increase of the angle in a joint by moving bones apart, usually in the sagittal plane. Using the elbow, an example is when the hand moves away from the shoulder.

**Circumduction:** Circular movement of a limb that delineates an arc or describes a cone. It is a combination of flexion, extension, abduction, and adduction. Sometimes referred to as circumflexion. An example is when the shoulder joint or the hip joint moves in a circular fashion around a fixed point, either clockwise or counterclockwise.

**Diagonal abduction:** Movement by a limb through a diagonal plane away from the midline of the body, such as in the hip or glenohumeral joint.

**Diagonal adduction:** Movement by a limb through a diagonal plane toward and across the midline of the body, such as in the hip or glenohumeral joint.

**External rotation:** Rotary movement around the longitudinal axis of a bone away from the midline of the body. Occurs in the transverse plane and is also known as rotation laterally, outward rotation, and lateral rotation.

**Internal rotation:** Rotary movement around the longitudinal axis of a bone toward the midline of the body. Occurs in the transverse plane and is also known as rotation medially, inward rotation, and medial rotation.

## Terms describing ankle and foot movements

**Eversion:** Turning the subtalar and traverse tarsal joints of the foot outward or laterally in the frontal plane; abduction, characterized by the soles of the feet being turned outward. An example is standing with the weight on the inner edge of the foot.

**Inversion:** Turning the subtalar and traverse tarsal joints medially into adduction in the frontal plane; characterized by the soles of the feet being turned inward. An example is standing with the weight on the outer edge of the foot.

**Dorsal flexion (dorsiflexion):** Flexion movement of the ankle that results in the top of the foot moving toward the anterior tibia in the sagittal plane.

**Plantar flexion:** Extension movement of the ankle that results in the foot and/or toes moving away from the body in the sagittal plane.

**Pronation:** A position of the foot and ankle resulting from a combination of ankle dorsiflexion, subtalar eversion, and forefoot abduction (toe-out).

**Supination:** A position of the foot and ankle resulting from a combination of ankle plantar flexion, subtalar inversion, and forefoot adduction (toe-in).

## Terms describing radioulnar joint movements

**Pronation:** Internally rotating the radius in the transverse plane so that it lies diagonally across the ulna, resulting in the palm-down position of the forearm.

**Supination:** Externally rotating the radius in the transverse plane so that it lies parallel to the ulna, resulting in the palm-up position of the forearm.

## Terms describing shoulder girdle (scapulothoracic) movements

**Depression:** Inferior movement of the shoulder girdle in the frontal plane. An example is returning to the normal position from a shoulder shrug.

**Elevation:** Superior movement of the shoulder girdle in the frontal plane. An example is shrugging the shoulders.

**Protraction (abduction):** Forward movement of the shoulder girdle in the horizontal plane away from the spine. Abduction of the scapula.

**Retraction (adduction):** Backward movement of the shoulder girdle in the horizontal plane toward the spine. Adduction of the scapula.

**Rotation downward:** Rotary movement of the scapula in the frontal plane with the inferior angle of the scapula moving medially and downward. Occurs primarily in the return from upward rotation. The inferior angle may actually move upward slightly as the scapula continues in extreme downward rotation.

**Rotation upward:** Rotary movement of the scapula in the frontal plane with the inferior angle of the scapula moving laterally and upward.

## Terms describing shoulder joint (glenohumeral) movements

**Horizontal abduction:** Movement of the humerus or femur in the horizontal plane away from the midline of the body. Also known as horizontal extension or transverse abduction.

**Horizontal adduction:** Movement of the humerus or femur in the horizontal plane toward the midline of the body. Also known as horizontal flexion or transverse adduction.

**Scaption:** Movement of the humerus away from the body in the scapular plane. Glenohumeral abduction in a plane 30 to 45 degrees between the sagittal and frontal planes.

## Terms describing spine movements

**Lateral flexion (side bending):** Movement of the head and/or trunk in the frontal plane laterally away from the midline. Abduction of the spine.

**Reduction:** Return of the spinal column in the frontal plane to the anatomic position from lateral flexion. Adduction of the spine.

## Terms describing wrist and hand movements

**Dorsal flexion (dorsiflexion):** Extension movement of the wrist in the sagittal plane with the dorsal or posterior side of the hand moving toward the posterior side of the forearm.

**Palmar flexion:** Flexion movement of the wrist in the sagittal plane with the volar or anterior side of the hand moving toward the anterior side of the forearm.

**Radial flexion (radial deviation):** Abduction movement at the wrist in the frontal plane of the thumb side of the hand toward the lateral forearm.

**Ulnar flexion (ulnar deviation):** Adduction movement at the wrist in the frontal plane of the little finger side of the hand toward the medial forearm.

**Opposition of the thumb:** Diagonal movement of the thumb across the palmar surface of the hand to make contact with the fingers.

**Reposition of the thumb:** Diagonal movement of the thumb as it returns to the anatomical position from opposition with the hand and/or fingers.

These movements are considered in detail in the chapters that follow as they apply to the individual joints.

Combinations of movements can occur. Flexion or extension can occur with abduction, adduction, or rotation.

## Movement icons (pedagogical feature)

Throughout this text a series of movement icons will be utilized to represent different joint movements. These icons will be displayed in the page margins to indicate the joint actions of the muscles displayed on that page. As further explained in Chapter 2, the actions displayed represent the movements that occur when the muscle contracts concentrically. Table 1.7 provides a complete list of the icons. Refer to them as needed when reading Chapters 4–7 and 9–12.

TABLE 1.7 · **Movement icons representing joint actions**

TABLE 1.7 (continued) • **Movement icons representing joint actions**

| Thumb carpometacarpal joint | | | Thumb metacarpophalangeal joint | | Thumb interphalangeal joint | |
|---|---|---|---|---|---|---|
| Thumb CMC flexion | Thumb CMC extension | Thumb CMC abduction | Thumb MCP flexion | Thumb MCP extension | Thumb IP flexion | Thumb IP extension |
| 2nd, 3rd, 4th, and 5th MCP, PIP, and DIP joints | | 2nd, 3rd, 4th, and 5th MCP and PIP joints | 2nd, 3rd, 4th, and 5th metacarpophalangeal joints | | 2nd, 3rd, 4th, and 5th PIP joints | 2nd, 3rd, 4th, and 5th DIP joints |
| 2nd–5th MCP, PIP, and DIP flexion | 2nd–5th MCP, PIP, and DIP extension | 2nd–5th MCP and PIP flexion | 2nd–5th MCP flexion | 2nd–5th MCP extension | 2nd–5th PIP flexion | 2nd–5th DIP flexion |

Hip

| Hip flexion | Hip extension | Hip abduction | Hip adduction | Hip external rotation | Hip internal rotation |
|---|---|---|---|---|---|

Knee

| Knee flexion | Knee extension | Knee external rotation | Knee internal rotation |
|---|---|---|---|

| Ankle | | Transverse tarsal and subtalar joints | |
|---|---|---|---|
| Ankle plantar flexion | Ankle dorsal flexion | Transverse tarsal and subtalar inversion | Transverse tarsal and subtalar eversion |

TABLE 1.7 (continued) • **Movement icons representing joint actions**

| Great toe metatarsophalangeal and interphalangeal joints | | 2nd–5th metatarsophalangeal, proximal interphalangeal, and distal interphalangeal joints | |
|---|---|---|---|
| Great toe MTP and IP flexion | Great toe MTP and IP extension | 2nd–5th MTP, PIP, and DIP flexion | 2nd–5th MTP, PIP, and DIP extension |
| Cervical spine | | | |
| Cervical flexion | Cervical extension | Cervical lateral flexion | Cervical rotation unilaterally |
| Lumbar spine | | | |
| Lumbar flexion | Lumbar extension | Lumbar lateral flexion | Lumbar rotation unilaterally |

## Physiological movements versus accessory motions

Movements such as flexion, extension, abduction, adduction, and rotation occur by the bones moving through planes of motion about an axis of rotation at the joint. These movements may be referred to as physiological movements. The motion of the bones relative to the three cardinal planes resulting from these physiological movements is referred to as **osteokinematic motion**. In order for these osteokinematic motions to occur, there must be movement between the actual articular surfaces of the joint. This motion between the articular surfaces is known as **arthrokinematics**, and it includes three specific types of **accessory motions**. These accessory motions, named specifically to describe the actual change in relationship between the articular surface of one bone relative to another, are **spin**, **roll**, and **glide** (Fig. 1.22).

FIG. 1.22 • Joint arthrokinematics. **A**, Spin; **B**, Roll; **C**, Glide.

FIG. 1.23 • Knee joint arthrokinematics. **A**, Standing from squatting; **B**, Flexing from nonweight-bearing position.

Roll is sometimes referred to as rock or rocking, whereas glide is sometimes referred to as slide or translation. If accessory motion is prevented from occurring, then physiological motion cannot occur to any substantial degree other than by joint compression or distraction. Because most diarthrodial joints in the body are composed of a concave surface articulating with a convex surface, roll and glide must occur together to some degree. For example, as illustrated in Fig. 1.23, as a person stands from a squatting position, in order for the knee to extend, the femur must roll forward and simultaneously slide backward on the tibia. If not for the slide, the femur would roll off the front of the tibia, and if not for the roll, the femur would slide off the back of the tibia.

Spin may occur in isolation or in combination with roll and glide, depending upon the joint structure. To some degree, spin occurs at the knee as it flexes and extends. In the squatting to standing example, the femur spins medially or internally rotates as the knee reaches full extension. Table 1.8 provides examples of accessory motion.

**Roll (rock):** A series of points on one articular surface contacts a series of points on another articular surface.

**Glide (slide, translation):** A specific point on one articulating surface comes in contact with a series of points on another surface.

**Spin:** A single point on one articular surface rotates about a single point on another articular surface. Motion occurs around some stationary longitudinal mechanical axis in either a clockwise or a counterclockwise direction.

## TABLE 1.8 • Accessory motion

| Accessory motion | Anatomical joint example | Analogy | |
|---|---|---|---|
| Roll (rocking) | Knee extension occurring from femoral condyles rolling forward on tibia as a person stands from squatting position | Tire rolling across a road surface, as in normal driving with good traction | Combination of roll and glide: Tire spinning on slick ice (i.e., poor traction) but still resulting in movement across the road surface |
| Glide (slide or translation) | Knee extension occurring from femoral condyles sliding backward on tibia as a person stands from squatting position | Tire skidding across a slick surface with the brakes locked | |
| Spin | Radioulnar pronation/supination occurring from spinning of radial head against humeral capitulum | Point of a toy top spinning around in one spot on the floor | |

## REVIEW EXERCISES

1. Complete the blanks in the following paragraphs using each word from the list below only once except for the ones marked with two asterisks,**, which are used twice. The number of dashes indicates the number of letters of the word for each blank.

a. anterior**
b. anteroinferior
c. anterolateral
d. anteromedial
e. anteroposterior
f. anterosuperior
g. bilateral
h. caudal
i. cephalic
j. contralateral
k. deep
l. distal
m. dorsal
n. inferior
o. inferolateral
p. inferomedial
q. ipsilateral
r. lateral

s. medial
t. palmar
u. plantar
v. posterior**
w. posteroinferior
x. posterolateral
y. posteromedial
z. posterosuperior
aa. prone
bb. proximal
cc. superficial
dd. superior
ee. superolateral**
ff. superomedial
gg. supine
hh. ventral
ii. volar

When Jacob greeted Stephanie at the beach, he reached out with the _ _ _ _ _ _ surface of his hand to grasp the _ _ _ _ _ surface of her hand for a handshake. As the _ _ _ _ _ _ _ _ aspect of their bodies faced each other, Jacob noticed that the hair located on the most _ _ _ _ _ _ _ _ part of Stephanie's head appeared to be a different color than he remembered. He then asked her to turn around so that he could see it from a _ _ _ _ _ _ _ _ _ view. As she did so, it became obvious to him that she had blonde streaks running from her _ _ _ _ _ _ _ region in an _ _ _ _ _ _ _ _ direction all the way down to her _ _ _ _ _ _ region.

Stephanie then asked Jacob if the sunburn on the _ _ _ _ _ _ _ _ _ _ _ _ _ portions of his shoulders was due to the exposure that his tank-top shirt provided. He replied yes but that it was only an _ _ _ _ _ _ _ _ _ _ _ burn and did not go too _ _ _ _. He then said, "I wished I would have had my shirt off so that I would have gotten some more sun on the _ _ _ _ _ _ _ _ _ _ _ _ _ portion of my shoulders up to my neck."

Stephanie said that she recently got sunburned on her back while lying _ _ _ _ _ at the beach. She then flipped her hair around the _ _ _ _ _ _ _ side of her neck toward the _ _ _ _ _ _ portion of her trunk to expose her _ _ _ _ _ _ region. Jacob remarked, "Wow, instead of the bikini you have on today with straps over your shoulders running from your _ _ _ _ _ _

_ _ _ _ _ _ _ _ chest to your _ _ _ _ _ _ _ _ _ _ _ _ _ _ shoulders, you must have been wearing one with crossing straps as I see you have tan lines running in an _ _ _ _ _ _ _ _ _ _ _ _ _ direction to your _ _ _ _ _ low back from the _ _ _ _ _ _ _ _ _ _ _ aspect of your _ _ _ _ _ _ _ _ shoulders. You should have spent more time lying _ _ _ _ _ _ ." She replied, "Well, I did lie partially on my back and my right side for a while. See where the _ _ _ _ _ _ _ _ _ _ _ _ _ portion of my right thigh and the _ _ _ _ _ _ _ _ _ _ _ _ _ portion of my left thigh are tanned just right, but unfortunately in that position the _ _ _ _ _ _ _ _ _ _ _ _ _ right thigh and _ _ _ _ _ _ _ _ _ _ _ _ _ thigh received relatively little exposure." Jacob commented, "Yep, when you lie on one side most of the time, you get all the sun on the _ _ _ _ _ _ _ _ _ _ _ _ _ side and none on the _ _ _ _ _ _ _ _ _ _ _ side. It looks like you must have had a towel covering your feet and ankles since your _ _ _ _ _ _ lower extremities are not nearly as tan as your _ _ _ _ _ _ _ _ lower extremities." Stephanie replied, "You are correct. I kept the bottom of my lower legs covered almost all of the time while lying on both sides so that the sensitive skin on my _ _ _ _ _ _ _ _ _ _ _ _ and _ _ _ _ _ _ _ _ _ _ _ _ _ shins would not burn. But, I did get a good _ _ _ _ _ _ _ _ _ tan on my _ _ _ _ _ _ _ _ trunk, except for the _ _ _ _ _ _ _ _ _ _ _ _ _ _ aspect of my right elbow I was resting on." As Jacob slipped his sandals on to protect the _ _ _ _ _ _ _ aspect of his feet from the hot sand, he said, "Well, nice to see you. I have to go by the doctor's office and get an _ _ _ _ _ _ _ _ _ _ _ _ _ chest X-ray to make sure my pneumonia has cleared up."

## 2. Joint movement terminology chart

The specific body area joint movement terms arise from basic motions in the three specific planes: flexion/extension in the sagittal plane, abduction/adduction in the frontal plane, and rotation in the transverse plane. With this in mind, complete the chart by writing the basic motion in the right column for each specific motion listed in the left column by using either *flexion, extension, abduction, adduction,* or *rotation (external or internal).*

| Specific motion | Basic motion |
|---|---|
| Eversion | |
| Inversion | |
| Dorsal flexion | |
| Plantar flexion | |
| Pronation (radioulnar) | |
| Supination (radioulnar) | |
| Lateral flexion | |

| Specific motion | Basic motion |
|-----------------|--------------|
| Reduction | |
| Radial flexion | |
| Ulnar flexion | |

## 3. Bone typing chart

| Utilizing Fig. 1.8 and other resources, place an "X" in the appropriate column to indicate its classification. | | | | | |
|------|------|-------|------|-----------|----------|
| Bone | Long | Short | Flat | Irregular | Sesamoid |
| Frontal | | | | | |
| Zygomatic | | | | | |
| Parietal | | | | | |
| Temporal | | | | | |
| Occipital | | | | | |
| Maxilla | | | | | |
| Mandible | | | | | |
| Cervical vertebrae | | | | | |
| Clavicle | | | | | |
| Scapula | | | | | |
| Humerus | | | | | |
| Ulna | | | | | |
| Radius | | | | | |
| Carpal bones | | | | | |
| Metacarpals | | | | | |
| Phalanges | | | | | |
| Ribs | | | | | |
| Sternum | | | | | |
| Lumbar vertebrae | | | | | |
| Ilium | | | | | |
| Ischium | | | | | |
| Pubis | | | | | |
| Femur | | | | | |
| Patella | | | | | |
| Fabella | | | | | |
| Tibia | | | | | |
| Fibula | | | | | |
| Talus | | | | | |
| Calcaneus | | | | | |
| Navicular | | | | | |
| Cuneiforms | | | | | |
| Metatarsals | | | | | |

4. What are the five functions of the skeleton?
5. List the bones of the upper extremity.
6. List the bones of the lower extremity.
7. List the bones of the shoulder girdle.
8. List the bones of the pelvic girdle.
9. Describe and explain the differences and similarities between the radius and ulna.
10. Describe and explain the differences and similarities between the humerus and femur.
11. Using body landmarks, how would you suggest determining the length of each lower extremity for comparison to determine whether someone had a true total leg length discrepancy?
12. Explain why the fibula is more susceptible to fractures than the tibia.
13. Why is anatomical position so important in understanding anatomy and joint movements?
14. Label the parts of a long bone.

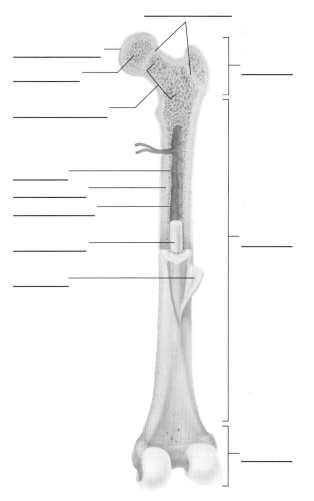

Femur

## 15. Joint type, movement, and plane of motion chart

Complete the chart by filling in the type of diarthrodial joint and then listing the movements of the joint under the plane of motion in which they occur.

| Joint | Type | Planes of motion | | |
| --- | --- | --- | --- | --- |
| | | Sagittal | Lateral | Transverse |
| Scapulothoracic joint | | | | |
| Sternoclavicular | | | | |
| Acromioclavicular | | | | |
| Glenohumeral joint | | | | |
| Elbow | | | | |
| Radioulnar joint | | | | |
| Wrist | | | | |
| 1st carpometacarpal joint | | | | |
| 1st metacarpophalangeal joint | | | | |
| Thumb interphalangeal joint | | | | |
| 2nd, 3rd, 4th, and 5th metacarpophalangeal joints | | | | |
| 2nd, 3rd, 4th, and 5th proximal interphalangeal joints | | | | |
| 2nd, 3rd, 4th, and 5th distal interphalangeal joints | | | | |
| Cervical spine C1–C2 | | | | |
| Cervical spine C2–C7 | | | | |
| Lumbar spine | | | | |
| Hip | | | | |
| Knee (tibiofemoral joint) | | | | |
| Knee (patellofemoral joint) | | | | |
| Ankle | | | | |
| Transverse tarsal and subtalar joints | | | | |
| Metatarsophalangeal joints | | | | |
| Great toe interphalangeal | | | | |
| 2nd, 3rd, 4th, and 5th proximal interphalangeal joints | | | | |
| 2nd, 3rd, 4th, and 5th distal interphalangeal joints | | | | |

## 16. Joint position chart

| Using proper terminology, complete the chart by listing the name of each joint involved and its position upon completion of the multiple joint movement. In all cases, begin in a sitting, upright position. | |
| --- | --- |
| Multiple joint movement | Joints and respective position of each |
| Reach straight over the superior aspect of your head to touch the contralateral ear | |
| Place the toe of one foot against the posterior aspect of the contralateral calf | |
| Reach behind the back and use your thumb to touch a spinous process | |
| Actively pull the knee as far as possible to the the shoulder | |
| Place the plantar aspect of both feet against against each other | |

## 17. Plane of motion and axis of rotation chart

| For each joint motion in the chart, list the plane in which the motion occurs and its axis of rotation. | | |
| --- | --- | --- |
| Motion | Plane of motion | Axis of rotation |
| Cervical rotation | | |
| Shoulder girdle elevation | | |
| Glenohumeral horizontal adduction | | |
| Elbow flexion | | |
| Radioulnar pronation | | |
| Wrist radial deviation | | |
| Metacarpophalangeal abduction | | |
| Lumbar lateral flexion | | |
| Hip internal rotation | | |
| Knee extension | | |
| Ankle inversion | | |
| Great toe extension | | |

18. List two sport skills that involve movements more clearly seen from the side. List the primary movements that occur in the ankle, knee, hip, spine, glenohumeral joint, elbow, and wrist. In which plane are these movements predominantly occurring? What axis of rotation is primarily involved?

19. List two sport skills that involve movements more clearly seen from the front or rear. List the primary movements that occur in the transverse tarsal/subtalar joint, hip, spine, glenohumeral joint, and wrist. In which plane are these movements predominantly occurring? What axis of rotation is primarily involved?

20. List the similarities between the ankle/foot/toes and the wrist/hand/fingers regarding the bones, joint structures, and movements. What are the differences?

21. Compare and contrast the glenohumeral and acetabulofemoral joints. Which one is more susceptible to dislocations and why?

22. Compare and contrast the elbow and knee joints. Considering the bone and joint structures and their functions, what are the similarities and differences?

## LABORATORY EXERCISES

1. Choose several different locations on your body at random and label as 1., 2., 3., etc. Then use correct anatomical terminology to describe the path you take as you move your index finger from one point to another without leaving the skin surface.

2. Determine which joints have movements possible in each of the following planes:
   a. Sagittal
   b. Frontal
   c. Transverse

3. List all the diarthrodial joints of the body that are capable of the following paired movements:
   a. Flexion/extension
   b. Abduction/adduction
   c. Rotation (left and right)
   d. Rotation (internal and external)

4. Determine the planes in which the following activities occur. Also, use a pencil to visualize the axis for each of the following activities.
   a. Walking up stairs
   b. Turning a knob to open a door
   c. Nodding the head to agree

d. Shaking the head to disagree

e. Shuffling the body from side to side (side-stepping)

f. Looking over your shoulder to see behind you

5. Individually practice the various joint movements, on yourself or with another subject.

6. Locate the various types of joints on a human skeleton and palpate their movements on a living subject.

7. Stand in the anatomical position facing a closed door. Reach out and grasp the knob with your right hand. Turn it and open the door widely toward you. Determine all of the joints involved in this activity and list the movements for each joint.

8. Utilize a goniometer to measure the joint ranges of motion for several students in your class for each of the following movements. Compare your results with the average ranges provided in Appendix 1.

a. External and internal rotation of the shoulder with the shoulder in 90 degrees of abduction while supine

b. Elbow flexion in the supine position

c. Wrist extension with the forearm in neutral and the elbow in 90 degrees of flexion

d. Hip external and internal rotation in the sitting position with the hip and knee each in 90 degrees of flexion

e. Knee flexion in the prone position

f. Ankle dorsiflexion with the knee in 90 degrees of flexion versus knee in full extension

9. Discuss the following joints among your classmates and place them in order from the least total range of motion in the sagittal plane to the most. Be prepared to defend your answer.

a. Ankle
b. Elbow
c. Glenohumeral

d. Hip
e. Knee
f. Wrist

10. Is there more inversion or more eversion possible in the transverse tarsal and subtalar joints? Explain this occurrence based on anatomy.

11. Is there more abduction or more adduction possible in the wrist joint? Explain this occurrence based on anatomy.

# References

Anthony C, Thibodeau G: *Textbook of anatomy and physiology*, ed 10, St. Louis, 1979, Mosby.

Booher JM, Thibodeau GA: *Athletic injury assessment*, ed 4, New York, 2000, McGraw-Hill.

Goss CM: *Gray's anatomy of the human body*, ed 29, Philadelphia, 1973, Lea & Febiger.

Hamilton N, Weimar W, Luttgens K: *Kinesiology: scientific basis of human motion*, ed 12, New York, 2012, McGraw-Hill.

Lindsay DT: *Functional human anatomy*, St. Louis, 1996, Mosby.

Logan GA, McKinney WC: *Anatomic kinesiology*, ed 3, Dubuque, IA, 1982, Brown.

National Strength and Conditioning Association; Baechle TR, Earle RW: *Essentials of strength training and conditioning*, ed 2, Champaign, IL, 2000, Human Kinetics.

Neumann, DA: *Kinesiology of the musculoskeletal system: foundations for physical rehabilitation*, ed 2, St. Louis, 2010, Mosby.

Northrip JW, Logan GA, McKinney WC: *Analysis of sport motion: anatomic and biomechanic perspectives*, ed 3, Dubuque, IA, 1983, Brown.

Nwanka ON, Hayashi D, et al.,: Sesamoids and accessory ossicles of the foot: anatomical variability and related pathology. *Insights Imaging.* 4(5): 581–593, 2013.

Prentice WE: *Principles of athletic training: a competency based approach*, ed 15, New York, 2014, McGraw-Hill.

Prentice WE: *Rehabilitation techniques in sports medicine*, ed 5, New York, 2011, McGraw-Hill.

Seeley RR, Stephens TD, Tate P: *Anatomy & physiology*, ed 8, New York, 2008, McGraw-Hill.

Shier D, Butler J, Lewis R: *Hole's essentials of human anatomy and physiology*, ed 10, New York, 2009, McGraw-Hill.

Stedman TL: *Stedman's medical dictionary*, ed 28, Baltimore, 2005, Lippincott Williams & Wilkins.

Steindler A: *Kinesiology of the human body*, Springfield, IL, 1970, Thomas.

Van De Graaff KM: *Human anatomy*, ed 6, New York, 2002, McGraw-Hill.

Van De Graaff KM, Fox SI, LaFleur KM: *Synopsis of human anatomy & physiology*, Dubuque, IA, 1997, Brown.

# Worksheet Exercises

For in- or out-of-class assignments, or for testing, utilize this tear-out worksheet.

## Anterior skeletal worksheet

On the anterior skeletal worksheet, label the bones and their prominent features by filling in the blanks.

1 _____
2 _____
3 _____
4 _____
5 _____
6 _____
7 _____
8 _____
9 _____
10 _____
11 _____
12 _____
13 _____
14 _____
15 _____
16 _____
17 _____
18 _____
19 _____
20 _____
21 _____

22 _____
23 _____
24 _____
25 _____
26 _____
27 _____
28 _____
29 _____
30 _____
31 _____
32 _____
33 _____
34 _____
35 _____
36 _____
37 _____
38 _____
39 _____
40 _____
41 _____
42 _____
43 _____
44 _____
45 _____
46 _____
47 _____
48 _____
49 _____
50 _____
51 _____
52 _____

# Worksheet Exercises

For in- or out-of-class assignments, or for testing, utilize this tear-out worksheet.

## Posterior skeletal worksheet

On the posterior skeletal worksheet, label the bones and their prominent features by filling in the blanks.

1 _____
2 _____
3 _____
4 _____
5 _____
6 _____
7 _____
8 _____
9 _____
10 _____
11 _____
12 _____
13 _____
14 _____
15 _____
16 _____
17 _____
18 _____
19 _____
20 _____
21 _____
22 _____
23 _____
24 _____
25 _____

26 _____
27 _____
28 _____
29 _____
30 _____
31 _____
32 _____
33 _____
34 _____
35 _____
36 _____
37 _____
38 _____
39 _____
40 _____
41 _____
42 _____

# CHAPTER 2

# NEUROMUSCULAR FUNDAMENTALS

## Objectives

- To review the basic anatomy and function of the muscular and nervous systems

- To review and understand the basic terminology used to describe muscular locations, arrangements, characteristics, and roles, as well as neuromuscular functions

- To learn and understand the different types of muscle contraction and the factors involved in each

- To learn and understand basic neuromuscular concepts in relation to how muscles function in joint movement and work together in effecting motion

- To develop a basic understanding of the neural control mechanisms for movement

Skeletal muscles are responsible for movement of the body and all its joints. Muscle contraction produces the force that causes joint movement in the human body. In addition to the function of movement, muscles also provide both dynamic stability of joints and protection, contribute to posture and support, and produce a major portion of total body heat. There are over 600 skeletal muscles, which constitute approximately 40% to 50% of body weight. Of these, there are 215 pairs of skeletal muscles. These pairs of muscles usually work in cooperation with each other to perform opposite actions at the joints they cross. In most cases, muscles work in groups rather than independently to achieve a given joint motion. This is known as **aggregate muscle action**.

## Muscle nomenclature

In attempting to learn the skeletal muscles, it is helpful to have an understanding of how they are named. Muscles are usually named because of one or more distinctive characteristics, such as their visual appearance, anatomical location, or function. Examples of skeletal muscle naming are as follows:

*Shape*—deltoid, rhomboid
*Size*—gluteus maximus, teres minor
*Number of divisions*—triceps brachii
*Direction of fibers*—external abdominal oblique
*Location*—rectus femoris, palmaris longus
*Points of attachment*—coracobrachialis, extensor hallucis longus, flexor digitorum longus
*Action*—erector spinae, supinator, extensor digiti minimi
*Action and shape*—pronator quadratus
*Action and size*—adductor magnus
*Shape and location*—serratus anterior
*Location and attachment*—brachioradialis
*Location and number of divisions*—biceps femoris

In discussions regarding the muscles, they are often grouped together for brevity of conversation and clearer understanding. The naming of muscle groups follows a similar pattern. Here are some muscle groups assembled according to different naming rationales:

*Shape*—hamstrings
*Number of divisions*—quadriceps, triceps surae
*Location*—peroneals, abdominal, shoulder girdle
*Action*—hip flexors, rotator cuff

Figs. 2.1 and 2.2 depict the muscular system from both a superficial and a deep point of view.

Muscles shown in these figures, and many other muscles, will be studied in more detail as each joint of the body is considered in later chapters.

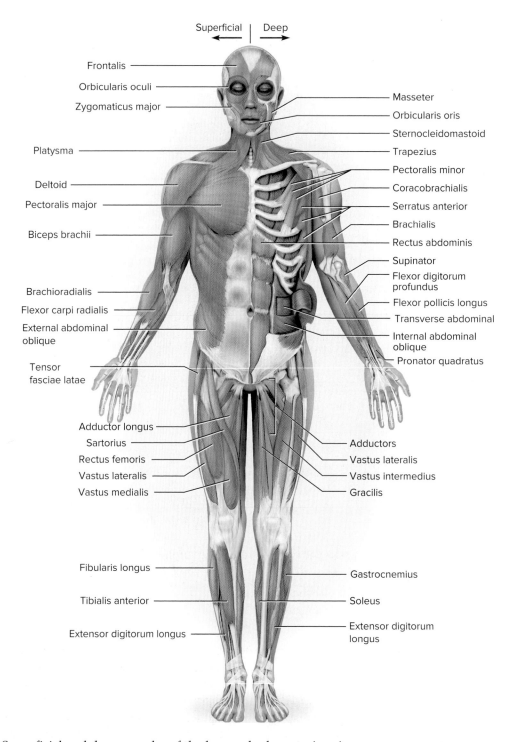

Superficial | Deep

Frontalis

Orbicularis oculi

Zygomaticus major

Masseter

Orbicularis oris

Sternocleidomastoid

Platysma

Trapezius

Pectoralis minor

Deltoid

Coracobrachialis

Pectoralis major

Serratus anterior

Biceps brachii

Brachialis

Rectus abdominis

Supinator

Brachioradialis

Flexor digitorum profundus

Flexor carpi radialis

Flexor pollicis longus

External abdominal oblique

Transverse abdominal

Internal abdominal oblique

Tensor fasciae latae

Pronator quadratus

Adductor longus

Sartorius

Adductors

Rectus femoris

Vastus lateralis

Vastus lateralis

Vastus intermedius

Vastus medialis

Gracilis

Fibularis longus

Gastrocnemius

Tibialis anterior

Soleus

Extensor digitorum longus

Extensor digitorum longus

FIG. 2.1 • Superficial and deep muscles of the human body, anterior view.

Deep | Superficial

Occipitalis

Semispinalis capitis
Sternocleidomastoid
Splenius capitis
Levator scapulae
Supraspinatus
Rhomboideus minor
Rhomboideus major
Deltoid (cut)
Infraspinatus
Serratus anterior
Triceps brachii (cut)
Serratus posterior inferior
External abdominal oblique
Internal abdominal oblique
Erector spinae
Flexor carpi ulnaris
Extensor digitorum (cut)
Gluteus minimus
Lateral rotators
Adductor
magnus

Iliotibial band
Semimembranosus
Biceps femoris

Gastrocnemius (cut)
Soleus
Tibialis posterior
Flexor digitorum longus
Extensor hallucis longus
Fibularis longus
Calcaneal tendon

Trapezius

Infraspinatus
Teres minor
Teres major
Triceps brachii

Latissimus dorsi

External abdominal
oblique
Gluteus medius
Gluteus maximus

Gracilis
Semitendinosus
Iliotibial band
Biceps femoris

Gastrocnemius

Soleus

FIG. 2.2 • Superficial and deep muscles of the human body, posterior view.

## Shape of muscles and fiber arrangement

Various muscles have different shapes, and their fibers may be arranged differently in relation to each other and to the tendons that connect them to bone. The shape and fiber arrangement play a role in the muscle's ability to exert force and in the range through which it can effectively exert force on the bones to which it is attached. A factor in the ability of a muscle to exert force is its cross-section diameter. Keeping all other factors constant, a muscle with a greater cross-section diameter will be able to exert a greater force. A factor in the ability of a muscle to move a joint through a large range of motion is its ability to shorten. Generally, longer muscles can shorten through a greater range and therefore are more effective in moving joints through large ranges of motion.

Essentially, all skeletal muscles may be grouped into two major types of fiber arrangements: parallel and pennate. Each may be subdivided further according to shape.

Parallel muscles have their fibers arranged parallel to the length of the muscle. Generally, parallel muscles will produce a greater range of movement than similar-size muscles with a pennate arrangement. Parallel muscles are categorized into the following shapes:

*Flat* muscles are usually thin and broad, originating from broad, fibrous, sheetlike aponeuroses that allow them to spread their forces over a broad area. Examples include the rectus abdominis and external oblique.

*Fusiform* muscles are spindle-shaped with a central belly that tapers to tendons on each end; this allows them to focus their power on small, bony targets. Examples are the brachialis and the brachioradialis.

*Strap muscles* are more uniform in diameter with essentially all their fibers arranged in a long parallel manner. This also enables a focusing of power on small, bony targets. The sartorius is an example.

*Radiate* muscles are also sometimes described as being triangular, fan-shaped, or convergent. They have the combined arrangement of flat and fusiform muscles, in that they originate on a broad surface or an aponeurosis and converge onto a tendon. Examples include the pectoralis major and trapezius.

*Sphincter* or circular muscles are technically endless strap muscles that surround openings and function to close them upon contraction. An example is the orbicularis oris, surrounding the mouth.

Pennate muscles have shorter fibers that are arranged obliquely to their tendons in a structure similar to that of a feather. This arrangement increases the cross-sectional area of the muscle, thereby increasing its force production capability. Pennate muscles are categorized on the basis of the exact arrangement between the fibers and the tendon, as follows:

*Unipennate* muscle fibers run obliquely from a tendon on one side only. Examples are seen in the biceps femoris, extensor digitorum longus, and tibialis posterior.

*Bipennate* muscle fibers run obliquely from a central tendon on both sides, as in the rectus femoris and flexor hallucis longus.

*Multipennate* muscles have several tendons with fibers running diagonally between them, as in the deltoid.

Bipennate and unipennate muscles produce the strongest contractions. Review Table 2.1 regarding muscle shapes and fiber arrangements.

## Muscle tissue properties

Skeletal muscle tissue has four properties related to its ability to produce force effecting movement about joints. **Irritability** or **excitability** is the muscle property of being sensitive or responsive to chemical, electrical, or mechanical stimuli. When an appropriate stimulus is provided, muscle responds by developing tension. **Contractility** is the ability of muscle to contract and develop tension or internal force against resistance when stimulated. The ability of muscle tissue to develop tension or contract is unique in that other body tissues do not have this property. **Extensibility** is the ability of muscle to be passively stretched beyond its normal resting length. As an example, the triceps brachii displays extensibility when it is stretched beyond its normal resting length by the biceps brachii and other elbow flexors contracting to achieve full elbow flexion. **Elasticity** is the ability of muscle to return to its original resting length following stretching. To continue with the elbow example, the triceps brachii displays elasticity by returning to its original resting length when the elbow flexors cease contracting and relax.

## Muscle terminology

Locating the muscles, their proximal and distal attachments, and their relationship to the joints they cross is critical to determining the effects that muscles have on the joints. It is also necessary for understanding certain terms when considering body movement.

### Intrinsic

Pertaining usually to muscles within or belonging solely to the body part on which they act. The small intrinsic muscles found entirely within the hand are examples. See page 211.

### Extrinsic

Pertaining usually to muscles that arise or originate outside of (proximal to) the body part on which they act. The forearm muscles that attach proximally on the distal humerus and insert on

## TABLE 2.1 · **Muscle shape and fiber arrangement**

| Fiber arrangement | Advantage | Shape | Appearance | Characteristics/description | Examples |
|---|---|---|---|---|---|
| Parallel (fibers arranged parallel to the length of the muscle) | Produces greater range of movement than similar-size pennate muscles; long excursion (contract over a great distance); good endurance | Flat | | Usually thin and broad, originating from broad, fibrous, sheetlike aponeuroses that allow them to spread their forces over a broad area | Rectus abdominis, external oblique |
| | | Fusiform | Tendon — Belly — Tendon | Spindle-shaped with central belly that tapers to tendons on each end; can focus their power on small, bony targets | Biceps brachii, brachialis |
| | | Strap | | More uniform in diameter with essentially all their fibers arranged in a long parallel manner; can focus their power on small, bony targets | Sartorius |
| | | Radiate (triangular, fan-shaped, convergent) | | Combined arrangement of flat and fusiform muscles; originate on broad aponeuroses and converge to a single point of attachment via a tendon | Pectoralis major, trapezius |
| | | Sphincter (circular) | | Fibers concentrically arranged around a body opening; technically endless strap muscles surround openings and function to close them upon contraction | Orbicularis oris, orbicularis oculi |
| Pennate (shorter fibers, arranged obliquely to their tendons) | Produces greater power than similar-size parallel muscles due to increased cross-sectional area; strong muscles; short excursion | Unipennate | | Run obliquely from a tendon on one side only | Biceps femoris, extensor digitorum longus, tibialis posterior |
| | | Bipennate | | Run obliquely from a central tendon on both sides | Rectus femoris, flexor hallucis longus |
| | | Multipennate | | Several tendons with fibers running diagonally between them | Deltoid |

Modified from Saladin, KS: *Anatomy & physiology: the unity of form and function*, ed 4, New York, 2007, McGraw-Hill; and Seeley RR, Stephens TD, Tate P: *Anatomy & physiology*, ed 7, New York, 2008, McGraw-Hill.

the fingers are examples of extrinsic muscles of the hand. See Chapter 7.

## Action

Action is the specific movement of the joint resulting from a concentric contraction of a muscle that crosses the joint. An example is the biceps brachii, which has the action of flexion at the elbow. In most cases a particular action is caused by a group of muscles working together. Any of the muscles in the group can be said to cause the action, even though it is usually an effort of the entire group. A particular muscle may cause more than one action either at the same joint or at a different joint, depending upon the characteristics of the joints crossed by the muscle and the exact location of the muscle and its attachments in relation to the joint(s).

## Innervation

Innervation occurs in the segment of the nervous system responsible for providing a stimulus to muscle fibers within a specific muscle or portion of a muscle. A particular muscle may be innervated by more than one nerve, and a particular nerve may innervate more than one muscle or portion of a muscle.

## Amplitude

The amplitude is the range of muscle fiber length between maximal and minimal lengthening.

## Gaster (belly or body)

The **gaster** is the central, fleshy portion of the muscle. This contractile portion of the muscle generally increases in diameter as the muscle contracts.

When a particular muscle contracts, it tends to pull both ends toward the gaster, or middle, of the muscle. Consequently, if neither of the bones to which a muscle is attached were stabilized, both bones would move toward each other upon contraction. The more common case, however, is that one bone is more stabilized by a variety of factors, and as a result the less stabilized bone usually moves toward the more stabilized bone upon contraction.

## Tendon

Tendons are tough yet flexible bands of fibrous connective tissue, often cordlike in appearance, that connect muscles to bones and other structures. By providing this connection, tendons transmit the force generated by the contracting muscle to the bone. In some cases, two muscles may share a common tendon, such as the Achilles tendon of the gastrocnemius and soleus muscles. In other cases a muscle may have multiple tendons connecting it to one or more bones, such as the three proximal attachments of the triceps brachii.

## Aponeurosis

An aponeurosis is a tendinous expansion of dense fibrous connective tissue that is sheet- or ribbon-like in appearance and resembles a flattened tendon. Aponeuroses serve as a fascia to bind muscles together or as a means of connecting muscle to bone, such as the palmar aponeurosis for the palmaris longus.

## Fascia

Fascia is a sheet or band of fibrous connective tissue that envelopes, separates, or binds together parts of the body such as muscles, organs, and other soft-tissue structures of the body. In certain places throughout the body, such as around joints like the wrist and ankle, fascial tissue forms a **retinaculum** to retain tendons close to the body.

## Origin

From a structural perspective, the proximal attachment of a muscle or the part that attaches closest to the midline or center of the body is usually considered to be the origin. From a functional and historical perspective, the least movable part or attachment of the muscle has generally been considered to be the origin.

## Insertion

Structurally, the distal attachment, or the part that attaches farthest from the midline or center of the body, is considered the insertion. Functionally and historically, the most movable part is generally considered the insertion.

As an example, in the biceps curl exercise, the biceps brachii muscle in the arm has its origin on the scapula (least movable bone) and its insertion on the radius (most movable bone). In some movements this process can be reversed. An example of this reversal can be seen in the pull-up, where the radius is relatively stable and the scapula moves up. Even though in this example the most movable bone is reversed, the proximal attachment of the biceps brachii is always on the scapula and is still considered to be the origin, and the insertion is still on the radius. The biceps brachii would be an extrinsic muscle of the elbow, whereas the brachialis would be intrinsic to the elbow. For each muscle studied, the origin and insertion are indicated.

## Types of muscle contraction (action)

When tension is developed in a muscle as a result of a stimulus, it is known as a contraction. The term *muscle contraction* may be confusing, because in some types of contractions the muscle does not shorten in length as the term *contraction* indicates. As a result, it has become increasingly common to refer to the various types of muscle contractions as *muscle actions* instead.

Muscle contractions can be used to *cause, control,* or *prevent* joint movement. To elaborate, muscle contractions can be used to initiate or accelerate the movement of a body segment, to slow down or decelerate the movement of a body segment, or to prevent movement of a body segment by external forces. All muscle contractions or actions can be classified as either isometric or isotonic. An isometric contraction occurs when tension is developed within the muscle but the joint angles remain constant. Isometric contractions may be thought of as **static** contractions, because a significant amount of active tension may be developed in the muscle to maintain the joint angle in a relatively static or stable position. Isometric contractions may be used to stabilize a body segment to prevent it from being moved by external forces.

Isotonic contractions involve the muscle developing tension to either cause or control joint movement. They may be thought of as **dynamic** contractions, because the varying degrees of active tension in the muscles are either causing the joint angles to change or controlling the joint angle change that is caused by external forces. The isotonic type of muscle contraction is classified further as either concentric or eccentric on the basis of whether shortening or lengthening occurs. Concentric contractions involve the muscle developing active tension as it shortens, whereas eccentric contractions involve the muscle lengthening under active tension. In Fig. 2.3, *A, B, E,* and *F* illustrate isotonic contractions, while *C* and *D* demonstrate isometric contractions.

It is also important to note that movement may occur at any given joint without any muscle contraction whatsoever. Such movement is referred to as passive and is due solely to external forces, such as those applied by another person, object, or resistance, or to the force of gravity in the presence of muscle relaxation.

### Concentric contraction

Concentric contractions involve the muscle developing active tension as it shortens and occur when the muscle develops enough force to overcome the applied resistance. Concentric contractions may be thought of as causing movement against gravity or resistance and are described as positive contractions. The force developed by the muscle is greater than that of the resistance. This results in the joint angle being changed in the direction of the applied muscular force and causes the body part to move against gravity or external forces. Concentric contractions are used to accelerate the movement of a body segment from a lower speed to a higher speed.

### Eccentric contraction (muscle action)

Eccentric contractions involve the muscle lengthening under active tension and occur when the muscle gradually lessens in tension to control the descent of the resistance. The weight or resistance may be thought of as overcoming the muscle contraction, but not to the point that the muscle cannot control the descending movement. Eccentric muscle actions control movement with gravity or resistance and are described as negative contractions. The force developed by the muscle is less than that of the resistance; this results in a change in the joint angle in the direction of the resistance or external force and allows the body part to move with gravity or external forces (resistance). Eccentric contractions are used to decelerate the movement of a body segment from a faster speed to a slower speed or stop the movement of a joint already in motion. Because the muscle is lengthening as opposed to shortening, the relatively recent change in terminology from muscle contraction to muscle action is becoming more commonly accepted.

### Movement differentiation

Some confusion exists regarding body movement and the factors affecting it. Joint movement may occur with muscle groups on either or both sides of the joint actively contracting or even without any muscles contracting. Similarly, when no movement is occurring there may or may not be muscle contraction present, depending on the external forces acting on the joint. To further add to the confusion, a variety of terms and descriptive phrases are used by different authorities to describe these phenomena. Table 2.2 attempts to provide an exhaustive explanation of the various types of contraction and resulting joint movements. The varying terminology utilized in defining and describing these

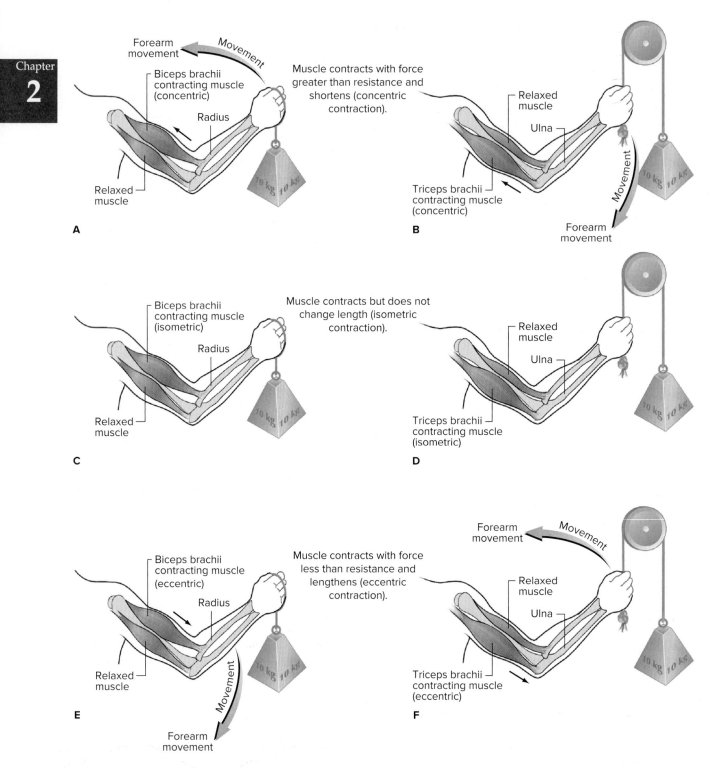

FIG. 2.3 • Agonist–antagonist relationship with isotonic and isometric contractions. **A,** Biceps brachii is agonist in flexing the elbow with a concentric contraction, and triceps brachii is antagonist; **B,** Triceps brachii is agonist in extending the elbow with a concentric contraction, and biceps brachii is antagonist; **C,** Biceps brachii is maintaining the elbow in a flexed position with an isometric contraction, and triceps brachii is antagonist; **D,** Triceps brachii is maintaining the elbow in a flexed position with an isometric contraction, and biceps brachii is antagonist; **E,** Biceps brachii is controlling elbow extension with an eccentric contraction, and triceps brachii is antagonist; **F,** Triceps brachii is controlling elbow flexion with an eccentric contraction, and biceps brachii is antagonist.

## TABLE 2.2 · **Muscle contraction and movement matrix**

| Definitive and descriptive factors | Type of contraction (muscle action) | | | Movement without contraction |
| --- | --- | --- | --- | --- |
| | Isotonic | | | |
| | Isometric | Concentric | Eccentric | |
| Agonist muscle length | No appreciable change | Shortening ➡️⬅️ | Lengthening ⬅️➡️ | Dictated solely by gravity and/or external forces |
| Antagonist muscle length | No appreciable change | Lengthening ⬅️➡️ | Shortening ➡️⬅️ | Dictated solely by gravity and/or external forces |
| Joint angle changes | No appreciable change | In direction of applied muscular force | In direction of external force (resistance) | Dictated solely by gravity and/or external forces |
| Direction of body part | Against immovable object or matched external force (resistance) | Against gravity and/or other external force (resistance) | With gravity and/or other external force (resistance) | Consistent with gravity and/or other external forces |
| Motion | Prevents motion; pressure (force) applied, but no resulting motion | Causes motion | Controls motion | Either no motion or passive motion as a result of gravity and/or other external forces |
| Description | Static; fixating | Dynamic shortening; positive work | Dynamic lengthening; negative work | Passive; relaxation |
| Applied muscle force versus resistance | Force 5 resistance | Force > resistance | Force < resistance | No force, all resistance |
| Speed relative to gravity or applied resistance including inertial forces | Equal to speed of applied resistance | Faster than the inertia of the resistance | Slower than the speed of gravity or applied inertial forces | Consistent with inertia of applied external forces or the speed of gravity |
| Acceleration/ deceleration | Zero acceleration | Acceleration ↗️ | Deceleration ↘️ | Either zero or acceleration consistent with applied external forces |
| Descriptive symbol | (=) | (+) | (–) | (0) |
| Practical application | Prevents external forces from causing movement | Initiates movement or speeds up the rate of movement | Slows down the rate of movement or stops movement, "braking action" | Passive motion by force from gravity and/or other external forces |

actions is included. Appendix 4 provides an algorithm for determining if a muscle or muscle group is contracting and, if so, the type of contraction.

Various exercises may use any one or all of these contraction types for muscle development. Development of exercise machines has resulted in another type of muscle exercise known as **isokinetics**. Isokinetics is not another type of contraction, as some authorities have mistakenly described; rather, it is a specific technique that may use any or all of the different types of contractions. Isokinetics is a type of dynamic exercise usually using concentric and/or eccentric muscle contractions in which the speed (or velocity) of movement is constant and muscular contraction (ideally, maximum contraction) occurs throughout the movement. Biodex, Cybex, and other types of apparatuses are engineered to allow this type of exercise.

Students well educated in kinesiology should be qualified to prescribe exercises and activities for the development of large muscles and of muscle groups in the human body. They should be able to read the description of an exercise or observe an exercise and immediately know the most important muscles being used. Terms describing how muscles function in joint movements follow.

## Role of muscles

When a muscle contracts, it simply attempts to pull the bones to which both of its ends are attached toward each other. Usually this does not happen, however, because one of the bones is usually more stable than the other. As a result, the less stable bone moves toward the more stable bone. When a muscle that is capable of performing multiple actions contracts, it attempts to perform all of its actions unless other forces, such as those provided by other muscles, prevent the undesired actions.

### Agonist    FIG. 2.3

Agonist muscles, when contracting concentrically, cause joint motion through a specified plane of motion. Any concentrically contracting muscle that causes the same joint motion is an agonist for the motion. However, some muscles, because of their relative location, size, length, or force generation capacity, are able to contribute significantly more to the joint movement than other agonists. These muscles are known as prime or **primary movers** or as muscles most involved. Agonist muscles that contribute significantly less to the joint motion are commonly referred to as assisters or assistant movers. Consensus among all authorities regarding which muscles are primary movers and which are weak assistants does not exist in every case. This text will emphasize the primary movers. The remaining agonists or assistants, when listed, will be referred to as weak contributors to the motion involved. As an example, the hamstrings (semitendinosus, semimembranosus, biceps femoris), sartorius, gracilis, popliteus, and gastrocnemius are all agonists in knee flexion, but most kinesiologists regard only the hamstrings as the prime movers.

### Antagonist    FIG. 2.3

Antagonist muscles have the opposite concentric action from agonists. Referred to as contralateral muscles, antagonists are located on the opposite side of the joint from the agonist and work in cooperation with agonist muscles by relaxing and allowing movement; but when contracting concentrically, they perform the joint motion opposite to that of the agonist. Using the previous example, the quadriceps muscles are antagonists to the hamstrings in knee flexion.

### Stabilizers

Stabilizers surround the joint or body part and contract to fixate or stabilize the area to enable another limb or body segment to exert force and move. Known as fixators, they are essential in establishing a relatively firm base for the more distal joints to work from when carrying out movements. In a biceps curling example, the muscles of the scapula and glenohumeral joint must contract in order to maintain the shoulder complex and humerus in a relatively static position so that the biceps brachii can more effectively perform the curls. The antagonists for each motion of the proximal joint co-contract or contract against each other to prevent motion. This is an example of proximal stabilization to enhance the effectiveness of distal joint motion, which occurs commonly with the upper extremity.

### Synergist

Muscles that assist in the action of an agonist but are not necessarily prime movers for the action, known as guiding muscles, assist in refined movement and rule out undesired motion. Synergist muscles may be either helping synergists or true synergists. **Helping synergists** have an action in common but also have actions antagonistic to each other. They help another muscle move the joint in the desired manner and simultaneously prevent undesired actions. An example involves the anterior and posterior deltoid. The anterior deltoid acts as an agonist in glenohumeral flexion, while the posterior deltoid acts as an extensor. Helping each other, they work in synergy with the middle deltoid to accomplish abduction. **True synergists** contract to prevent an undesired joint action of the agonist and have no direct effect on the agonist action. The finger flexors are provided true synergy by the wrist extensors when one is grasping an object. The finger flexors originating on the forearm and humerus are agonists in both wrist flexion and finger flexion. The wrist extensors contract to prevent wrist flexion by the finger flexors. This allows the finger flexors to maintain more of their length and therefore utilize more of their force in flexing the fingers.

### Neutralizers

Neutralizers counteract or neutralize the action of other muscles to prevent undesirable movements such as inappropriate muscle substitutions. They contract to resist specific actions of other muscles. As an example, when only the supination action of the biceps brachii is desired, the triceps brachii contracts to neutralize the flexion action of the biceps brachii. Another example may be seen in the biceps curl, when only the flexion force of the biceps brachii is desired. When the biceps brachii contracts, it normally attempts to both flex the elbow and supinate the forearm. In this case the pronator teres contracts to neutralize the supination component of the biceps.

### Force couples

Force couples occur when two or more forces are pulling in different directions on an object, causing the object to rotate about its axis. Fig. 2.4, *A* depicts a force couple consisting of one hand on each side of a steering wheel. One hand pulls the wheel up and to the right, and the other hand pulls it down and to the left. Coupling of muscular forces in the body can result in a more efficient movement. Figure 2.4, *B* illustrates a force couple in which the middle trapezius, lower trapezius, and serratus anterior all attach at different points on the scapula. Each muscle pulls on the scapula from a different direction to produce the combined result of upward rotation. Another example of muscular force couples is seen in standing, when

the hip flexors (iliopsoas and rectus femoris) are used to pull the front of the pelvis downward and the erector spinae are used to pull the posterior pelvis upward, resulting in anterior pelvic rotation.

## Tying the roles of muscles together

When a muscle with multiple agonist actions contracts, it attempts to perform all its actions. Muscles cannot determine which actions are appropriate for the task at hand. The resulting actions actually performed depend upon several factors, such as the motor units activated, joint position at the time of contraction, planes of motion allowed in the joint, axis of rotation possible in the joint, muscle length, and the relative contraction or relaxation of other muscles acting on the joint. In certain instances, two muscles may work in synergy by counteracting their opposing actions to accomplish a common action.

As discussed, agonist muscles are primarily responsible for a given movement, such as those of hip flexion and knee extension while kicking a ball. In this example, the hamstrings are antagonistic and relax to allow the kick to occur. This does not mean that all other muscles in the hip area are

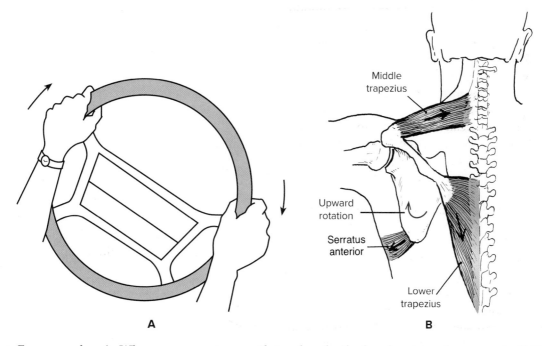

A                                    B

FIG. 2.4 • Force couples. **A,** When a person steers with two hands, the hands act as a force couple; **B,** Two force couples act on the scapula to rotate it upward. The middle trapezius and lower serratus anterior are excellent examples. The middle trapezius and lower trapezius also tend to act as a force couple, although their pulls are not in opposite directions.

uninvolved. The preciseness of the kick depends on the involvement of many other muscles. As the lower extremity swings forward, its route and subsequent angle at the point of contact depend on a certain amount of relative contraction or relaxation in the hip abductors, adductors, internal rotators, and external rotators. These muscles act in a synergistic fashion to guide the lower extremity in a precise manner. That is, they are not primarily responsible for knee extension and hip flexion, but they do contribute to the accuracy of the total movement. These guiding muscles assist in refining the kick and preventing extraneous motions. Additionally, the muscles in the contralateral hip and pelvic area must be under relative tension to help fixate or stabilize the pelvis on that side in order to provide a relatively stable pelvis for the hip flexors on the involved side to contract against. In kicking the ball, the pectineus and tensor fascia latae are adductors and abductors, respectively, in addition to flexors. The actions of adduction and abduction are neutralized by each other, and the common action of the two muscles results in hip flexion.

From a practical point of view, it is not essential that individuals know the exact force exerted by each of the elbow flexors—biceps, brachialis, and brachioradialis—in chinning. It is important to understand that this muscle group is the agonist or primary mover responsible for elbow joint flexion. Similarly, it is important to understand that these muscles contract concentrically when the chin is pulled up to the bar and that they contract eccentrically when the body is lowered slowly. Antagonistic muscles produce actions opposite those of the agonist. For example, the muscles that produce extension of the elbow joint are antagonistic to the muscles that produce flexion of the elbow joint. It is important to understand that specific exercises need to be prescribed for the development of each antagonistic muscle group. The return movement to the hanging position at the elbow joint after chinning is elbow joint extension, but the triceps and anconeus are not being strengthened. A concentric contraction of the elbow joint flexors occurs, followed by an eccentric contraction of the same muscles.

### Reversal of muscle function

A muscle group that is described to perform a given function can contract to control the exact opposite motion. Fig. 2.3, *A* illustrates how the biceps brachii is an agonist by contracting concentrically to flex the elbow. The triceps brachii is an antagonist to elbow flexion, and the pronator teres is considered to be a synergist to the biceps brachii in this example. If the biceps brachii were to slowly lengthen and control elbow extension, as in Fig. 2.3, *E*, it would still be the agonist, but it would be contracting eccentrically. Fig. 2.3, *B* illustrates how the triceps brachii is an agonist by contracting concentrically to extend the elbow. The biceps brachii is an antagonist to elbow extension in this example. If the triceps brachii were to slowly lengthen and control elbow flexion, as in Fig. 2.3, *F*, it would still be the agonist, but it would be eccentrically contracting. In both of these examples, the deltoid, trapezius, and various other shoulder muscles are serving as stabilizers of the shoulder area.

## Determination of muscle action

The specific action of a muscle may be determined through a variety of methods. These include considering anatomical lines of pull, anatomical dissection, palpation, models, electromyography, and electrical stimulation.

With an understanding of a muscle's line of pull relative to a joint, one may determine the muscle's action at the joint. (See lines of pull below.) Although not available to all students, cadaver dissection of muscles and joints is an excellent way to further understand muscle action.

For most of the skeletal muscles, **palpation** is a very useful way to determine muscle action. It is done through using the sense of touch to feel or examine a muscle as it contracts. Palpation is limited to superficial muscles but is helpful in furthering an understanding of joint mechanics. Models such as long rubber bands may be used to facilitate understanding of lines of pull and to simulate muscle lengthening or shortening as joints move through various ranges of motion.

Electromyography (EMG) utilizes either surface electrodes that are placed over the muscle or fine wire/needle electrodes placed into the muscle. As the subject then moves the joint and contracts the muscles, the EMG unit detects the action potentials of the muscles and provides an electronic readout of the contraction intensity and duration. EMG is the most accurate way of detecting the presence and extent of muscle activity.

Electrical muscle stimulation is somewhat a reverse approach of electromyography. Instead of electricity being used to detect muscle action, it is used to cause muscle activity. Surface electrodes are placed over a muscle, and then the stimulator causes the muscle to contract. The joint's actions may then be observed to see the effect of the muscle's contraction on it.

## Lines of pull   FIG. 2.5

Combining the knowledge of a particular joint's functional design and diarthrodial classification with an understanding of the specific location of a musculotendinous unit as it crosses a joint is extremely helpful in understanding its action on the joint. For example, knowing that the rectus femoris has its origin on the anterior inferior iliac spine and its insertion on the tibial tuberosity via the patella, you can then determine that the muscle must have an anterior relationship to the knee and hip. Combining this knowledge with the knowledge that both joints are capable of sagittal plane movements such as flexion/extension, you can then determine that when the rectus femoris contracts concentrically, it should cause the knee to extend and the hip to flex.

Furthermore, knowing that the semitendinosus, semimembranosus, and biceps femoris all originate on the ischial tuberosity and that the semitendinosus and semimembranosus cross the knee posteromedially before inserting on the tibia, but that the biceps femoris crosses the knee posterolaterally before inserting on the fibula head, you may determine that all three muscles have posterior relationships to the hip and knee, which would enable them to be hip extensors and knee flexors upon concentric contraction. The specific knowledge related to their distal attachments and the knee's ability to rotate when flexed would allow you to determine that the semitendinosus and semimembranosus will cause internal rotation, whereas the biceps femoris will cause external rotation. Knowledge that the knee's axes of rotation are only frontal and vertical, but not sagittal, enables you to determine that even though the semitendinosus and semimembranosus have a posteromedial line of pull and the biceps femoris has a posterolateral line of pull, they are not capable of causing knee adduction and abduction, respectively.

You can also apply this concept in reverse. For example, if the only action of a muscle such as the brachialis is known to be elbow flexion, then you should be able to determine that its line of pull must be anterior to the joint. Additionally, you would know that the origin of the brachialis must be somewhere on the anterior humerus and the insertion must be somewhere on the anterior ulna.

Consider all the following factors and their relationships as you study movements of the body to gain a more thorough understanding.

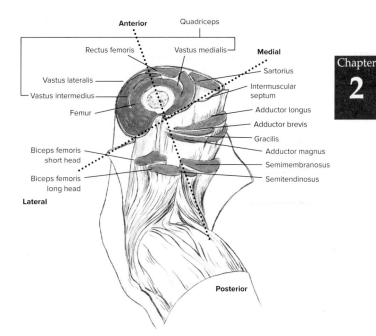

FIG. 2.5 • Lines of pull in relation to the left knee. Biceps femoris with a posterolateral relationship enables it to externally rotate the knee; semitendinosus and semimembranosus have a posteromedial relationship enabling them to internally rotate the knee; hamstrings (biceps femoris, semitendinosus, and semimembranosus) all have a posterior relationship enabling them to flex the knee; quadriceps muscles have an anterior relationship enabling them to extend the knee.

1. Exact locations of bony landmarks to which muscles attach proximally and distally and their relationship to joints
2. The planes of motion through which a joint is capable of moving
3. The muscle's relationship or line of pull relative to the joint's axes of rotation
4. As a joint moves through a particular range of motion, the ability of the line of pull of a particular muscle to change and even result in the muscle having a different or opposite action than in the original position
5. The potential effect of other muscles' relative contraction or relaxation on a particular muscle's ability to cause motion
6. The effect of a muscle's relative length on its ability to generate force (See muscle length–tension relationship, p. 59, and active and passive insufficiency, p. 63.)
7. The effect of the position of other joints on the ability of a biarticular or multiarticular

muscle to generate force or allow lengthening (See uniarticular, biarticular, and multiarticular muscles, p. 62.)

(See uniarticular, biarticular, and multiarticular muscles, p. 62.)

# Neural control of voluntary movement

When we discuss muscular activity, we should really state it as neuromuscular activity, since muscle cannot be active without nervous innervation. All voluntary movement is a result of the muscular and the nervous systems working together. All muscle contraction occurs as a result of stimulation from the nervous system. Ultimately, every muscle fiber is innervated by a somatic motor neuron, which, when an appropriate stimulus is provided, results in a muscle contraction. Depending upon a variety of factors, this stimulus may be processed in varying degrees at different levels of the **central nervous system (CNS)**. The CNS, for the purposes of this discussion, may be divided into five levels of control. Listed in order from the most general level of control and the most superiorly located to the most specific level of control and the most inferiorly located, these levels are the cerebral cortex, the basal ganglia, the cerebellum, the brain stem, and the spinal cord.

The **cerebral cortex**, the highest level of control, provides for the creation of voluntary movement as aggregate muscle action but not as specific muscle activity. Sensory stimuli from the body also are interpreted here, to a degree, for the determination of needed responses.

At the next level, the **basal ganglia** control the maintenance of postures and equilibrium and learned movements such as driving a car. Sensory integration for balance and rhythmic activities is controlled here.

**Cervical plexus (C1–C4)**
Ansa cervicalis
Lesser occipital nerve
Transverse cervical nerve
Supraclavicular nerve
Phrenic nerve

**Brachial plexus (C5–T1)**
Axillary nerve
Radial nerve
Musculocutaneous nerve
Median nerve
Ulnar nerve

**Intercostal (thoracic) nerves**

Dura mater of spinal cord

**Lumbar plexus (L1–L4)**
Iliohypogastric nerve
Ilioinguinal nerve
Genitofemoral nerve
Lateral femoral cutaneous nerve
Femoral nerve
Obturator nerve

**Sacral plexus (L5–S4)**
Sciatic nerve — Common peroneal nerve / Tibial nerve
Posterior cutaneous femoral nerve
Pudendal nerve

**Lumbosacral plexus**

Atlas (first cervical vertebra)
**Cervical nerves (8 pairs)**
Cervical enlargement
First thoracic vertebra
**Thoracic nerves (12 pairs)**
Lumbar enlargement
First lumbar vertebra
Conus medullaris
**Lumbar nerves (5 pairs)**
Cauda equina
Ilium
Sacrum
**Sacral nerves (5 pairs)**
**Coccygeal nerves (1 pair)**
Filum terminale

FIG. 2.6 • Spinal nerve roots and plexuses.

The **cerebellum** is a major integrator of sensory impulses and provides feedback relative to motion. It controls the timing and intensity of muscle activity to assist in the refinement of movements.

Next, the **brain stem** integrates all central nervous system activity through excitation and inhibition of desired neuromuscular actions and functions in arousal or maintaining a wakeful state.

Finally, the **spinal cord** is the common pathway between the CNS and the **peripheral nervous system (PNS)**, which contains all the remaining nerves throughout the body. It has the most specific control and integrates various simple and complex spinal reflexes, as well as cortical and basal ganglia activity.

Functionally, the PNS can be divided into sensory and motor divisions. The sensory or **afferent nerves** bring impulses from receptors in the skin,

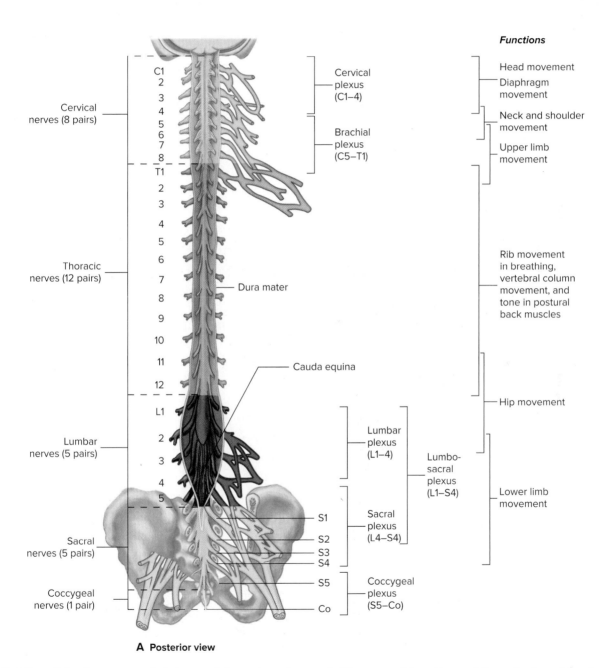

**A** Posterior view

FIG. 2.7 • Spinal nerves and plexus and dermatomal map. **A,** Nerves and functions of the spinal cord (regions color-coded). The regional designations and the numbers of the spinal nerves are shown on the left. The plexuses formed by the spinal nerves and function are both shown on the right. **B,** Letters and numbers indicate the spinal nerves innervating a given region.

joints, muscles, and other peripheral aspects of the body to the CNS, while the motor or **efferent nerves** carry impulses to the outlying regions of the body.

The spinal nerves, illustrated in Fig. 2.6, also provide both motor and sensory function for their respective portions of the body and are named for the locations from which they exit the vertebral column. From each side of the spinal column, there are 8 cervical nerves, 12 thoracic nerves, 5 lumbar nerves, 5 sacral nerves, and 1 coccygeal nerve. Cervical nerves 1 through 4 form the cervical plexus, which is generally responsible for sensation from the upper part of the shoulders to the back of the head and front of the neck. The cervical plexus supplies motor innervation to several muscles of the neck. Cervical nerves 5 through 8, along with thoracic

nerve 1, form the brachial plexus, which supplies motor and sensory function to the upper extremity and most of the scapula. Thoracic nerves 2 through 12 run directly to specific anatomical locations in the thorax. All of the lumbar, sacral, and coccygeal nerves form the lumbosacral plexus, which supplies sensation and motor function to the lower trunk and the entire lower extremity and perineum.

One aspect of the sensory function of spinal nerves is to provide feedback to the CNS regarding skin sensation. A defined area of skin supplied by a specific spinal nerve is known as a **dermatome** (Fig. 2.7). Regarding motor function of spinal nerves, a **myotome** is defined as a muscle or group of muscles supplied by a specific spinal nerve. Certain spinal nerves are also responsible for reflexes. Table 2.3 summarizes the specific spinal nerve functions.

FIG. 2.7 (continued) • Spinal nerves and plexus dermatomal map. **A,** Nerves and functions of the spinal cord (regions color-coded). The regional designations and the numbers of the spinal nerves are shown on the left. The plexuses formed by the spinal nerves and function are both shown on the right. **B,** Letters and numbers indicate the spinal nerves innervating a given region.

TABLE 2.3 · **Spinal nerve root dermatomes, myotomes, reflexes, and functional applications**

| | Nerve root | Dermatome afferent (sensory) | Myotome efferent (motor) | Reflexes | Functional application |
|---|---|---|---|---|---|
| Cervical plexus | C1 | Touch: Vertex of skull | Upper neck muscles | None | Capital flexion and extension |
| | C2 | Touch: Temple, forehead, occiput | Upper neck muscles | None | Sensation behind the ear and posterior skull<br>Capital and upper cervical movements |
| | C3 | Touch: Entire neck, posterior cheek, temporal area, under mandible | Trapezius, splenius, capitis | None | Scapula retraction, neck extension<br>Sensation to cheek and side of neck |
| | C4 | Touch: Shoulder area, clavicular area, upper scapular area | Trapezius, levator scapulae | None | Scapula retraction and elevation<br>Sensation to clavicle and upper scapula |
| Brachial plexus | C5 | Touch: Deltoid area, anterior aspect of entire arm to base of thumb | Supraspinatus, infraspinatus, deltoid, biceps brachii | Biceps brachii | Shoulder abduction<br>Sensation to lateral side of arm and elbow |
| | C6 | Touch: Anterior arm, radial side of hand to thumb and index finger | Biceps brachii, supinator, wrist extensors | Biceps brachii, brachioradialis | Elbow flexion, wrist extension<br>Sensation to lateral side of forearm including thumb and index fingers |
| | C7 | Touch: Lateral arm and forearm to index, long, and ring fingers | Triceps brachii, wrist flexors | Triceps brachii | Elbow extension, wrist flexion<br>Sensation to middle of anterior forearm and long finger |
| | C8 | Touch: Medial side of forearm to ring and little fingers | Ulnar deviators, thumb extensors, thumb adductors (rarely triceps brachii) | None | Wrist ulnar deviation, thumb extension<br>Sensation to posterior elbow and medial forearm to little fingers |
| | T1 | Touch: Medial arm and forearm to wrist | Intrinsic muscles of the hand except for opponens pollicis and abductor pollicis brevis | None | Abduction and adduction of fingers<br>Sensation to medial arm and elbow |
| Thoracic | T2 | Touch: Medial side of upper arm to medial elbow, pectoral, and midscapular areas | Intercostal muscles | None | Sensation to medial upper arm, upper chest, and midscapular area |
| | T3–T12 | Touch: T3–T6, upper thorax; T5–T7, coastal margin; T8–T12, abdomen and lumbar region | Intercostal muscles, abdominal muscles | None | Sensation to chest, abdomen, and low back |

The basic functional units of the nervous system responsible for generating and transmitting impulses are nerve cells known as **neurons**. Neurons consist of a **neuron cell body**; one or more branching projections known as **dendrites**, which transmit impulses to the neuron and cell body; and an **axon**, an elongated projection that transmits impulses away from neuron cell bodies. As shown in Fig. 2.8, neurons are classified into three types, according to the direction in which they transmit impulses. **Sensory neurons** transmit impulses to the spinal cord and brain from all parts of the body, whereas **motor neurons** transmit impulses away from the brain and spinal cord to muscle and glandular tissue. **Interneurons** are central or connecting neurons that conduct impulses from sensory neurons to motor neurons.

**TABLE 2.3** (continued) • **Spinal nerve root dermatomes, myotomes, reflexes, and functional applications**

| | Nerve root | Dermatome afferent (sensory) | Myotome efferent (motor) | Reflexes | Functional application |
|---|---|---|---|---|---|
| Lumbosacral plexus | L1 | Touch: Lower abdomen, groin, lumbar region from 2nd to 4th vertebrae, upper and outer aspect of buttocks | Quadratus lumborum | None | Sensation to low back, over trochanter and groin |
| | L2 | Touch: Lower lumbar region, upper buttock, anterior aspect of thigh | Iliopsoas, quadriceps | None | Hip flexion<br>Sensation to back, front of thigh to knee |
| | L3 | Touch: Medial aspect of thigh to knee, anterior aspect of lower 1/3 of the thigh to just below patella | Psoas, quadriceps | Patella or knee extensors | Hip flexion and knee extension<br>Sensation to back, upper buttock, anterior thigh and knee, medial lower leg |
| | L4 | Touch: Medial aspect of lower leg and foot, inner border of foot, great toe | Tibialis anterior, extensor hallucis and digitorum longus, peroneals | Patella or knee extensors | Ankle dorsiflexion, transverse tarsal/subtalar inversion<br>Sensation to medial buttock, lateral thigh, medial leg, dorsum of foot, great toe |
| | L5 | Touch: Lateral border of leg, anterior surface of lower leg, top of foot to middle three toes | Extensor hallucis and digitorum longus, peroneals, gluteus maximus and medius, dorsiflexors | None | Great toe extension, transverse tarsal/subtalar eversion<br>Sensation to upper lateral leg, anterior surface of the lower leg, middle three toes |
| | S1 | Touch: Posterior aspect of the lower 1/4 of the leg, posterior aspect of the foot, including the heel, lateral border of the foot and sole | Gastrocnemius, soleus, gluteus maximus and medius, hamstrings, peroneals | Achilles reflex | Ankle plantar flexion, knee flexion, transverse tarsal/subtalar eversion<br>Sensation to lateral leg, lateral foot, lateral two toes, plantar aspect of foot |
| | S2 | Touch: Posterior central strip of the leg from below the gluteal fold to 3/4 of the way down the leg | Gastrocnemius, soleus, gluteus maximus, hamstrings | None | Ankle plantar flexion and toe flexion<br>Sensation to posterior thigh and upper posterior leg |
| | S3 | Touch: Groin, medial thigh to knee | Intrinsic foot muscles | None | Sensation to groin and adductor region |
| | S4 | Touch: Perineum, genitals, lower sacrum | Bladder, rectum | None | Urinary and bowel control<br>Sensation to saddle area, genitals, anus |
| Coccygeal plexus | S5 | Touch: Perianal | Coccygeus muscle | None | Pelvic floor muscles, bladder, sphincters and external genitalia |
| | Co | Skin over the back of the coccyx | None | None | Sensation to skin over the coccyx |

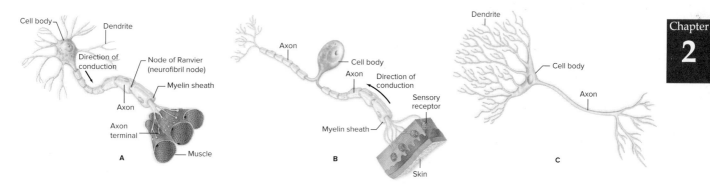

FIG. 2.8 • Neuron anatomy. **A,** Motor neuron. Note the branched dendrites and the single long axon, which branches only near its tip; **B,** Sensory neuron with dendritelike structures projecting from the peripheral end of the axon; **C,** Interneuron (from the cortex of the cerebellum) with very highly branched dendrites.

## Proprioception and kinesthesis

The performance of various activities is significantly dependent upon neurological feedback from the body. Very simply, we use the various senses to determine a response to our environment, as when we use sight to know when to lift our hand to catch a fly ball. We are familiar with the senses of smell, touch, sight, hearing, and taste. We are also aware of other sensations, such as pain, pressure, heat, and cold, but we often take for granted the sensory feedback provided by proprioceptors during neuromuscular activity. Proprioceptors are internal receptors located in the skin, joints, muscles, and tendons that provide feedback relative to the tension, length, and contraction state of muscle, the position of the body and limbs, and movements of the joints. These proprioceptors in combination with the other sense organs of the body are vital in **kinesthesis**, the conscious awareness of the position and movement of the body in space. For example, if standing on one leg with the other knee flexed, you do not have to look at your non-weight-bearing leg to know the approximate number of degrees that you may have it flexed. The proprioceptors in and around the knee provide information so that you are kinesthetically aware of your knee position. Muscle spindles and Golgi tendon organs (GTO) are proprioceptors specific to the muscles, whereas Meissner's corpuscles, Ruffini's corpuscles, Pacinian corpuscles, and Krause's end-bulbs are proprioceptors specific to the joints and skin.

While kinesthesis is concerned with the conscious awareness of the body's position, **proprioception** is the subconscious mechanism by which the body is able to regulate posture and movement by responding to stimuli originating in the proprioceptors embedded in the joints, tendons, muscles, and inner ear. When we unexpectedly step on an unlevel or unstable surface, if we have good proprioception the muscles in and about our lower extremity may respond very quickly by contracting appropriately to prevent a fall or injury. This protective response of the body occurs without our having time to make a conscious decision about how to respond.

Muscle spindles (Fig. 2.9), concentrated primarily in the muscle belly between the fibers, are sensitive to stretch and rate of stretch. Specifically, they insert into the connective tissue within the muscle and run parallel with the muscle fibers. The number of spindles in a particular muscle varies depending upon the level of control needed for the area. Consequently, the concentration of muscle spindles in the hands is much greater than in the thigh.

When rapid stretch occurs, an impulse is sent to the CNS. The CNS then activates the motor neurons of the muscle and causes it to contract. All muscles possess this **myotatic** or **stretch reflex**, but it is most remarkable in the extensor muscles of the extremities. The knee jerk or patellar tendon reflex is an example, as shown in Fig. 2.10. When the reflex hammer strikes the patellar tendon, it causes a quick stretch of the musculotendinous unit of the quadriceps. In response, the quadriceps fires and the knee extends. To an extent, the more sudden the tap of the hammer, the more significant the reflexive contraction. A more practical example is seen in maintaining posture, as when a student begins to doze off in class. As the head starts to nod forward, a sudden stretch is placed on the neck

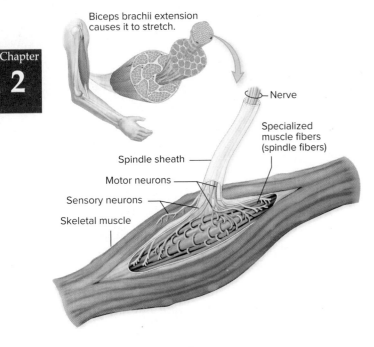

Biceps brachii extension causes it to stretch.

Nerve

Specialized muscle fibers (spindle fibers)

Spindle sheath

Motor neurons

Sensory neurons

Skeletal muscle

FIG. 2.9 • Muscle spindles.

extensors, which activates the muscle spindles and ultimately results in a sudden jerk back to an extended position.

The stretch reflex provided by the muscle spindle may be utilized to facilitate a greater response, as in the case of a quick, short squat before attempting a jump. The quick stretch placed upon the muscles in the squat enables the same muscles to generate more force in the subsequent jump off the floor.

The Golgi tendon organs (Fig. 2.11), serially located in the tendon close to the muscle–tendon junction, are continuously sensitive to both muscle tension and active contraction. The GTO is much less sensitive to stretch than muscle spindles are and requires a greater stretch to be activated. Tension in tendons and consequently in the GTO increases as the muscle contracts, which in turn activates the GTO. When the GTO stretch threshold is reached, an impulse is sent to the CNS, which in turn causes the muscle to relax and facilitates activation of the antagonists as a protective mechanism. That is, the GTO, through this inverse stretch reflex, protects us from excessive contraction by causing the muscle it supplies to relax. As an example, when a weight lifter attempts a very heavy resistance in the biceps

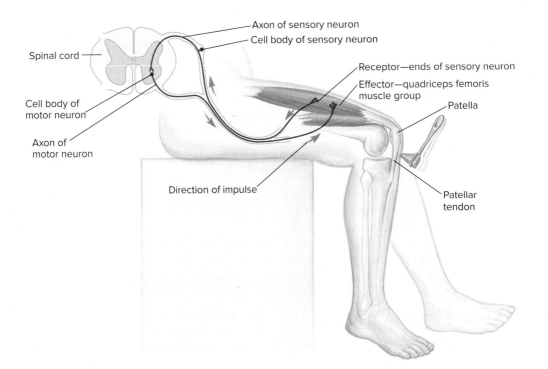

Axon of sensory neuron

Cell body of sensory neuron

Spinal cord

Receptor—ends of sensory neuron

Effector—quadriceps femoris muscle group

Patella

Cell body of motor neuron

Axon of motor neuron

Direction of impulse

Patellar tendon

FIG. 2.10 • Knee jerk, or patellar tendon reflex. A sudden tap on the patellar tendon causes a quick stretch of the quadriceps, which activates the muscle spindle. The information regarding the stretch is sent via the axon of the sensory neuron to the spinal cord, where it synapses with a motor neuron that, in turn, carries via its axon a motor response for the quadriceps to contract.

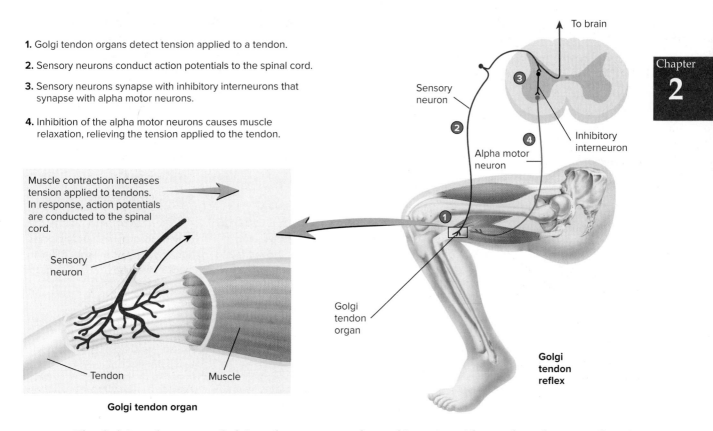

1. Golgi tendon organs detect tension applied to a tendon.

2. Sensory neurons conduct action potentials to the spinal cord.

3. Sensory neurons synapse with inhibitory interneurons that synapse with alpha motor neurons.

4. Inhibition of the alpha motor neurons causes muscle relaxation, relieving the tension applied to the tendon.

Muscle contraction increases tension applied to tendons. In response, action potentials are conducted to the spinal cord.

Sensory neuron

Tendon

Muscle

**Golgi tendon organ**

To brain

Sensory neuron

Inhibitory interneuron

Alpha motor neuron

Golgi tendon organ

**Golgi tendon reflex**

FIG. 2.11 • The Golgi tendon organ. Golgi tendon organs are located in series with muscle and serve as "tension monitors" that act as a protective device for muscle.

curl and reaches the point of extreme overload, the GTO is activated, the biceps suddenly relaxes, and the triceps brachii contracts. This is why it appears as if the lifter is throwing the weight down.

Pacinian corpuscles, concentrated around joint capsules, ligaments, and tendon sheaths and beneath the skin, are activated by rapid changes in the joint angle and by pressure changes affecting the capsule. This activation lasts only briefly and is not effective in detecting constant pressure. Pacinian corpuscles are helpful in providing feedback regarding the location of a body part in space following quick movements such as running or jumping.

Ruffini's corpuscles, located in deep layers of the skin and the joint capsule, are activated by strong and sudden joint movements as well as by pressure changes. Compared to Pacinian corpuscles, their reaction to pressure changes is slower to develop, but their activation is continued as long as pressure is maintained. They are essential in detecting even minute joint position changes and providing information as to the exact joint angle.

Meissner's corpuscles and Krause's end-bulbs are located in the skin and in subcutaneous tissues. They are important in receiving stimuli from touch, but they are not so relevant to our discussion of kinesthesis. See Table 2.4 for further comparisons of sensory receptors.

The quality of movement and how we react to position change are significantly dependent upon proprioceptive feedback from the muscles and joints. Like the other factors involving body movement, proprioception may be enhanced through specific training that utilizes the proprioceptors to a high degree, such as balancing and functional activities. Attempting to maintain your balance on one leg, first with the eyes open on a level surface, may serve as an initial low-level proprioceptive activity, which may eventually progress to a much higher level such as balancing on an unlevel, unstable surface with your eyes closed. There are numerous additional proprioceptive training activities that are limited only by the imagination and the level of proprioception.

TABLE 2.4 • **Sensory receptors**

| Receptors | Sensitivity | Location | Response |
|---|---|---|---|
| Muscle spindles | Subconscious muscle sense, muscle length changes | In skeletal muscles among muscle fibers in parallel with fibers | Initiate rapid contraction of stretched muscle<br><br>Inhibit development of tension in antagonistic muscles |
| Golgi tendon organs | Subconscious muscle sense, muscle tension changes | In tendons, near muscle–tendon junction in series with muscle fibers | Inhibit development of tension in stretched muscles<br><br>Initiate development of tension in antagonistic muscles |
| Pacinian corpuscles | Rapid changes in joint angles, pressure, vibration | Subcutaneous, submucosa, and subserous tissues around joints and external genitals, mammary glands | Provide feedback regarding location of body part in space following quick movements |
| Ruffini's corpuscles | Strong, sudden joint movements, touch, pressure | Skin and subcutaneous tissue of fingers, collagenous fibers of the joint capsule | Provide feedback as to exact joint angle |
| Meissner's corpuscles | Fine touch, vibration | In skin | Provide feedback regarding touch, two-point discrimination |
| Krause's end-bulbs | Touch, thermal change | Skin, subcutaneous tissue, lip and eyelid mucosa, external genitals | Provide feedback regarding touch |

## Neuromuscular concepts

### Motor units and the all or none principle

Each muscle cell is connected to a motor neuron at the **neuromuscular junction**. A **motor unit**, shown in Fig. 2.12, consists of a single motor neuron and all the muscle fibers it innervates. Motor units function as a single unit. When a particular muscle contracts, the contraction actually occurs at the muscle fiber level within a particular motor unit. In a typical muscle contraction, the number of motor units responding and consequently the number of muscle fibers contracting within the muscle may vary significantly, from relatively few to virtually all of the muscle fibers, depending on the number of muscle fibers within each activated motor unit and the number of motor units activated (Fig. 2.13). Regardless of the number involved, the individual muscle fibers within a given motor unit will fire and contract either maximally or not at all. This is referred to as the **all or none principle**.

### Muscle fiber type

Most agree that humans have three types of muscle fiber—two subtypes of fast fibers known as type IIa and type IIb (more recently referred to as type

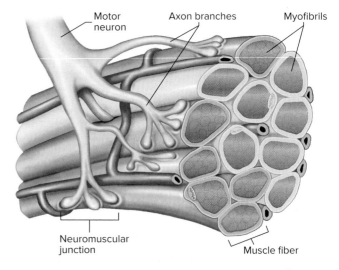

FIG. 2.12 • Motor unit. A motor unit consists of a single motor neuron and all the muscle fibers its branches innervate.

IIx), and a slow fiber known as type I. Fast fibers can produce greater forces due to a greater shortening velocity but fatigue more quickly than slow fibers. Slow muscle fibers have a higher resistance to fatigue, but generally produce less tension than fast fibers.

**Factors affecting muscle tension development**

The difference between a particular muscle contracting to lift a minimal resistance and the same muscle contracting to lift a maximal resistance lies in the number of muscle fibers recruited. The number of muscle fibers recruited may be increased by activating those motor units containing a greater number of muscle fibers, by activating more motor units, or by increasing the frequency of motor unit activation. The number of muscle fibers per motor unit varies significantly from fewer than 10 in muscles requiring a very precise and detailed response, such as the muscles of the eye, to as many as a few thousand in large muscle groups, such as the quadriceps, that perform less complex activities.

Additionally, recruitment of motor units containing fast muscle fibers is helpful in developing greater tension. Finally, recruitment of muscle fibers that are at optimal length can help generate greater muscle tension. Tension development is minimized in muscle fibers that are shortened to around 60% of their resting length, and muscle fibers stretched beyond 130% of their resting length are significantly compromised in their tension development capability. (See Muscle length–tension relationship, p. 59.)

FIG. 2.13 • The number and size of motor units. **A,** Precise muscle contractions require smaller motor units; **B,** Large muscle movements require larger motor units.

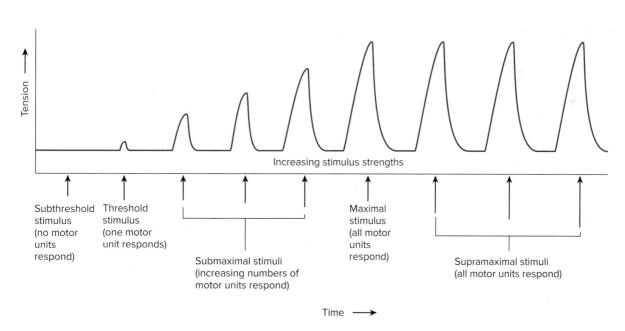

FIG. 2.14 • Achieving threshold stimulus and the effect on increasing tension of recruiting more motor units. If the stimulus does not reach threshold, there is no motor unit response. As the stimulus strength increases, more motor units are recruited until eventually all motor units are recruited and maximal tension of the muscle is generated. Increasing stimulus strength beyond this point has no effect.

For the muscle fibers in a particular motor unit to contract, the motor unit must first receive a stimulus via an electrical signal known as an **action potential** from the brain and spinal cord through its axons. If the stimulus is not strong enough to cause an action potential, it is known as a **subthreshold stimulus** and does not result in a contraction. When the stimulus becomes strong enough to produce an action potential in a single motor unit axon, it is known as a **threshold stimulus**, and all the muscle fibers in the motor unit contract. Stimuli that are stronger to the point of producing action potentials in additional motor units are known as **submaximal stimuli**. For action potentials to be produced in all the motor units of a particular muscle, a **maximal stimulus** is required. As the strength of the stimulus increases from threshold up to maximal, more motor units are recruited, and the overall force of the muscle contraction increases in a graded fashion. Increasing the stimulus beyond maximal has no effect. The effect of increasing the number of motor units activated is detailed in Fig. 2.14.

Greater contraction forces may also be achieved by increasing the frequency of motor unit activation. To simplify the phases of a single muscle fiber contraction or twitch, a stimulus is provided and followed by a brief **latent period** of a few milliseconds. Then the second phase, known as the **contraction phase**, begins and the muscle fiber starts shortening. The contraction phase lasts about 40 milliseconds and is followed by the **relaxation phase**, which lasts approximately 50 milliseconds. This sequence is illustrated in Fig. 2.15. When successive stimuli are provided before the relaxation phase of the first twitch is complete, the subsequent twitches combine with the first to produce a sustained contraction. This **summation** of contractions generates greater tension than a single contraction would produce on its own. As the frequency of stimuli increases, the resultant summation increases accordingly, producing increasingly greater total muscle tension. If the stimuli are provided at a frequency high enough that no relaxation can occur between contractions, then **tetanus** results. Fig. 2.16 illustrates the effect of increasing the rate of stimulation to gain increased muscle tension.

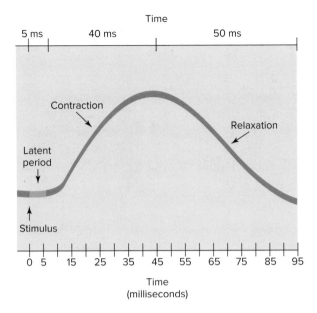

FIG. 2.15 • A recording of a simple twitch. Note the three time periods (latent period, contraction, and relaxation) following the stimulus.

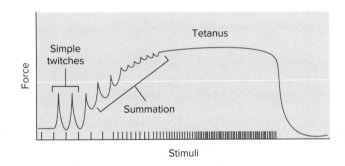

FIG. 2.16 • Recording showing the change from simple twitches to summation and finally tetanus. Peaks to the left represent simple twitches. Increasing the frequency of the stimulus results in a summation of the twitches and finally tetanus.

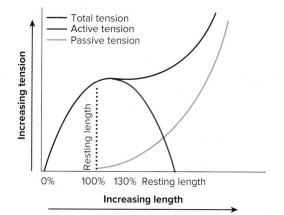

FIG. 2.17 • Treppe. When a rested muscle is stimulated repeatedly with a maximal stimulus at a frequency that allows complete relaxation between stimuli, the second contraction produces a slightly greater tension than the first, and the third contraction produces greater tension than the second. After a few contractions, the tension produced by all contractions is equal.

FIG. 2.18 • Muscle length–tension relationship. As the length increases, the amount of active tension that can be developed increases until approximately 130% of the muscle's resting length is reached. After this point, further increases in length result in decreased ability to generate active tension. Passive tension begins to increase on the muscle lengthened beyond its resting length.

**Treppe**, another phenomenon of muscle contraction, occurs when multiple maximal stimuli are provided to rested muscle at a low enough frequency to allow complete relaxation between contractions. Slightly greater tension is produced by the second stimulus than by the first. A third stimulus produces even greater tension than the second. This staircase effect, illustrated in Fig. 2.17, occurs only with the first few stimuli, with the resultant contractions after the initial ones resulting in equal tension being produced.

### Muscle length–tension relationship

Tension in a muscle can be thought of as a pulling force. Tension may be either passive through external applied forces or active via muscle contraction. **Passive tension** is developed as a muscle is stretched beyond its normal resting length. As it becomes stretched further, its passive tension increases similarly to that of a rubber band being stretched. **Active tension** is dependent on the number of motor units and their respective muscle fibers recruited in a given contraction. However, the length of the muscle during the contraction is a factor in the amount of active tension the muscle may be able to generate.

Generally, depending on the particular muscle involved, the greatest amount of active tension can be developed when a muscle is stretched to between 100% and 130% of its resting length. As a muscle is stretched beyond this point, the amount

of active tension it can generate decreases significantly. Likewise, there is a proportional decrease in the ability to develop tension as a muscle is shortened. When a muscle is shortened to around 50% to 60% of its resting length, its ability to develop contractile tension is essentially reduced to zero.

In the preparatory phase of most sporting activities, we generally place an optimum stretch on the muscles we intend to contract forcefully in the subsequent movement or action phase of the skill. The various phases of performing a movement skill are discussed in much greater detail in Chapter 12. This principle may be seen at work when we squat slightly to stretch the calf, hamstrings, and quadriceps before contracting them concentrically to jump. If we do not first lengthen these muscles through squatting slightly, they are unable to generate enough contractile force to allow us to jump very high. If we squat fully and lengthen the muscles too much, we lose the ability to generate as much force and cannot jump as high.

We can take advantage of this principle by effectively reducing the contribution of some muscles in a group by placing them in a shortened state so that we can isolate the work to those muscle(s) remaining in the lengthened state. For example, in hip extension, we may isolate the work of the gluteus maximus as a hip extensor by maximally shortening the hamstrings with flexion of the knee to reduce their ability to act as hip extensors. See Figs. 2.18 and 2.19.

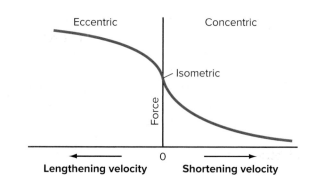

A    B    C

Courtesy of R.T. Floyd (all)

FIG. 2.19 • Practical application of the muscle length–tension relationship involving the calf, hamstrings, and quadriceps muscle groups in jumping. **A,** The muscles are in a relatively shortened position and consequently are not able to generate much tension upon contraction; **B,** The muscles are in a more optimally lengthened position to generate significant tension to jump high; **C,** The muscles are lengthened too much and are not able to generate as much force as in B.

FIG. 2.20 • Muscle–force velocity relationship. From right to left: When little force is needed to move relatively light loads, the muscles may contract concentrically at a relatively high velocity. As the amount of force needed increases with greater loads, the velocity of the concentric contraction decreases proportionally. The amount of force needed continues to increase as the load increases until eventually the load cannot be moved, resulting in zero velocity and an isometric contraction. When the muscle can no longer generate the amount of force needed to maintain the load in a static position, the muscle begins eccentrically contracting to control the velocity, and it can do so at a relatively slow velocity. As the amount of force needed increases to control greater loads, the velocity increases proportionally.

### Muscle force–velocity relationship

When the muscle is either concentrically or eccentrically contracting, the rate of length change is significantly related to the amount of force potential. When contracting concentrically against a light resistance, the muscle is able to contract at a high velocity. As the resistance increases, the maximal velocity at which the muscle is able to contract decreases. Eventually, as the load increases, the velocity decreases to zero, resulting in an isometric contraction.

As the load increases even further beyond that which the muscle can maintain with an isometric contraction, the muscle begins to lengthen, resulting in an eccentric contraction or action. A slight increase in the load will result in a relatively low velocity of lengthening. As the load increases even further, the velocity of lengthening will increase as well. Eventually, the load may increase to the point where the muscle can no longer resist. This will result in uncontrollable lengthening or, more likely, dropping of the load.

From this explanation you can see that there is an inverse relationship between concentric velocity and force production. As the force needed to cause movement of an object increases, the

velocity of concentric contraction decreases. Furthermore, there is a somewhat proportional relationship between eccentric velocity and force production. As the force needed to control the movement of an object increases, the velocity of eccentric lengthening increases, at least until the point at which control is lost. This is illustrated in Fig. 2.20.

### Stretch-shortening cycle

In addition to the previously discussed factors affecting the force generation capabilities of muscle, the sequencing and timing of contractions can enhance the total amount of force produced. When a muscle is suddenly stretched, resulting in an eccentric contraction that is followed by a concentric contraction of the same muscle, the total force generated in the concentric contraction is greater than that of an isolated concentric contraction. This is often referred to as the **stretch-shortening cycle** and functions by integration of the Golgi tendon organ (GTO) and the muscle spindle. Elastic energy is stored during the eccentric stretch phase, transitioned, and utilized in the concentric contraction phase. A stretch reflex is elicited in the eccentric phase of the motion, which subsequently increases the activation of the muscle that was stretched, resulting in a more forceful concentric contraction. For this to be effective, the transition phase must be immediate or the potential energy gained in the eccentric phase will be lost as heat. The shorter the transition phase, the more effective the force production. This is the basis of plyometric training. An example may be seen when a jumper moves quickly downward immediately prior to jumping upward, resulting in a greater jumping height.

### Reciprocal inhibition or innervation

As stated earlier, antagonist muscle groups must relax and lengthen when the agonist muscle group contracts. This effect, called reciprocal innervation, occurs through reciprocal inhibition of the antagonists. Activation of the motor units of the agonists causes a reciprocal neural inhibition of the motor units of the antagonists. This reduction in neural activity of the antagonists allows them to subsequently lengthen under less tension. This may be demonstrated by comparing the ease with which one can stretch the hamstrings while simultaneously contracting the quadriceps with the difficulty of attempting to stretch the hamstrings without the quadriceps contracted. See Fig. 2.21.

### Angle of pull

Another factor of considerable importance in using the leverage system is the angle of pull of the muscles on the bone. The angle of pull may be defined as the angle between the line of pull of the muscle and the bone on which it inserts. For the sake of clarity and consistency, we need to specify that the actual angle referred to is the angle toward the joint. With every degree of joint motion, the angle of pull changes. Joint movements and insertion angles involve mostly small angles of pull. The angle of pull decreases as the bone moves away from its anatomical position through the contraction of the local muscle group. This range of movement depends on the type of joint and bony structure.

Most muscles work at a small angle of pull, generally less than 50 degrees. The amount of muscular force needed to cause joint movement is affected by the angle of pull. Three components of muscular force are involved. The **rotary component**, also referred to as the vertical

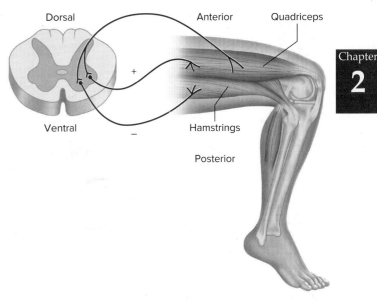

FIG. 2.21 • Reciprocal inhibition. A contraction of the agonist (quadriceps) will produce relaxation in the antagonist (hamstrings).

component, acts perpendicular to the long axis of the bone (lever). When the line of muscular force is at 90 degrees to the bone on which it attaches, all of the muscular force is rotary force; therefore, 100% of the force is contributing to the movement. That is, all of the force is being used to rotate the lever about its axis. The closer the angle of pull to 90 degrees, the greater the rotary component. At all other degrees of the angle of pull, one of the other two components of force is operating in addition to the rotary component. The same rotary component is continuing, although with less force, to rotate the lever about its axis. The horizontal or **nonrotary component** is either a **stabilizing component** or a **dislocating component** depending on whether the angle of pull is less than or greater than 90 degrees. If the angle is less than 90 degrees, the force is a stabilizing force because its pull directs the bone toward the joint axis. This increases the compressive forces within the joint and overall joint stability. If the angle is greater than 90 degrees, the force is dislocating because its pull directs the bone away from the joint axis (Fig. 2.22). Angles of pull greater than 90 degrees tend to decrease joint compressive forces and increase the distractive forces, thereby putting more stress on the joint's ligamentous structures. There is quite

FIG. 2.22 • **A** to **C,** Components of force due to the angle of pull.

a bit of variance in both cases, depending on the actual joint structure.

In some activities, it is desirable to have a person begin a movement when the angle of pull is at 90 degrees. Many boys and girls are unable to do a chin-up (pull-up) unless they start with the elbow in a position to allow the elbow flexor muscle group to approximate a 90-degree angle with the forearm.

This angle makes the chin-up easier because of the more advantageous angle of pull. The application of this fact can compensate for a lack of sufficient strength. In its range of motion, a muscle pulls a lever through a range characteristic of itself, but it is most effective when approaching and going beyond 90 degrees. An increase in strength is the only solution for muscles that operate at disadvantageous angles of pull and require a greater force to operate efficiently.

### Uniarticular, biarticular, and multiarticular muscles

Uniarticular muscles are those that cross and act directly only on the joint that they cross. The brachialis of the elbow is an example in that it can only pull the humerus and ulna closer to each other upon concentric contraction. When the humerus is relatively stabilized, as in an elbow curl, the brachialis contracts to flex the elbow and pulls the ulna closer to the humerus. However, when the ulna is relatively stabilized, as in a pull-up, the brachialis indirectly causes motion at the shoulder even though it does not cross it. In this example the brachialis contracts and pulls the humerus closer to the ulna as an elbow flexor. Correspondingly, the shoulder has to move from flexion to extension for the pull-up to be accomplished.

Biarticular muscles are those that cross and act directly on two different joints. Depending on a variety of factors, a biarticular muscle may contract to cause, control, or prevent motion at either one or both of its joints. Biarticular muscles have two advantages over uniarticular muscles. They can cause, control, and/or prevent motion at more than one joint, and they may be able to maintain a relatively constant length due to "shortening" at one joint and "lengthening" at another joint. The muscle does not actually shorten at one joint and lengthen at the other; rather, the concentric shortening of the muscle to move one joint is off-set by motion of the other joint, which moves its attachment of the muscle farther away. This maintenance of a relatively constant length results in the muscle's ability to continue to exert force. In the pull-up example, the biceps brachii acts as a flexor at the elbow. In the initial stage of the pull-up, the biceps brachii is in a relatively lengthened state at the elbow due to its extended position and in a relatively shortened state at the shoulder due to its flexed position. To accomplish the pull-up, the biceps brachii contracts concentrically to flex the elbow, so it effectively "shortens" at the elbow. Simultaneously, the shoulder is extending during the pull-up, which effectively "lengthens" the biceps brachii at the shoulder.

The biarticular muscles of the hip and knee provide excellent examples of two different patterns of action. **Concurrent** movement patterns allow the involved biarticular muscle to maintain a relatively consistent length because of the same action (extension) at both its joints. An example occurs within the rectus femoris (and also the hamstrings) when both the knee and the hip extend at the same time, as in standing from a squatting position as shown in Fig. 2.23. If only the knee were to extend,

the rectus femoris would shorten and its ability to exert force similar to the other quadriceps muscles would decrease, but its relative length and subsequent force production capability are maintained due to its relative lengthening at the hip joint during extension.

Due to opposite actions occurring simultaneously at both joints of a biarticular muscle, **countercurrent** movement patterns result in substantial shortening of the biarticular muscle. Substantial lengthening of its biarticular antagonist also occurs. This may be observed in the rectus femoris when kicking a ball. During the forward movement phase of the lower extremity, the rectus femoris is concentrically contracted to both flex the hip and extend the knee. These two movements, when combined, result in decreased force production capability in the rectus femoris and increased passive tension or stretch on the hamstring muscles at both the knee and the hip as the kick nears completion. Countercurrent movement patterns result in active insufficiency in the contracting agonist muscles and passive insufficiency in the antagonist muscles. See Fig. 2.24, *B*.

Multiarticular muscles act on three or more joints due to the line of pull between their origin and insertion crossing multiple joints. The principles discussed relative to biarticular muscles also apply to multiarticular muscles.

### Active and passive insufficiency

As a muscle shortens, its ability to exert force diminishes, as discussed earlier. When the muscle becomes shortened to the point where it cannot generate or maintain active tension, **active insufficiency** is reached. As a result, the muscle cannot shorten any further. If the opposing muscle becomes stretched to the point where it can no longer lengthen and allow movement, **passive insufficiency** is reached. These principles are most easily observed in either biarticular or multiarticular muscles when the full range of motion is attempted in all the joints crossed by the muscle.

An example is when the rectus femoris contracts concentrically to both flex the hip and extend the knee. It may completely perform either action one at a time, as shown in Fig. 2.24, *A*, but is actively insufficient to obtain full range at both joints simultaneously, as shown in Fig. 2.24, *B*. Likewise, the hamstrings will not usually stretch enough to allow both maximal hip flexion and maximal knee extension; hence, they are passively insufficient. It is virtually impossible to actively extend the knee fully when beginning with the hip fully flexed, or vice versa.

Courtesy of Lisa Floyd (left, right)

A          B

FIG. 2.23 • Example of concurrent movement pattern. When moving from a squatted position to a standing position, the concurrent movement pattern of extension at the hip and extension at the knee allow the biarticular agonist muscles (hamstrings and rectus femoris, respectively) to maintain a relatively consistent length.

Courtesy of R.T. Floyd (left, right)

A          B

FIG. 2.24 • Active and passive insufficiency.
**A,** The rectus femoris is easily able to actively flex the hip or extend the knee through their respective full ranges of motion individually without fully stretching the hamstrings; **B,** However, when one tries to actively flex the hip and simultaneously extend the knee (countercurrent movement pattern), active insufficiency is reached in the rectus femoris and passive insufficiency is reached in the hamstrings, resulting in the inability to reach full range of motion in both joints.

## REVIEW EXERCISES

### 1. Muscle nomenclature chart

Complete the chart by writing in the distinctive characteristics for which each of the muscles is named, such as shape, size, number of divisions, fiber direction, location, and/or action. Some muscles have more than one. Refer to Chapters 4, 5, 6, 7, 9, 10, 11, and 12 if needed.

| Muscle name | Distinctive characteristic(s) for which it is named |
|---|---|
| Adductor magnus | |
| Biceps brachii | |
| Biceps femoris | |
| Brachialis | |
| Brachioradialis | |
| Coracobrachialis | |
| Deltoid | |
| Extensor carpi radialis brevis | |
| Extensor carpi ulnaris | |
| Extensor digiti minimi | |
| Extensor digitorum | |

| Extensor hallucis longus | |
|---|---|
| Extensor indicis | |
| Extensor pollicis brevis | |
| External oblique | |
| Fibularis brevis | |
| Flexor carpi radialis | |
| Flexor digitorum longus | |
| Flexor digitorum profundus | |
| Flexor digitorum superficialis | |
| Flexor pollicus longus | |
| Gastrocnemius | |
| Gluteus maximus | |
| Gluteus medius | |
| Iliacus | |
| Iliocostalis thoracis | |
| Infraspinatus | |
| Latissimus dorsi | |
| Levator scapulae | |
| Longissimus lumborum | |
| Obturator externus | |
| Palmaris longus | |
| Pectoralis minor | |
| Peroneus tertius | |
| Plantaris | |
| Pronator quadratus | |
| Pronator teres | |
| Psoas major | |
| Rectus abdominis | |
| Rectus femoris | |
| Rhomboid | |
| Semimembranous | |
| Semitendinosus | |
| Serratus anterior | |
| Spinalis cervicis | |
| Sternocleidomastoid | |
| Subclavius | |
| Subscapularis | |
| Supinator | |
| Supraspinatus | |
| Tensor fasciae latae | |
| Teres major | |
| Tibialis posterior | |
| Transversus abdominis | |
| Trapezius | |
| Triceps brachii | |
| Vastus intermedius | |
| Vastus lateralis | |
| Vastus medialis | |

## 2. Muscle shape and fiber arrangement chart

For each muscle listed, determine first whether it should be classified as parallel or pennate. Complete the chart by writing in *flat, fusiform, strap, radiate,* or *sphincter* under those you classify as parallel. Write in *unipennate, bipennate,* or *multipennate* for those you classify as pennate.

| Muscle | Parallel | Pennate |
|---|---|---|
| Adductor longus | | |
| Adductor magnus | | |
| Brachioradialis | | |
| Extensor digitorum | | |
| Flexor carpi ulnaris | | |
| Flexor digitorum longus | | |
| Gastrocnemius | | |
| Gluteus maximus | | |
| Iliopsoas | | |
| Infraspinatus | | |
| Latissimus dorsi | | |
| Levator scapulae | | |
| Palmaris longus | | |
| Pronator quadratus | | |
| Pronator teres | | |
| Rhomboid | | |
| Serratus anterior | | |
| Subscapularis | | |
| Triceps brachii | | |
| Vastus intermedius | | |
| Vastus medialis | | |

3. Choose a particular sport skill and determine the types of muscle contractions occurring in various major muscle groups throughout the body at different phases of the skill.

## 4. Muscle contraction typing chart

For each of the following exercises, write the type of contraction (isometric, concentric, or eccentric), if any, in the cell of the muscle group that is contracting. Place a dash in the cell if there is no contraction occurring. *Hint:* In some instances you may have more than one type of contraction in the same muscle groups throughout various portions of the exercises. If so, list them in the order of occurrence.

| Exercise | Quadriceps | Hamstrings |
|---|---|---|
| a. Lie prone on a table with your knee in full extension. | | |
| Maintain your knee in full extension. | | |
| Very slowly flex your knee maximally. | | |
| Maintain your knee in full flexion. | | |
| From the fully flexed position, extend your knee fully as fast as possible but stop immediately before reaching maximal extension. | | |
| From the fully flexed position, very slowly extend your knee fully. | | |
| b. Begin sitting on the edge of the table with your knee in full extension. | | |
| Maintain your knee in full extension. | | |
| Very slowly flex your knee maximally. | | |
| Maintain your knee in full flexion. | | |
| Maintain your knee in approximately 90 degrees of flexion. | | |
| From the fully flexed position, slowly extend your knee fully. | | |
| c. Stand on one leg and move the other knee as directed. | | |
| Maintain your knee in full extension. | | |
| Very slowly flex your knee maximally. | | |
| From the fully flexed position, slowly extend your knee fully. | | |
| From the fully flexed position, extend your knee fully as fast as possible. | | |

5. With the wrist in neutral, extend the fingers maximally and attempt to maintain the position and then extend the wrist maximally. What happens to the fingers and why?

6. Maximally flex your fingers around a pencil with your wrist in neutral. Maintain the maximal finger flexion while you allow a partner to grasp your forearm with one hand and use his or her other hand to push your wrist into maximal flexion. Can you maintain control of the pencil? Explain.

7. You are walking in a straight line down the street when a stranger bumps into you. You stumble but "catch" your balance. Using the information from this chapter and other resources, explain what happened.

8. Drinking a glass of water is a normal daily activity in which the mind and body are involved in the controlled task. Explain how the movements happen once you decide to drink, in terms of the nerve roots, muscle contractions, and angle of pull.

## LABORATORY EXERCISES

1. Observe on a fellow student some of the muscles shown in Figs. 2.1 and 2.2.

2. With a partner, choose a diarthrodial joint on the body and carry out each of the following exercises:
   a. Familiarize yourself with all of the joint's various movements and list them.
   b. Determine which muscles or muscle groups are responsible for each movement you listed in 2a.
   c. For the muscles or muscle groups you listed for each movement in 2b, determine the type of contraction occurring.
   d. Determine how to change the parameters of gravity and/or resistance so that the opposite muscles contract to control the same movements in 2c. Name the type of contraction occurring.
   e. Determine how to change the parameters of movement, gravity, and/or resistance so that the same muscles listed in 2c contract differently to control the opposite movement.

3. Utilizing a reflex hammer or the flexed knuckle of your long finger PIP joint, compare the patellar reflex among several subjects.

4. Request a partner to stand with eyes closed while you position his or her arms in an odd position at the shoulders, elbows, and wrists. Ask your partner to describe the exact position of each joint while keeping the eyes closed. Then have your partner begin in the anatomical position, close the eyes, and subsequently reassume the position in which you had previously placed him or her. Explain the neuromechanisms involved in your partner's being able both to sense the joint position in which you placed him or her and then to reassume the same position.

5. Stand up straight on one leg on a flat surface with the other knee flexed slightly and not in contact with anything. Look straight ahead and attempt to maintain your balance in this position for up to 5 minutes. What do you notice happening in terms of the muscles in your lower leg? Try this again with the knee of the leg you are standing on slightly flexed. What differences do you notice? Try it again standing on a piece of thick foam. Try it in the original position with your eyes closed. Elaborate on the differences among the various attempts.

6. Hold a heavy book in your hand with your forearm supinated and your elbow flexed approximately 90 degrees while standing. Have a partner suddenly place another heavy book atop the one you are holding. What is the immediate result regarding the angle of flexion in your elbow? Explain why this result occurs.

7. Sit up very straight on a table with the knees flexed 90 degrees and the feet hanging free. Maintain this position while flexing the right hip and attempting to cross your legs to place the right leg across the left knee. Is this difficult? What tends to happen to the low back and trunk? How can you modify this activity to make it easier?

8. Determine your one-repetition maximum for a biceps curl beginning in full extension and ending in full flexion. Carry out each of the following exercises with adequate periods for recovery in between:
   a. Begin with your elbow flexed 45 degrees, then have a partner hand you a weight slightly heavier than your one-repetition maximum (about 5 pounds). Attempt to lift this weight through the remaining range of flexion. Can you reach full flexion? Explain.
   b. Begin with your elbow in 90 degrees of flexion. Have your partner hand you a slightly heavier weight than in 8a. Attempt to hold the elbow flexed in this position for 10 seconds. Can you do this? Explain.
   c. Begin with your elbow in full flexion. Have your partner hand you an even slightly heavier weight than in 8b. Attempt to slowly lower the weight under control until you reach full extension. Can you do this? Explain.

# References

Bernier MR: Perturbation and agility training in the rehabilitation for soccer athletes, *Athletic Therapy Today* 8(3):20–22, 2003.

Blackburn T, Guskiewicz KM, Petschauer MA, Prentice WE: Balance and joint stability: the relative contributions of proprioception and muscular strength, *Journal of Sport Rehabilitation* 9(4):315–328, 2000.

Carter AM, Kinzey SJ, Chitwood LF, Cole JL: Proprioceptive neuromuscular facilitation decreases muscle activity during the stretch reflex in selected posterior thigh muscles, *Journal of Sport Rehabilitation* 9(4):269–278, 2000.

Chimera N, Swanik K, Swanik C: Effects of plyometric training on muscle activation strategies and performance in female athletes. *Journal of Athletic Training* 39(1):24, 2004.

Dover G, Powers ME: Reliability of joint position sense and force reproduction measures during internal and external rotation of the shoulder, *Journal of Athletic Training* 38(4):304–310, 2003.

Hall SJ: *Basic biomechanics,* ed 7, New York, 2015, McGraw-Hill.

Hamill J, Knutzen KM: *Biomechanical basis of human movement,* ed 3, Baltimore, 2008, Lippincott Williams & Wilkins.

Hamilton N, Weimar W, Luttgens K: *Kinesiology: scientific basis of human motion,* ed 12, New York, 2012, McGraw-Hill.

Knight KL, Ingersoll CD, Bartholomew J: Isotonic contractions might be more effective than isokinetic contractions in developing muscle strength, *Journal of Sport Rehabilitation* 10(2):124–131, 2001.

Kreighbaum E, Barthels KM: *Biomechanics: a qualitative approach for studying human movement,* ed 4, Boston, 1996, Allyn & Bacon.

Lindsay DT: *Functional human anatomy,* St. Louis, 1996, Mosby.

Logan GA, McKinney WC: *Anatomic kinesiology,* ed 3, Dubuque, IA, 1982, Brown.

Mader SS: *Biology,* ed 9, New York, 2007, McGraw-Hill.

McArdle WD, Katch FI, Katch VI: *Exercise physiology: nutrition, energy, and human performance,* ed 7, Baltimore, 2009, Lippincott Williams & Wilkins.

McCrady BJ, Amato HK: Functional strength and proprioception testing of the lower extremity, *Athletic Therapy Today* 9(5):60–61, 2005.

Myers JB, Guskiewicz KM, Schneider, RA, Prentice WE: Proprioception and neuromuscular control of the shoulder after muscle fatigue, *Journal of Athletic Training* 34(4):362–367, 1999.

National Strength and Conditioning Association; Baechle TR, Earle RW: *Essentials of strength training and conditioning,* ed 2, Champaign, IL, 2000, Human Kinetics.

Neumann DA: *Kinesiology of the musculoskeletal system: foundations for physical rehabilitation,* ed 2, St. Louis, 2010, Mosby.

Norkin CC, Levangie PK: *Joint structure and function—a comprehensive analysis,* ed 5, Philadelphia, 2011, Davis.

Northrip JW, Logan GA, McKinney WC: *Analysis of sport motion,* ed 3, Dubuque, IA, 1983, Brown.

Olmsted-Kramer LC, Hertel J: Preventing recurrent lateral ankle sprains: an evidence-based approach, *Athletic Therapy Today* 9(2):19–22, 2004.

Powers ME, Buckley BD, Kaminski TW, Hubbard TJ, Ortiz C: Six weeks of strength and proprioception training does not affect muscle fatigue and static balance in functional ankle stability, *Journal of Sport Rehabilitation* 13(3):201–227, 2004.

Powers SK, Howley ET: *Exercise physiology: theory ed. 9, 2015, and application of fitness and performance,* ed 8, New York, 2012, McGraw-Hill.

Rasch PJ: *Kinesiology and applied anatomy,* ed 7, Philadelphia, 1989, Lea & Febiger.

Raven PH, Johnson GB, Losos JB, Mason KA, Singer SR: *Biology,* ed 8, New York, 2008, McGraw-Hill.

Riemann BL, Lephart SM: The sensorimotor system, part I: the physiological basis of functional joint stability, *Journal of Athletic Training* 37(1):71–79, 2002.

Riemann BL, Lephart SM: The sensorimotor system, part II: the role of proprioception in motor control and functional joint stability, *Journal of Athletic Training* 37(1):80–84, 2002.

Riemann BL, Myers JB, Lephart SM: Sensorimotor system measurement techniques, *Journal of Athletic Training* 37(1): 85–98, 2002.

Riemann BL, Tray NC, Lephart SM: Unilateral multiaxial coordination training and ankle kinesthesia, muscle strength, and postural control, *Journal of Sport Rehabilitation* 12(1):13–30, 2003.

Ross S, Guskiewicz K, Prentice W, Schneider R, Yu B: Comparison of biomechanical factors between the kicking and stance limbs, *Journal of Sport Rehabilitation* 13(2):135–150, 2004.

Saladin, KS: *Anatomy & physiology: the unity of form and function,* ed 5, New York, 2010, McGraw-Hill.

Sandrey MA: Using eccentric exercise in the treatment of lower extremity tendinopathies, *Athletic Therapy Today* 9(1):58–59, 2004.

Seeley RR, Stephens TD, Tate P: *Anatomy & physiology,* ed 7, New York, 2006, McGraw-Hill.

Seeley RR, VanPutte CL, Regan JL, Russo AF: *Seeley's Anatomy & Physiology,* ed 9, New York, 2011, McGraw-Hill.

Shier D, Butler J, Lewis R: *Hole's essentials of human anatomy and physiology,* ed 10, New York, 2009, McGraw-Hill.

Shier D, Butler J, Lewis R: *Hole's human anatomy and physiology,* ed 12, New York, 2010, McGraw-Hill.

Van De Graaff KM: *Human anatomy,* ed 6, New York, 2002, McGraw-Hill.

Van De Graaf KM, Fox SI, LaFleur KM: *Synopsis of human anatomy & physiology,* Dubuque, IA, 1997, Brown.

Yaggie J, Armstrong WJ: Effects on lower extremity fatigue on indices of balance, *Journal of Sport Rehabilitation* 10(2):124–131, 2004.

# Worksheet Exercises

For in- or out-of-class assignments, or for testing, utilize this tear-out worksheet.

## Anterior muscular system worksheet

On the anterior muscular system worksheet, label the major superficial muscles on the right and the deeper muscles on the left.

Superficial | Deep

1 _____
2 _____
3 _____
4 _____
5 _____
6 _____
7 _____
8 _____
9 _____
10 _____
11 _____
12 _____
13 _____
14 _____
15 _____
16 _____
17 _____
18 _____
19 _____

20 _____
21 _____
22 _____
23 _____
24 _____
25 _____
26 _____
27 _____
28 _____
29 _____
30 _____
31 _____
32 _____
33 _____
34 _____
35 _____
36 _____
37 _____
38 _____
39 _____
40 _____
41 _____

# Worksheet Exercises

For in- or out-of-class assignments, or for testing, utilize this tear-out worksheet.

## Posterior muscular system worksheet

On the posterior muscular system worksheet, label the major superficial muscles on the right and the deeper muscles on the left.

Deep | Superficial

1 _____
2 _____
3 _____
4 _____
5 _____
6 _____
7 _____
8 _____
9 _____
10 _____
11 _____
12 _____
13 _____
14 _____
15 _____
16 _____
17 _____
18 _____
19 _____
20 _____
21 _____
22 _____
23 _____
24 _____
25 _____
26 _____
27 _____
28 _____
29 _____
30 _____

31 _____
32 _____
33 _____
34 _____
35 _____
36 _____
37 _____
38 _____
39 _____
40 _____
41 _____
42 _____
43 _____
44 _____
45 _____
46 _____

# BASIC BIOMECHANICAL FACTORS AND CONCEPTS

## Objectives

- To know and understand how the proper application of levers can help improve physical performance

- To know and understand how the musculoskeletal system functions as a series of simple machines

- To know and understand how knowledge of torque and lever arm lengths can help improve physical performance

- To know and understand how the proper application of Newton's laws of motion can help improve physical performance

- To know and understand how knowledge of balance, equilibrium, and stability can help improve physical performance

- To know and underst and how knowledge of force and momentum can help improve physical performance

- To know and understand the basic effects of mechanical loading on body tissues

In Chapter 1 we defined kinesiology, very simply, as the study of muscles, bones, and joints as they are involved in the science of movement. From this general definition we can go into greater depth in exploring the science of body movement, which primarily includes anatomy, physiology, and mechanics. For a true understanding of movement, a vast amount of knowledge is needed in all three areas. The focus of this text is primarily structural and functional anatomy. We have only very minimally touched on some physiology in the first two chapters. A much greater study of physiology as it relates to movement should be addressed in an exercise physiology course, for which there are many excellent texts and resources. Likewise, the study of mechanics as it relates to the functional and anatomical analysis of biological systems, known as **biomechanics**, should be addressed to a greater degree in a separate course. Human movement is quite complex. In order to make recommendations for its improvement, we need to study movements from a biomechanical perspective, both qualitatively and quantitatively. This chapter introduces some basic biomechanical factors and concepts, with the understanding that many readers will subsequently study these in more depth in a dedicated course utilizing much more thorough resources.

Many students in kinesiology classes have some knowledge, from a college or high school physics course, of the laws that affect motion. These principles and others are discussed briefly in this chapter, which should prepare you as you begin to apply them to motion in the human body. The more you can put these principles and concepts into practical application, the easier it will be to understand them.

**Mechanics**, the study of physical actions of forces, can be subdivided into **statics** and **dynamics**. Statics involves the study of systems that are in a constant state of motion, whether at rest with no motion or moving at a constant velocity without acceleration. In statics all forces acting on the body are in balance, resulting in the body being in equilibrium. Dynamics involves the study of systems in motion with acceleration. A system in acceleration is unbalanced due to unequal forces acting on the body. Additional components of biomechanical study include **kinematics** and **kinetics**. Kinematics is concerned with the description of motion and includes consideration of time, displacement, velocity, acceleration, and space factors of a system's motion. Kinetics is the study of forces associated with the motion of a body.

## Types of machines found in the body

As discussed in Chapter 2, we utilize muscles to apply force to the bones on which they attach to cause, control, or prevent movement in the joints they cross. As is often the case, we utilize bones such as those in the hand to either hold, push, or pull on an object while simultaneously using a series of bones and joints throughout the body to apply force via the muscles to affect the position of the object. In doing so we are using a series of simple machines to accomplish the tasks. Machines are used to increase or multiply the applied force in performing a task or to provide a **mechanical advantage**. The mechanical advantage provided by machines enables us to apply a relatively small force, or effort, to move a much greater resistance or to move one point of an object a relatively small distance to result in a relatively large amount of movement of another point of the same object. We can determine mechanical advantage by dividing the load by the effort. The mechanical aspect of each component should be considered with respect to the component's machinelike function.

Another way of thinking about machines is to note that they convert smaller amounts of force exerted over a longer distance to larger amounts of force exerted over a shorter distance. This may be turned around so that a larger amount of force exerted over a shorter distance is converted to a smaller amount of force over a greater distance. Machines function in four ways:

1. To balance multiple forces
2. To enhance force in an attempt to reduce the total force needed to overcome a resistance

3. To enhance range of motion and speed of movement so that resistance can be moved farther or faster than the applied force
4. To alter the resulting direction of the applied force

Simple machines are the lever, wheel and axle, pulley, inclined plane, screw, and wedge. The arrangement of the musculoskeletal system provides three types of machines in producing movement: levers, wheel/axles, and pulleys. Each of these involves a balancing of rotational forces about an axis. The lever is the most common form of simple machine found in the human body.

## Levers

It may be difficult for a person to visualize his or her body as a system of levers, but this is actually the case. Human movement occurs through the organized use of a system of levers. While the anatomical levers of the body cannot be changed, when the system is properly understood they can be used more efficiently to maximize the muscular efforts of the body.

A lever is defined as a rigid bar that turns about an **axis of rotation**, or fulcrum. The axis is the point of rotation about which the lever moves. The lever rotates about the axis as a result of **force** (sometimes referred to as effort, $E$) being applied to it to cause its movement against a **resistance** (sometimes referred to as load or weight). In the body, the bones represent the bars, the joints are the axes, and the muscles contract to apply the force. The amount of resistance can vary from maximal to minimal. In fact, the bones themselves or the weight of the body segment may be the only resistance applied. All lever systems have each of these three components in one of three possible arrangements.

The arrangement or location of three points in relation to one another determines the type of lever and the application for which it is best suited. These points are the axis, the point of force application (usually the muscle insertion), and the point of resistance application (sometimes the center of gravity of the lever and sometimes the location of an external resistance). When the axis ($A$) is placed anywhere between the force ($F$) and the resistance ($R$), a first-class lever is produced (Fig. 3.1). In second-class levers, the resistance is somewhere between the axis and the force (Fig. 3.2). When the force is placed somewhere between the axis and the resistance, a third-class lever is created (Fig. 3.3). Table 3.1 provides a summary of the three classes of levers and the characteristics of each.

The mechanical advantage of levers may be determined using the following equations:

$$\text{Mechanical advantage} = \frac{\text{resistance}}{\text{force}}$$

or

$$\text{Mechanical advantage} = \frac{\text{length of force arm}}{\text{length of resistance arm}}$$

## First-class levers

Typical examples of a first-class lever are the crowbar, the seesaw, pliers, oars, and the triceps in overhead elbow extension. In the body an example is when the triceps applies the force to the olecranon (*F*) in extending the nonsupported forearm (*R*) at the elbow (*A*). Other examples are when the agonist and antagonist muscle groups on either side of a joint axis are contracting simultaneously, with the agonist producing the force and the antagonist supplying the resistance. A first-class lever (see Fig. 3.1) is designed basically to produce

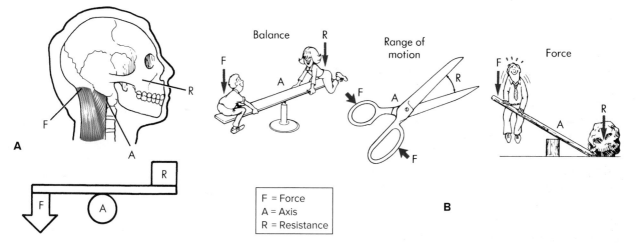

FIG. 3.1 • **A** and **B**, First-class levers.

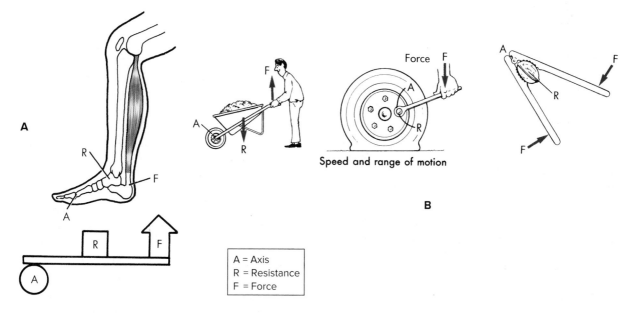

FIG. 3.2 • **A** and **B**, Second-class levers.

R = Resistance
F = Force
A = Axis

FIG. 3.3 • **A** and **B**, Third-class levers. Note that the paddle and shovel function as third-class levers only when the top hand does not apply force but serves as a fixed axis of rotation. If the top hand applied force and the lower hand acted as the axis, then these would represent first-class levers.

## TABLE 3.1 • **Classification of levers and characteristics of each**

| Class | Illustration | Arrangement | Arm movement | Direction of force versus resistance | Functional design | Relationship to axis | Mechanical advantage | Practical example | Human example |
|---|---|---|---|---|---|---|---|---|---|
| 1st | F    R First class A | **F–A–R** Axis between force and resistance | Resistance arm and force arm move in opposite directions | Resistance and force are applied in same direction | Balanced movements | Axis in middle | Equal to 1 | Seesaw | Erector spinae extending the head on cervical spine |
| | | | | | Speed and range of motion | Axis near force | Less than 1 | Scissors | Triceps brachii in extending the elbow |
| | | | | | Force motion | Axis near resistance | Greater than 1 | Crow bar | |
| 2nd | R    F A Second class | **A–R–F** Resistance between axis and force | Resistance arm and force arm move in the same direction | Resistance and force are applied in opposite directions | Force motion (large resistance can be moved with relatively small force) | Axis near resistance | Always greater than 1 | Wheel barrow, nut-cracker | Gastroc-nemius and soleus in plantar flexing the foot to raise the body on the toes |
| 3rd | F    R A Third class | **A–F–R** Force between axis and resistance | Resistance arm and force arm move in the same direction | Resistance and force are applied in opposite directions | Speed and range of motion (requires large force to move a relatively small resistance) | Axis near force | Always less than 1 | Shovel-ing dirt, catapult | Biceps brachii and brachialis in flexing the elbow |

balanced movements when the axis is midway between the force and the resistance (e.g., a seesaw). When the axis is close to the force, the lever produces speed and range of motion (e.g., the triceps in elbow extension). When the axis is close to the resistance, the lever produces force motion (e.g., a crowbar).

In applying the principle of levers to the body, it is important to remember that the force is applied where muscle inserts in bone, not in the belly of the muscle. For example, in elbow extension with the shoulder fully flexed and the arm beside the ear, the triceps applies the force to the olecranon of the ulna behind the axis of the elbow joint. As the applied force exceeds the amount of forearm resistance, the elbow extends.

The type of lever may be changed for a given joint and muscle depending on whether the body segment is in contact with a surface such as a floor or wall. For example, we have demonstrated that the triceps in elbow extension is a first-class lever with the hand free in space and the arm pushed away from the body. If the hand is placed in contact with the floor, as in performing a push-up to push the body away from the floor, the same muscle action at this joint now changes the lever to second class, because the axis is at the hand and the resistance is the body weight at the elbow joint.

### Second-class levers

A second-class lever (see Fig. 3.2) is designed to produce force movements, since a large resistance can be moved by a relatively small force. Examples of second-class levers include a bottle opener, a wheelbarrow, and a nutcracker. We have just noted the example of the triceps extending the elbow in a push-up. A similar example of a second-class lever in the body is plantar flexion of the ankle to raise the body on the toes. The ball (A) of the foot serves as the axis of rotation as the ankle plantar flexors apply force to the calcaneus (F) to lift the resistance of the body at the tibiofibular articulation (R) with the talus. Opening the mouth against resistance provides another example of a second-class lever. There are relatively few other examples of second-class levers in the body.

### Third-class levers

Third-class levers (see Fig. 3.3), with the force being applied between the axis and the resistance, are designed to produce speed and range of motion. Most of the levers in the human body are of this type, which requires a great deal of force to

move even a small resistance. Examples include a catapult, a screen door operated by a short spring, and the application of lifting force to a shovel handle with the lower hand while the upper hand on the shovel handle serves as the axis of rotation. The biceps brachii is a typical example in the body. Using the elbow joint (A) as the axis, the biceps brachii applies force at its insertion on the radial tuberosity (F) to rotate the forearm up, with its center of gravity (R) serving as the point of resistance application.

The brachialis is an example of true third-class leverage. It pulls on the ulna just below the elbow, and, since the ulna cannot rotate, the pull is direct and true. The biceps brachii, on the other hand, supinates the forearm (applying the rotational force of a first-class lever as in a wheel and axle to the radius) as it flexes, so the third-class leverage applies to flexion only.

Other examples include the hamstrings contracting to flex the leg at the knee in a standing position and the iliopsoas being used to flex the thigh at the hip.

## Factors in use of anatomical levers

Our anatomical leverage system can be used to gain a mechanical advantage that will improve simple or complex physical movements. Some individuals unconsciously develop habits of using human levers properly, but frequently this is not the case.

### Torque and length of lever arms

To understand the leverage system, the concept of torque must be understood. **Torque**, or moment of force, is the turning effect of an eccentric force. **Eccentric force** is a force that is applied off center or in a direction not in line with the center of rotation of an object with a fixed axis. In objects without a fixed axis, it is an applied force that is not in line with the object's center of gravity; for rotation to occur, an eccentric force must be applied. In the human body, the contracting muscle applies an eccentric force (not to be confused with eccentric contraction) to the bone on which it attaches and causes the bone to rotate about an axis at the joint. The amount of torque can be determined by multiplying the **force magnitude** (amount of force) by the **force arm**. The perpendicular distance between the location of force application and the axis is known as the force arm, moment arm, or torque arm. The force arm may be best understood as the shortest distance from the axis of rotation to the line of action of the force. The greater the distance of the force arm, the more torque produced by the force. A frequent practical application of torque and levers occurs when we purposely increase the force arm length in order to increase the torque so that we can more easily move a relatively large resistance. This is commonly referred to as increasing our leverage.

It is also important to note the **resistance arm**, which may be defined as the distance between the axis and the point of resistance application. In discussing the application of levers, it is necessary to understand the length relationship between the two lever arms. There is an inverse relationship between force and the force arm, just as there is between resistance and the resistance arm. The longer the force arm, the less force required to move the lever if the resistance and resistance arm remain constant, as shown graphically in Fig. 3.4. In addition, if the force and force arm remain constant, a greater resistance may be moved by shortening the resistance arm. Because the muscular

force is applied internally, in musculoskeletal discussions the force arm may also be referred to as the internal moment arm; and because the load is applied externally, the resistance arm may be referred to as the external moment arm.

Also, there is a proportional relationship between the force components and the resistance components. That is, for movement to occur when either of the resistance components increases, there must be an increase in one or both of the force components. See Figs. 3.5, 3.6, and 3.7 to see how these relationships apply to first-, second-,

FIG. 3.5 • First-class levers. **A**, If the force arm and resistance arm are equal in length, a force equal to the resistance is required to balance it; **B**, As the force arm becomes longer, a decreasing amount of force is required to move a relatively larger resistance; **C**, As the force arm becomes shorter, an increasing amount of force is required to move a relatively smaller resistance, but the speed and range of motion that the resistance can be moved are increased. *Forces (moments) are calculated to balance the lever system. The effort and resistance forces sum to zero. If any of the components are moved in relation to one another, then either a greater force or a greater resistance will be required.*

FIG. 3.4 • Relationships among forces, force arms, and resistance arms. (The graph assumes a constant resistance of 20 kilograms, and as a result the graphical representations of the resistance arm and force arm lie directly over each other.) With the resistance held constant at 20 kilograms and a resistance arm of 1 meter, the product of the (force) × (force arm) must equal 20 newtons. Thus there is an inverse relationship between the force and the force arm. As the force increases in newtons, the force arm length decreases in meters, and vice versa.

FIG. 3.6 • Second-class levers have a positive mechanical advantage due to the force arm always being longer than the resistance arm and are well suited for moving larger resistances with smaller forces. **A**, Placing the resistance halfway between the axis and the point of force application provides a mechanical advantage of 2; **B**, Moving the resistance closer to the axis increases the mechanical advantage but decreases the distance the resistance is moved; **C**, The closer the resistance is positioned to the point of force application, the less the mechanical advantage but the greater the distance the resistance is moved. *Forces (moments) are calculated to balance the lever system. The effort and resistance forces sum to zero. If any of the components are moved in relation to one another, then either a greater force or a greater resistance will be required.*

FIG. 3.7 • Third-class levers. **A**, A force greater than the resistance, regardless of the point of force application, is required due to the resistance arm always being longer; **B**, Moving the point of force application closer to the axis increases the range of motion and speed but requires more force; **C**, Moving the point of force application closer to the resistance decreases the force needed but also decreases the speed and range of motion. *Forces (moments) are calculated to balance the lever system. The effort and resistance forces sum to zero. If any of the components are moved in relation to one another, then either a greater force or a greater resistance will be required.*

and third-class levers, respectively. Even slight variations in the location of the force and the resistance are important in determining the mechanical advantage (MA) and effective force of the muscle. This point can be illustrated with the simple formula shown in Fig. 3.8, using the biceps brachii muscle in each example.

In Example A, the only way to move the insertion of the biceps brachii is with surgery, so this is not practical. In some orthopedic conditions, the attachments of tendons are surgically relocated in an attempt to change the dynamic forces

| Lever equation for a child performing biceps curls | | | | | |
| --- | --- | --- | --- | --- | --- |
| **Lever equation** | | | | | |
| F | x | FA | = | R | x | RA |
| (Force) | x | (Force arm) | = | (Resistance) | x | (Resistance arm) |

**Initial Example**

$$F \times 0.1 = 45 \text{ newtons} \times 0.25 \text{ meter}$$
$$F \times 0.1 = 11.25 \text{ newton-meters}$$
$$F = 112.5 \text{ newtons}$$

**Example A – Lengthening the force arm**

Increase FA by moving the insertion distally 0.05 meter:

$$F \times 0.15 = 45 \text{ newtons} \times 0.25 \text{ meter}$$
$$F \times 0.15 = 11.25 \text{ newton-meters}$$
$$F = 75 \text{ newtons}$$

An increase in the insertion from the axis by 0.05 meter results in a substantial reduction in the force necessary to move the resistance.

**Example B – Shortening the resistance arm**

Reduce RA by moving the point of resistance application proximally 0.05 meter:

$$F \times 0.1 = 45 \text{ newtons} \times 0.2 \text{ meter}$$
$$F \times 0.1 = 9 \text{ newton-meters}$$
$$F = 90 \text{ newtons}$$

A decrease in the resistance application from the axis by 0.05 meter results in a considerable reduction in the force necessary to move the resistance.

**Example C – Reducing the resistance**

Reduce R by reducing the resistance 1 newton:

$$F \times 0.1 = 44 \text{ newtons} \times 0.25 \text{ meter}$$
$$F \times 0.1 = 11 \text{ newton-meters}$$
$$F = 110 \text{ newtons}$$

Decreasing the amount of resistance can decrease the amount of force needed to move the lever.

FIG. 3.8 • Torque calculations with examples of modifications in the force arm, resistance arm, and resistance.

the muscles exert on the joints. In Example B, we can and often do shorten the resistance arm to enhance our ability to move an object. When attempting a maximal weight in a biceps curl exercise, we may flex our wrist to move the weight just a little closer, which shortens the resistance arm. Example C is straightforward in that we obviously can reduce the force needed by reducing the resistance.

The system of leverage in the human body is built for speed and range of motion at the expense of force. Short force arms and long resistance arms require great muscular strength to produce movement. In the forearm, the attachments of the biceps brachii and triceps brachii muscles clearly illustrate this point, since the force arm of the biceps brachii is 1 to 2 inches and that of the triceps brachii is less than 1 inch. Many similar examples are found all over the body. From a practical point of view, this means that the muscular system should be strong to supply the necessary force for body movements, especially in strenuous sports activities.

When we speak of human leverage in relation to sport skills, we are generally referring to several levers. For example, throwing a ball involves levers at the shoulder, elbow, and wrist joints as well as from the ground up through the lower extremities and the trunk. In fact, it can be said that there is one long lever from the feet to the hand.

The longer the lever, the more effective it is in imparting velocity. A tennis player can hit a tennis ball harder (deliver more force to it) with a straight-arm drive than with a bent elbow, because the lever (including the racket) is longer and moves faster.

Fig. 3.9 indicates that a longer lever (Z1) travels faster than a shorter lever (S1) in traveling the same number of degrees. In sports activities in which it is possible to increase the length of a lever with a racket or bat, the same principle applies.

In baseball, hockey, golf, field hockey, and other sports, long levers produce more linear force and thus better performance. However, to be able to fully execute the movement in as short a time as possible, it is sometimes desirable to have a short lever arm. For example, a baseball catcher attempting to throw a runner out at second base does not have to throw the ball so that it travels as fast as when the pitcher is attempting to throw a strike. In the catcher's case, it is more important to initiate and complete the throw as soon as possible than to deliver as much velocity to the ball as possible. The pitcher, when attempting to throw a ball

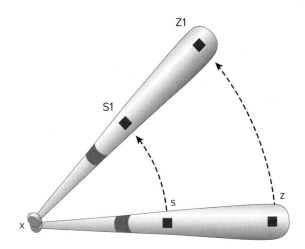

FIG. 3.9 • Length of levers. The end of a longer lever (Z1) travels faster than a shorter lever (S1) when moved over the same number of degrees in the same amount of time. The farther the resistance is from the axis, the farther it is moved and the more force is delivered to it, resulting in the resistance moving at a greater velocity. The same principle applies in sports in which it is possible to increase the lever length with a racket or bat.

at 90-plus miles per hour, will utilize his body as a much longer lever system throughout a greater range of motion to impart velocity to the ball.

## Wheels and axles

Wheels and axles are used primarily to enhance range of motion and speed of movement in the musculoskeletal system. A wheel and an axle essentially function as a form of a first-class lever. When either the wheel or the axle turns, the other must turn as well. Both complete one turn at the same time. The centers of the wheel and the axle both correspond to the fulcrum. Both the radius of the wheel and the radius of the axle correspond to the force arms. If the radius of the wheel is greater than the radius of the axle, then the wheel has a mechanical advantage over the axle due to the longer force arm. That is, a relatively smaller force may be applied to the wheel to move a relatively greater resistance applied to the axle. Very simply, if the radius of the wheel is five times the radius of the axle, then the wheel has a 5 to 1 mechanical advantage over the axle, as shown in Fig. 3.10. The mechanical advantage of a wheel and an axle for this scenario may be calculated by considering the radius of the wheel over the radius of the axle. This

application enables the wheel and axle to act as a second-class lever to gain force motion.

$$\text{Mechanical advantage} = \frac{\text{radius of the wheel}}{\text{radius of the axle}}$$

In this case the mechanical advantage is always more than 1. An application of this example is using the outer portion of an automobile steering wheel to turn the steering mechanism. Before the development of power steering, steering wheels had a much larger diameter than today in order to give the driver more of a mechanical advantage. An example in the body of applying force to the wheel occurs when we attempt to manually force a person's shoulder into internal rotation while she or he holds it in external rotation isometrically. The humerus acts as the axle, and the person's hand and wrist are located near the outside of the wheel when the elbow is flexed approximately 90 degrees. If we unsuccessfully attempt to break the force of the contraction of the external rotators by pushing internally at the midforearm, we can increase our leverage or mechanical advantage and our likelihood of success by applying the force nearer to the hand and wrist.

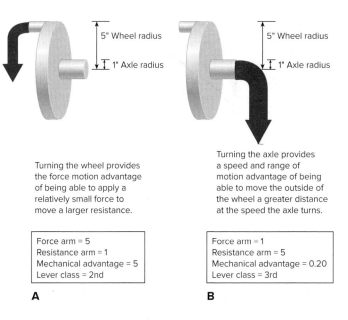

FIG. 3.10 • Wheel and axle. **A,** Mechanical advantage is gained by applying force to the outside of the wheel to more easily move a large resistance; **B,** The mechanical advantage in applying force to the axle is always less than 1 and requires a relatively large force, but the advantage is being able to move the larger wheel a relatively greater distance at a relatively greater speed than the axle.

If the application of force is reversed so that it is applied to the axle, then the mechanical advantage results from the wheel's turning a greater distance at greater speed. Using the same example, if the wheel radius is five times greater than the radius of the axle, the outside of the wheel will turn at a speed five times that of the axle. Additionally, the distance the outside of the wheel turns will be five times that of the outside of the axle. This application enables the wheel and axle to act as a third-class lever to gain speed and range of motion. The mechanical advantage of a wheel and axle for this scenario may be calculated by considering the radius of the axle over the radius of the wheel.

$$\text{Mechanical advantage} = \frac{\text{radius of the axle}}{\text{radius of the wheel}}$$

In this case the mechanical advantage is always less than 1. This is the principle utilized in the drivetrain of an automobile to turn the axle, which subsequently turns the tire one revolution for every turn of the axle. We utilize the powerful engine of the automobile to supply the force to increase the speed of the tire and subsequently carry us great distances. An example of the muscles applying force to the axle to result in greater range of motion and speed may again be seen in the upper extremity, in the case of the internal rotators attaching to the humerus. With the humerus acting as the axle and the hand and wrist located at the outside of the wheel (when the elbow is flexed approximately 90 degrees), the internal rotators apply force to the humerus. With the internal rotators concentrically internally rotating the humerus a relatively small amount, the hand and wrist travel a great distance. Using the wheel and axle in this manner enables us to significantly increase the speed at which we can throw objects.

## Pulleys

Single pulleys have a fixed axle and function to change the effective direction of force application. Single pulleys have a mechanical advantage of 1, as shown in Fig. 3.11, *A*. Numerous weight machines utilize pulleys to alter the direction of the resistive force. Pulleys may be movable and can be combined to form compound pulleys to further increase the mechanical advantage. Every additional rope connected to movable pulleys increases the mechanical advantage by 1, as shown in Fig. 3.11, *B*.

In the human body, an excellent example is provided by the lateral malleolus, acting as a pulley around which the tendon of the peroneus longus runs. As this muscle contracts, it pulls toward its belly, which is toward the knee. Due to its use of the lateral malleolus as a pulley (Fig. 3.12), the force is transmitted to the plantar aspect of the foot, resulting in downward and outward movement of the foot. Other examples in the human body include pulleys on the volar aspect of the phalanges to redirect the force of the flexor tendons.

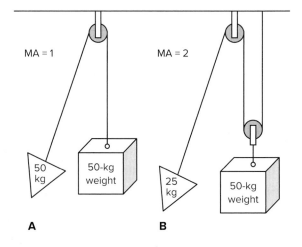

FIG. 3.11 • **A**, Single pulley; **B**, Compound movable pulley.

FIG. 3.12 • Pulley. The lateral malleolus serves as a pulley for the peroneus longus tendon.

# Laws of motion and physical activities

Motion is fundamental in physical education and sports activity. Body motion is generally produced, or at least started, by some action of the muscular system. Motion cannot occur without a force, and the muscular system is the source of force in the human body. Thus, development of the muscular system is indispensable to movement.

Basically, there are two types of motion: **linear motion** and **angular motion**. Linear motion, also referred to as translatory motion, is motion along a line. If the motion is along a straight line, it is **rectilinear** motion, whereas motion along a curved line is known as **curvilinear** motion. Angular motion, also known as rotary motion, involves rotation around an axis. In the human body, the axis of rotation is provided by the various joints. In a sense, these two types of motion are related, since angular motion of the joints can produce the linear motion of walking. In many sports activities, the cumulative angular motion of the joints of the body imparts linear motion to a thrown object (ball, shot) or to an object struck with an instrument (bat, racket). It is also important to consider the **center of rotation**, which is the point or line around which all other points in the body move. In a door hinge, the axis of rotation is fixed and all points of the door have equal arcs of rotation around the center of the hinge. But in joints of the body the axis is not usually fixed, due to their accessory motion as discussed in Chapter 1. As a result, the location of the exact center of rotation changes with changes in the joint angle. So we have to consider the **instantaneous center of rotation**, which is the center of rotation at a specific instant in time during movement. See Fig. 3.13.

## Quantity measurements—scalars versus vectors

In order to discuss measures of motion, we assign them a quantity of measurement. These mathematical quantities used to describe motion can be divided into either scalars or vectors. **Scalar** quantities are described by a magnitude (or numerical value) alone, such as speed as in miles per hour or meters per second. Other scalar quantities are length, area, volume, mass, time, density, temperature, pressure, energy, work, and power. **Vector** quantities are described by both a magnitude and a direction, such as velocity as in miles per hour in an eastward direction. Other vector quantities are acceleration, direction, displacement, force, drag, momentum, lift, weight, and thrust.

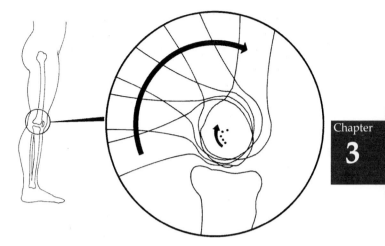

FIG. 3.13 • The path of the instantaneous center of rotation for the knee during extension.

**Displacement** is a change in the position or location of an object from its original point of reference, whereas **distance**, or the path of movement, is the actual sum length it is measured to have traveled. Thus an object may have traveled a distance of 10 meters along a linear path in two or more directions but be displaced from its original reference point by only 6 meters. Fig. 3.14 provides an example. **Angular displacement** is the change in location of a rotating body. **Linear displacement** is the distance a system moves in a straight line.

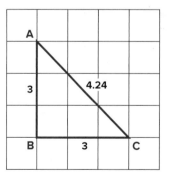

FIG. 3.14 • Displacement. If the path of movement is from A to B and then from B to C, the distance covered is AB + BC, but the displacement is the distance from A to C, or AC. If each cell is 1 square meter, then AB is 3 meters and BC is 3 meters, so the distance covered would be 6 meters. Using the Pythagorean Theorem (in a right triangle the square of the measure of the hypotenuse is equal to the sum of the squares of the measures of the legs, or $a^2 + b^2 = c^2$), we can then determine the displacement (AC) to be 4.24 meters with $AB^2 + BC^2 = AC^2$.

We are sometimes concerned about the time it takes for displacement to occur. **Speed** is how fast an object is moving, or the distance an object travels in a specific amount of time. **Velocity**, the rate at which an object changes its position, includes the direction and describes the rate of displacement.

A brief review of Newton's laws of motion will indicate the many applications of these laws to physical education activities and sports. Newton's laws explain all the characteristics of motion, and they are fundamental to understanding human movement.

### Law of inertia

**A body in motion tends to remain in motion at the same speed in a straight line unless acted on by a force; a body at rest tends to remain at rest unless acted on by a force.**

**Inertia** can be described as the resistance to action or change. In terms of human movement, *inertia* refers to resistance to acceleration or deceleration. Inertia is the tendency for the current state of motion to be maintained, whether the body segment is moving at a particular velocity or is motionless.

Muscles produce the force necessary to start motion, stop motion, accelerate motion, decelerate motion, or change the direction of motion. Put another way, inertia is the reluctance to change status; only force can do so. The greater the mass of an object, the greater its inertia. Therefore, the greater the mass, the more force needed to significantly change an object's inertia. Numerous examples of this law are found in physical education activities. A sprinter in the starting blocks must apply considerable force to overcome resting inertia. A runner on an indoor track must apply considerable force to overcome moving inertia and stop before hitting the wall. Fig. 3.15 provides an example of how a skier in motion remains in motion even though airborne after skiing off a hill. We routinely experience inertial forces when our upper body tends to move forward if we are driving a car at the speed limit and then suddenly have to slow down. Balls and other objects that are thrown or struck require force to stop them. Starting, stopping, and changing direction—a part of many physical activities—provide many examples of the law of inertia applied to body motion.

Because force is required to change inertia, it is obvious that any activity that is carried out at a steady pace in a consistent direction will conserve energy and that any irregularly paced or directed

FIG. 3.15 • An example of Newton's first law of motion. The skier continues airborne in space because of the previously established inertia.

activity will be very costly to energy reserves. This explains in part why activities such as handball and basketball are so much more fatiguing than jogging and dancing.

### Law of acceleration

**A change in the acceleration of a body occurs in the same direction as the force that caused it. The change in acceleration is directly proportional to the force causing it and inversely proportional to the mass of the body.**

**Acceleration** may be defined as the rate of change in velocity. To attain speed in moving the body, a strong muscular force is generally necessary. **Mass**, the amount of matter in a body, affects the speed and acceleration in physical movements. A much greater force is required from the muscles to accelerate an 80-kilogram man than to accelerate a 58-kilogram man to the same running speed. Also, it is possible to accelerate a baseball faster than a shot because of the difference in mass. The force required to run at half speed is less than the force required to run at top speed. To impart speed to a ball or an object, it is necessary to rapidly accelerate the part of the body holding the object. Football, basketball, track, and field hockey are a few sports that demand speed and acceleration.

### Law of reaction

**For every action there is an opposite and equal reaction.**

As we place force on a supporting surface by walking over it, the surface provides an equal resistance back in the opposite direction to the soles

FIG. 3.16 • An example of Newton's third law of motion. To accelerate forward, a walker must push backward. Note that the front part of the footprint in the sand is more deeply depressed than the rear part.

FIG. 3.17 • Friction. **A**, Static friction; **B**, Static friction also, but less than in A because there is less mass (weight); **C**, Kinetic friction is always less than static friction; **D**, Rolling friction is always less than kinetic friction.

of our feet. Our feet push down and back, while the surface pushes up and forward. The force of the surface reacting to the force we place on it is referred to as **ground reaction force**. We provide the action force, while the surface provides the reaction force. It is easier to run on a hard track than on a sandy beach because of the difference in the ground reaction forces of the two surfaces. The track resists the runner's propulsion force, and the reaction drives the runner ahead. The sand dissipates the runner's force, and the reaction force is correspondingly reduced, with an apparent loss in forward force and speed (Fig. 3.16). A sprinter applies a force in excess of 1335 Newtons on the starting blocks, which resist with an equal force. When a body is in flight, as in jumping, movement of one part of the body produces a reaction in another part because there is no resistive surface to supply a reaction force.

## Friction

**Friction** is the force that results from the resistance between the surfaces of two objects moving on each other. Depending on the activity involved, we may desire increased or decreased friction. In running, we depend on friction forces between our feet and the ground so that we may exert force against the ground and propel ourselves forward. When friction is reduced due to a slick ground or shoe surface, we are more likely to slip. In skating, we desire decreased friction so that we may slide across the ice with less resistance. Friction may be further characterized as either static or kinetic. See Fig. 3.17, *A, B,* and *C.* **Static friction** is the amount

of friction between two objects that have not yet begun to move, whereas **kinetic friction** is the friction between two objects that are sliding along each other. Static friction is always greater than kinetic friction. As a result, it is always more difficult to initiate dragging an object across a surface than it is to continue dragging it. Static friction may be increased by increasing the normal or perpendicular forces pressing the two objects together, as by adding more weight to one object sitting on another object. To determine the amount of friction forces, we must consider both the forces pressing the two objects together and the **coefficient of friction**, which depends on the hardness and roughness of the surface textures. The coefficient of friction is the ratio of the force needed to overcome the friction to the force holding the surfaces together. **Rolling friction** (Fig. 3.17, *D*) is the resistance to an object rolling across a surface, such as a ball rolling across a court or a tire rolling across the ground. Rolling friction is always much less than static or kinetic friction.

## Balance, equilibrium, and stability

**Balance** is the ability to control equilibrium, either static or dynamic. In relation to human movement, **equilibrium** refers to a state of zero acceleration, where there is no change in the speed or direction of the body. Equilibrium may be either static or dynamic. If the body is at rest or completely motionless, it is in **static equilibrium**. **Dynamic equilibrium** occurs when all the applied and inertial forces acting on the moving body are in balance, resulting in movement with unchanging speed or direction. For us to control equilibrium and hence achieve balance, we need to maximize **stability**. Stability is the resistance to a change in the body's acceleration or, more appropriately, the resistance to a disturbance of the body's equilibrium. Stability may be enhanced by determining the body's **center of gravity** and changing it appropriately. The center of gravity is the point at which all of the body's mass and weight is equally balanced or equally distributed in all directions. Very generally, the center of gravity for humans is located in the vicinity of the umbilicus.

Balance is important for the resting body as well as for the moving body. Generally, balance is to be desired, but there are circumstances in which movement is improved when the body tends to be unbalanced. Following are certain general factors that apply toward enhancing equilibrium, maximizing stability, and ultimately achieving balance.

1. A person has balance when the center of gravity falls within the base of support (Fig. 3.18).
2. A person has balance in direct proportion to the size of the base. The larger the base of support, the more balance.
3. A person has balance depending on the weight (mass). The greater the weight, the more balance.
4. A person has balance depending on the height of the center of gravity. The lower the center of gravity, the more balance.

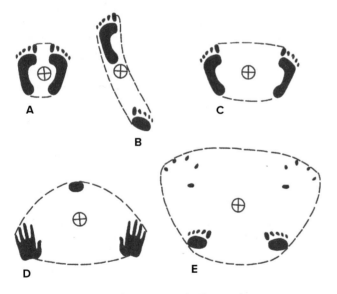

**A**  **B**  **C**  **D**  **E**

FIG. 3.18 • Base of support. The base of support includes the part(s) of the body in contact with the supporting surface and the intervening area. **A, B,** and **C,** The weight is supported by the feet; **D,** The weight is supported by the forehead and hands during a headstand; **E,** The weight is supported by the hands and feet while the body is in a squat position. Circled crosses indicate the point of intersection of the line of gravity with the base of support.

5. A person has balance depending on where the center of gravity is in relation to the base of support. The balance is less if the center of gravity is near the edge of the base. However, when anticipating an oncoming force, stability may be improved by placing the center of gravity nearer the side of the base of support expected to receive the force.
6. In anticipation of an oncoming force, stability may be increased by enlarging the size of the base of support in the direction of the anticipated force.
7. Equilibrium may be enhanced by increasing the friction between the body and the surfaces it contacts.
8. Rotation about an axis aids balance. A moving bike is easier to balance than a stationary bike.
9. Kinesthetic physiological functions contribute to balance. The semicircular canals of the inner ear, vision, touch (pressure), and kinesthetic sense all provide balance information to the performer. Balance and its components of equilibrium and stability are essential in all movements. All are affected by the constant force of gravity, as well as by inertia. Walking has been described as an activity in which a person throws the body in and out of balance with each step. In rapid running movements in which moving inertia is high, the individual has to lower the center of gravity to maintain balance when stopping or changing direction. Conversely, in jumping activities, the individual attempts to raise the center of gravity as high as possible.

## Force

Muscles are the main source of force that produces or changes movement of a body segment, the entire body, or an object thrown, struck, or stopped. As discussed previously, a variety of factors affect the ability of a muscle to exert force. We obviously need to understand these various factors. And we must utilize this knowledge in properly managing the factors to condition our muscles appropriately to achieve the desired response in dealing with both internal and external forces. As a result, we usually desire stronger muscles in order to be able to produce more force for both maximum and sustained exertion.

**Forces** either push or pull on an object in an attempt to affect motion or shape. Without forces acting on an object, there is no motion. Force is the product of mass times acceleration. The mass of a body segment or the entire body times the speed of acceleration determines the force. Obviously, in football this is very important, yet it is just as important in other activities that use only a part of the human body. In throwing a ball, the force applied to the ball is equal to the mass of the arm times the arm's speed of acceleration. Also, as previously discussed, leverage is important.

$$Force = mass \times acceleration$$
$$F = m \times a$$

The quality of motion, or, more scientifically stated, the **momentum**, which is equal to mass times velocity, is important in skill activities. The greater the momentum, the greater the resistance to change in the inertia or state of motion. In other words, a larger person with greater mass moving at the same velocity as a smaller person will have more momentum. On the other hand, a person with less mass moving at a higher velocity may have more momentum than a person with greater mass moving at a lower velocity. Momentum may be altered by **impulse**, which is the product of force and time.

It is not necessary to apply maximal force and thereby increase the momentum of a ball or an object being struck in all situations. In skillful performance, regulation of the amount of force is necessary. Judgment as to the amount of force required to throw a softball a given distance, hit a golf ball 200 yards, or hit a tennis ball across the net and into the court is important.

In activities involving movement of various joints, as in throwing a ball or putting a shot, there should be a summation of forces from the beginning of movement in the lower segment of the body to the twisting of the trunk and movement at the shoulder, elbow, and wrist joints. The speed at which a golf club strikes the ball is the result of a summation of forces of the lower extremities, trunk, shoulders, arms, and wrists. Shot-putting and discus and javelin throwing are other good examples that show that a summation of forces is essential.

## Mechanical loading basics

As we utilize the musculoskeletal system to exert force on the body to move and to interact with the ground and other objects or people, significant mechanical loads are generated and absorbed by the tissues of the body. The forces causing these loads may be internal or external. Only muscles can actively generate internal force, but tension in

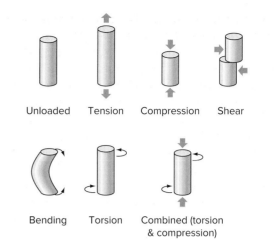

Unloaded  Tension  Compression  Shear

Bending  Torsion  Combined (torsion & compression)

FIG. 3.19 • Mechanical loading forces.

tendons, connective tissues, ligaments, and joint capsules may passively generate internal forces. External forces are produced from outside the body and originate from gravity, inertia, or direct contact. All tissues, in varying degrees, resist changes in their shape. Obviously, tissue deformation may result from external forces, but we also have the ability to generate internal forces large enough to fracture bones, dislocate joints, and disrupt muscles and connective tissues. To prevent injury or damage from tissue deformation, we must use the body to absorb energy from both internal and external forces. Along this line, it is to our advantage to absorb such force over larger aspects of our body rather than smaller ones, and to spread the absorption rate over a longer period of time. Additionally, the stronger and healthier we are, the more likely we are to be able to withstand excessive mechanical loading and the resultant excessive tissue deformation. Tension (stretching or strain), compression, shear, bending, and torsion (twisting) are all forces that act individually or in combination to provide mechanical loading that may result in excessive tissue deformation. Fig. 3.19 illustrates the mechanical forces that act on tissues of the body.

## Functional application

In the performance of various sport skills, many applications of the laws of leverage, motion, and balance may be found. A skill common to many activities is throwing. The object thrown may be some type of ball, but it is frequently an object of another size or shape, such as a rock, beanbag, Frisbee, discus, or javelin. A brief analysis of some of the basic mechanical principles involved in the skill of throwing will help indicate the importance of understanding the applications of these principles. Many activities involve these and often other mechanical principles. Motion is basic to throwing when the angular motion (Fig. 2.21) of the levers (bones) of the body (trunk, shoulder, elbow, and wrist) is used to give linear motion to the ball when it is released.

Newton's laws of motion apply in throwing because the individual's inertia and the ball's inertia (see p. 82) must be overcome by the application of force. The muscles of the body provide the force to move the body parts and the ball held in the hand. The **law of acceleration** (Newton's second law) comes into operation with the muscular force necessary to accelerate the arm, wrist, and hand. The greater the force (mass times acceleration) a person can produce, the faster the arm will move and, thus, the greater the speed that will be imparted to the ball. The reaction of the feet against the surface on which the person stands illustrates the application of the **law of reaction**.

The leverage factor is very important in throwing a ball or an object. For all practical purposes, the body from the feet to the fingers can be considered one long lever. The longer the lever, either from natural body length or from the movement of the body to the extended backward position (as in throwing a softball, with extension of the shoulder and elbow joints), the greater will be the arc through which it accelerates and thus the greater will be the speed imparted to the thrown object.

In certain circumstances, when the ball is to be thrown only a short distance, as in baseball when it is thrown by the catcher to the bases, the short lever is advantageous because it takes less total time to release the ball.

Balance, or equilibrium, is a factor in throwing when the body is rotated to the rear in the beginning of the throw. This motion moves the body nearly out of balance to the rear, and balance then changes again in the body with the forward movement. Balance is again established with the follow-through, when the feet are spread and the knees and trunk are flexed to lower the center of gravity.

## Summary

The preceding discussion has been a brief overview of some of the factors affecting motion. Analysis of human motion in light of the laws of physics poses a problem: How comprehensive is the analysis to be? It can become very complex, particularly when body motion is combined with the manipulation of a handheld object involved in throwing, kicking, striking, or catching.

These factors become involved when we attempt an analysis of the activities common to our physical education programs—football, baseball, basketball, track and field, field hockey, and swimming, to mention a few. However, to have a complete view of which factors control human movement, we must have a working knowledge of both the physiological and biomechanical principles of kinesiology.

It is beyond the scope of this book to make a detailed analysis of other activities. Some sources that consider these problems in detail are listed in the references.

# REVIEW AND LABORATORY EXERCISES

1. Lever component identification chart

Determine and list two practical examples of levers (in the body or in daily living) for each lever class. Do not use examples already discussed in this chapter. For each example identify the force, axis, and resistance. Also explain the advantage of using each lever—that is, whether it is to achieve balance, force, motion, speed, or range of motion.

| Lever class | Example | Force | Axis | Resistance | Advantage provided |
|---|---|---|---|---|---|
| 1st | | | | | |
| 1st | | | | | |
| 2nd | | | | | |
| 2nd | | | | | |
| 3rd | | | | | |
| 3rd | | | | | |

2. Anatomical levers can improve physical performance. Explain how this occurs using the information you have learned in relation to throwing.

3. If your biceps brachii muscle inserts to your forearm 2 inches below your elbow, the distance from the elbow to the palm of your hand is 18 inches, and you lift a 20-pound weight, how much force must your muscle exert to achieve elbow flexion?

4. If the weight of an object is 50 kilograms and your mechanical advantage is 4, how much force would you need to exert to lift the object with a lever system?

5. For the lever system component calculation chart, arrange the lever components as listed for each task a. through j. Determine the lever class and calculate the values for the force arm (FA), the resistance arm (RA), the force, and the mechanical advantage (MA). You may want to draw all the various arrangements of the components on a separate sheet of paper using the components below.

*Each vertical line on the lever bar represents the points at which the components are to be arranged, with the left endpoint representing 0 and the right endpoint representing 20.*

## Lever system component calculation chart

| Lever components | | | Variables | | | | |
|---|---|---|---|---|---|---|---|
| Force applied at | Axis placed at | 25 Newtons of resistance placed at | Lever class | FA | RA | Force | MA |
| a. | 0 | 2 | 20 | | | | |
| b. | 0 | 9 | 15 | | | | |
| c. | 3 | 17 | 13 | | | | |
| d. | 8 | 4 | 19 | | | | |
| e. | 12 | 0 | 18 | | | | |
| f. | 19 | 9 | 3 | | | | |
| g. | 16 | 2 | 7 | | | | |
| h. | 13 | 20 | 4 | | | | |
| i. | 8 | 17 | 1 | | | | |
| j. | 20 | 4 | 11 | | | | |

6. What class of lever is an automobile steering wheel?

7. List two different wheel and axles in which the force is applied to the wheel. From your observation, estimate which of the two has the greater mechanical advantage.

8. List two different wheel and axles in which the force is applied to the axle. From your observation, estimate which of the two has the greater mechanical advantage.

9. When attempting to remove a screw, is it easier when using a screwdriver with a larger grip on the handle? Why?

10. If a pulley setup has five supporting ropes, what is the MA of the setup?

11. What amount of force is needed to lift an object in a pulley system if the weight of the object being lifted is 200 kg and the number of supporting ropes is four?

12. Identify a practical example of Newton's law of inertia. Explain how the example illustrates the law.

13. Identify a practical example of Newton's law of acceleration. Explain how the example illustrates the law.

14. Identify a practical example of Newton's law of reaction. Explain how the example illustrates the law.

15. If a baseball player hit a triple and ran around the bases to third base, what would be his displacement? *Hint:* The distance from each base to the next is 90 feet.

16. Select a sporting activity and explain how the presence of too much friction becomes a problem in the activity.

17. Select a sporting activity and explain how the presence of too little friction becomes a problem in the activity.

18. Using the mechanical loading basic forces of compression, torsion, and shear, describe each force by using examples from soccer or volleyball.

19. Laws of motion task comparison chart

| | For this chart, assume that you possess the skill, strength, etc. to be able to perform each of the paired tasks. Circle the task that would be easier to perform based on Newton's laws of motion and explain why. | |
|---|---|---|
| | Paired tasks | Explanation |
| a. | Throw a baseball 60 mph OR throw a shot-put 60 mph. | |
| b. | Kick a bowling ball 40 yards OR kick a soccer ball 40 yards. | |
| c. | Bat a whiffle ball over a 320-yard fence OR bat a baseball over a 320-yard fence. | |
| d. | Catch a shot-put that was thrown at 60 mph OR catch a softball that was thrown at 60 mph. | |
| e. | Tackle a 240-pound running back sprinting toward you full speed OR tackle a 200-pound running back sprinting toward you full speed. | |
| f. | Run a 40-yard dash in 4.5 seconds on a wet field OR run a 40-yard dash in 4.5 seconds on a dry field. | |

20. Develop special projects and class reports by individuals or small groups of students on the mechanical analysis of all the skills involved in the following:
    a. Basketball
    b. Baseball
    c. Dancing
    d. Diving
    e. Football
    f. Field hockey
    g. Golf
    h. Gymnastics
    i. Soccer
    j. Swimming
    k. Tennis
    l. Wrestling

    q. Leverage
    r. Lift
    s. Linear displacement
    t. Mass
    u. Momentum
    v. Motion
    w. Projectiles
    x. Rebound angle
    y. Restitution
    z. Speed
    aa. Spin
    bb. Stability
    cc. Thrust
    dd. Torque
    ee. Velocity
    ff. Vector composition
    gg. Vector resolution
    hh. Weight
    ii. Work

21. Develop term projects and special class reports by individuals or small groups of students on the following factors in motion:
    a. Acceleration
    b. Aerodynamics
    c. Angular displacement
    d. Balance
    e. Base of support
    f. Buoyancy
    g. Center of gravity
    h. Drag
    i. Equilibrium
    j. Force
    k. Friction
    l. Gravity
    m. Hydrodynamics
    n. Impulse
    o. Inertia
    p. Instantaneous center of rotation

22. Develop demonstrations, term projects, or special reports by individuals or small groups of students on the following activities:
    a. Lifting
    b. Throwing
    c. Standing
    d. Walking
    e. Running
    f. Jumping
    g. Falling
    h. Sitting
    i. Pushing and pulling
    j. Striking

Chapter
**3**

# References

Adrian MJ, Cooper JM: *The biomechanics of human movement,* Indianapolis, IN, 1989, Benchmark.

American Academy of Orthopaedic Surgeons; Schenck RC, ed.: *Athletic training and sports medicine,* ed 3, Rosemont, IL, 1999, American Academy of Orthopaedic Surgeons.

Barham JN: *Mechanical kinesiology,* St. Louis, 1978, Mosby.

Broer MR: *An introduction to kinesiology,* Englewood Cliffs, NJ, 1968, Prentice-Hall.

Broer MR, Zernicke RF: *Efficiency of human movement,* ed 3, Philadelphia, 1979, Saunders.

Bunn JW: *Scientific principles of coaching,* ed 2, Englewood Cliffs, NJ, 1972, Prentice-Hall.

Cooper JM, Adrian M, Glassow RB: *Kinesiology,* ed 5, St. Louis, 1982, Mosby.

Donatelli R, Wolf SL: *The biomechanics of the foot and ankle,* Philadelphia, 1990, Davis.

Hall SJ: *Basic biomechanics,* ed 7, New York, 2015, McGraw-Hill.

Hamill J, Knutzen KM: *Biomechanical basis of human movement,* ed 3, Baltimore, 2008, Lippincott Williams & Wilkins.

Hamilton N, Weimar W, Luttgens K: *Kinesiology: scientific basis of human motion,* ed 12, New York, 2012, McGraw-Hill.

Hinson M: *Kinesiology,* ed 4, New York, 1981, McGraw-Hill.

Kegerreis S, Jenkins WL, Malone TR: Throwing injuries, *Sports Injury Management* 2:4, 1989.

Kelley DL: *Kinesiology: fundamentals of motion description,* Englewood Cliffs, NJ, 1971, Prentice-Hall.

Kreighbaum E, Barthels KM: *Biomechanics: a qualitative approach for studying human movement,* ed 4, New York, 1996, Allyn & Bacon.

Logan GA, McKinney WC: *Anatomic kinesiology,* ed 3, New York, 1982, McGraw-Hill.

McCreary EK, Kendall FP, Rodgers MM, Provance PG, Romani WA: *Muscles: testing and function with posture and pain,* ed 5, Philadelphia, 2005, Lippincott Williams & Wilkins.

McGinnis PM: *Biomechanics of sport and exercise,* ed 2, Champaign, IL, 2005, Human Kinetics.

Neumann, DA: *Kinesiology of the musculoskeletal system: foundations for physical rehabilitation,* ed 2, St. Louis, 2010, Mosby.

Nordin M, Frankel VH: *Basic biomechanics of the musculoskeletal system,* ed 3, Philadelphia, 2001, Lippincott Williams & Wilkins.

Norkin CC, Levangie PK: *Joint structure and function—a comprehensive analysis,* ed 5, Philadelphia, 2011, Davis.

Northrip JW, Logan GA, McKinney WC: *Analysis of sport motion: anatomic and biomechanic perspectives,* ed 3, New York, 1983, McGraw-Hill.

Piscopo J, Baley J: *Kinesiology: the science of movement,* New York, 1981, Wiley.

Prentice WE: *Principles of athletic training: a competency based approach,* ed 15, New York, 2014, McGraw-Hill.

Rasch PJ: *Kinesiology and applied anatomy,* ed 7, Philadelphia, 1989, Lea & Febiger.

Scott MG: *Analysis of human motion,* ed 2, New York, 1963, Appleton-Century-Crofts.

Segedy A: Braces' joint effects spur research surge, *Biomechanics,* February 2005.

Weineck J: *Functional anatomy in sports,* ed 2, St. Louis, 1990, Mosby.

Whiting WC, Zermicke R: *Biomechanics of musculoskeletal injury,* ed 2, Champaign, IL, 2008, Human Kinetics.

Wirhed R: *Athletic ability and the anatomy of motion,* ed 3, St. Louis, 2006, Mosby Elsevier.

Chapter
3

# THE SHOULDER GIRDLE

## Objectives

- To identify on the skeleton important bony features of the shoulder girdle

- To label on a skeletal chart the important bony features of the shoulder girdle

- To draw on a skeletal chart the muscles of the shoulder girdle and indicate shoulder girdle movements using arrows

- To demonstrate, using a human subject, all the movements of the shoulder girdle and list their respective planes of movement and axes of rotation

- To palpate the muscles of the shoulder girdle on a human subject and list their antagonists

- To palpate the joints of the shoulder girdle on a human subject during each movement through the full range of motion and appreciate the role of the ligaments in providing stability

- To learn and understand the innervation of the shoulder girdle muscles

- To determine, through analysis, the shoulder girdle movements and the muscles involved in selected skills and exercises

T he entire upper extremity depends upon the shoulder girdle to serve as a base from which to function. The only attachment of the upper extremity to the axial skeleton is via the scapula and its attachment through the clavicle at the sternoclavicular joint. To enhance understanding of how the shoulder joint and the rest of the upper extremity depend on the shoulder girdle, we will discuss it separately from the other structures.

Brief descriptions of the most important bones in the shoulder region will help you understand the skeletal structure and its relationship to the muscular system.

## Bones

Two bones are primarily involved in movements of the shoulder girdle. They are the scapula and the clavicle, which generally move as a unit. Their only bony link to the axial skeleton is provided by the clavicle's articulation with the sternum. Key bony landmarks for studying the shoulder girdle are the manubrium, clavicle, coracoid process, acromion process, glenoid fossa, lateral border, inferior angle, medial border, superior angle, and spine of the scapula (Figs. 4.1, 4.2, 4.3, and 4.4).

## Joints

When analyzing shoulder girdle (scapulothoracic) movements, it is important to realize that the scapula moves on the rib cage as a consequence of joint motion actually occurring at the sternoclavicular joint and to a much lesser extent at the acromioclavicular joint (see Figs. 4.1 and 4.3).

### Sternoclavicular (SC)

This is classified as a (multiaxial) arthrodial joint although there is anatomical justification to classify it as a sellar joint. In relation to the manubrium of the sternum, the clavicle moves anteriorly 15 degrees with protraction and moves posteriorly 15 degrees

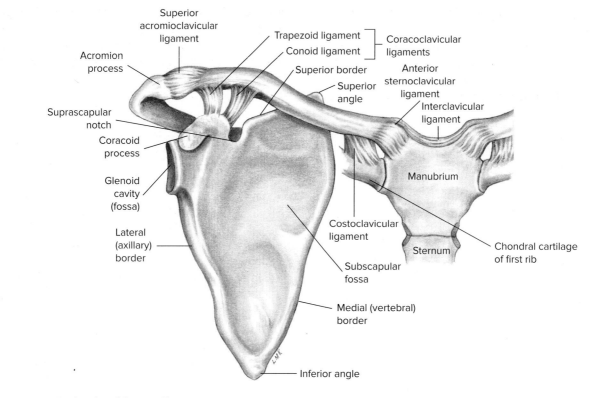

FIG. 4.1 • Right shoulder girdle, anterior view.

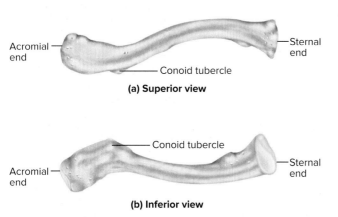

FIG. 4.2 • Right clavicle. **A**, Superior view; **B**, Inferior view.

with retraction. Some rotation of the clavicle along its axis during various movements of the shoulder girdle results in slight rotary gliding movement at the sternoclavicular joint. It is supported anteriorly by the anterior sternoclavicular ligament and posteriorly by the posterior sternoclavicular ligament. The costoclavicular and interclavicular ligaments also provide stability against superior displacement.

## Acromioclavicular (AC)

This joint is classified as an arthrodial joint. It has a 20- to 30-degree total gliding and rotational motion accompanying other shoulder girdle and shoulder joint motions. In addition to the strong support provided by the coracoclavicular ligaments (conoid and trapezoid), the superior and inferior acromioclavicular ligaments provide stability to this often-injured joint. The coracoclavicular joint, classified as a syndesmotic-type joint, functions through its ligaments to greatly increase the stability of the acromioclavicular joint. See Table 4.1 for further details on the ligaments supporting the shoulder girdle.

## Scapulothoracic

This joint is not a true synovial joint, due to its not having regular synovial features and to the fact that its movement is totally dependent on the sternoclavicular and acromioclavicular joints. Even though scapula movement occurs as a result of motion at the SC and AC joints, the scapula can be described as having a total range of 25-degree abduction–adduction movement, 60-degree upward–downward rotation, and 55-degree elevation–depression. The scapulothoracic joint is supported dynamically by its muscles and lacks ligamentous support, since it has no synovial features.

There is not a typical articulation between the anterior scapula and the posterior rib cage. Between these two osseous structures is the serratus anterior muscle originating off the upper nine ribs laterally and running just behind the rib cage posteriorly to insert on the medial border of the scapula. Immediately posterior to the serratus anterior is the subscapularis muscle (see Chapter 5) on the anterior scapula.

## TABLE 4.1 • Major ligaments of the shoulder girdle

| Joint | Ligament | | Proximal Attachments | Distal Attachments | Functional Application and Joint Stability |
|-------|----------|---|----------------------|--------------------|--------------------------------------------|
| Acromioclavicular (AC) joint | Acromio-clavicular | Superior AC ligament | Superior aspect of distal end of the clavicle | Superior surface of acromion process of the scapula | Functions as a capsular ligament that surrounds the AC joint, provides horizontal stability to the acromioclavicular joint |
| | | Inferior AC ligament | Inferior aspect of distal end of the clavicle | Inferior surface of acromion process of the scapula | |
| Coracoclavicular (CC) joint | Coraco-clavicular | Conoid ligament | Superior aspect of base of coracoid process | Conoid tubercle on inferior surface of distal clavicle | Posteromedial coracoclavicular ligament, cone shaped, major stabilizer of AC joint, limits anterior displacement of scapula from clavicle |
| | | Trapezoid ligament | Superior aspect of coracoid process | Inferior surface at trapezoid line of distal clavicle of the clavicle | Anterolateral coracoclavicular ligament, sheet like in appearance, major stabilizer of AC joint, limits posterior displacement of scapula from clavicle |
| Sternoclavicular (SC) joint | Costoclavicular | | Superior aspect of 1st costal cartilage | Inferior aspect of proximal clavicle just later to SC joint | Resists superior displacement of the proximal end of the clavicle and limits excessive shoulder girdle elevation |
| | Anterior sternoclavicular | | Anteriorly on superolateral border of the manubrium of the sternum | Anterior and superior aspect of proximal clavicle | Reinforces sternoclavicular joint anteriorly and assist in preventing anterior subluxation/dislocation |
| | Posterior sternoclavicular | | Posteriorly on superolateral border of the manubrium of the sternum | Posterior and superior aspect of proximal clavicle | Reinforces sternoclavicular joint posteriorly and assists in preventing posterior subluxation/dislocation |
| | Interclavicular | | Superior aspect of proximal clavicle via the superior margin of manubrium to superior aspect of contralateral proximal clavicle | Superior aspect of proximal clavicle via the superior margin of manubrium to superior aspect of contralateral proximal clavicle | Resists downward glide of proximal clavicle and excessive depression of shoulder girdle |

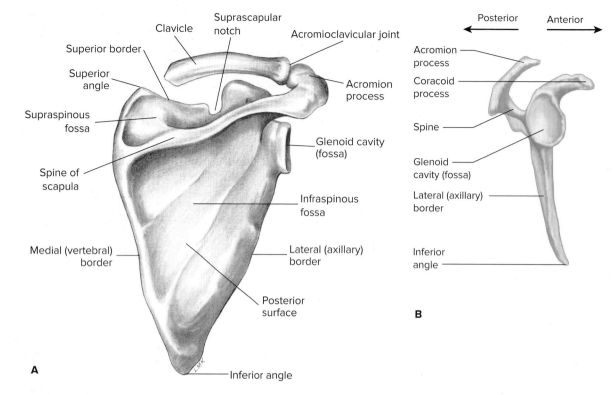

FIG. 4.3 • Right scapula. **A**, Posterior view; **B**, Lateral view.

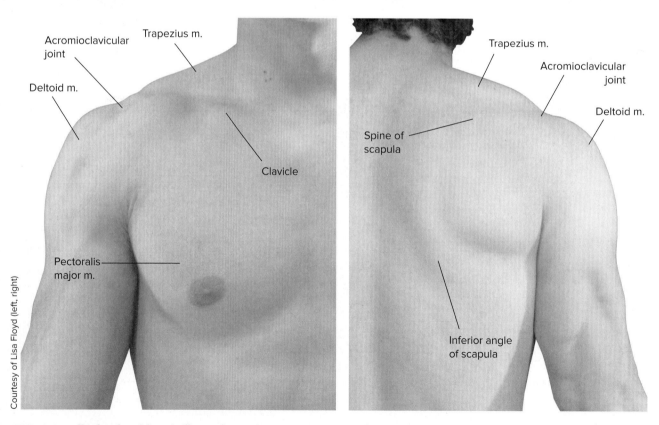

FIG. 4.4 • Right shoulder girdle surface anatomy, anterior and posterior views.

# Movements FIGS. 4.5, 4.6

In analyzing shoulder girdle movements, it is often helpful to focus on a specific scapular bony landmark, such as the inferior angle (posteriorly), the glenoid fossa (laterally), or the acromion process (anteriorly). All of these movements have their pivot point where the clavicle joins the sternum at the sternoclavicular joint.

Courtesy of Britt Jones (all)

**Abduction**
(protraction)
**A**

**Adduction**
(retraction)
**B**

**Elevation**
**C**

**Depression**
**D**

**Upward rotation**
**E**

**Downward rotation**
**F**

FIG. 4.5 • Movements of the shoulder girdle.

Scapula
abduction

Scapula
adduction

Scapula
elevation

Scapula
depression

Scapula
upward
rotation

Scapula
downward
rotation

Movements of the shoulder girdle can be described as movements of the scapula, but it is important to remember that wherever the scapula goes, the clavicle follows. Figs. 4.5 and 4.6 show movements of the shoulder girdle.

**Abduction (protraction):** Movement of the scapula laterally away from the spinal column, as in reaching for an object in front of the body

**Adduction (retraction):** Movement of the scapula medially toward the spinal column, as in pinching the shoulder blades together

**Elevation:** Upward or superior movement of the scapula, as in shrugging the shoulders

**Depression:** Downward or inferior movement of the scapula, as in returning to a normal position from a shoulder shrug

**Upward rotation:** Turning the glenoid fossa upward and moving the inferior angle superiorly and laterally away from the spinal column to assist in raising the arm out to the side

**Downward rotation:** Returning the inferior angle medially and inferiorly toward the spinal column and the glenoid fossa to its normal position, as in bringing the arm down to the side. (Once the scapula has returned to its anatomical position, further downward rotation actually results in the superior angle moving slightly superomedial.)

To accomplish some of the previously listed shoulder girdle movements, the scapula must rotate or tilt on its axis. While these are not primary movements of the shoulder girdle, they are necessary for the scapula to move normally throughout its range of motion during shoulder girdle movements.

**Lateral tilt (outward tilt):** Consequential movement during abduction in which the scapula rotates about its vertical axis, resulting in posterior movement of the medial border and anterior movement of the lateral border

**Medial tilt (inward tilt):** Return from lateral tilt; consequential movement during extreme adduction in which the scapula rotates about its vertical axis, resulting in anterior movement of the medial border and posterior movement of the lateral border

**Anterior tilt (upward tilt):** Consequential rotational movement of the scapula about the frontal axis occurring during hyperextension of the glenohumeral joint, resulting in the superior border moving anteroinferiorly and the inferior angle moving posterosuperiorly

**Posterior tilt (downward tilt):** Consequential rotational movement of the scapula about the frontal axis occurring during hyperflexion of the glenohumeral joint, resulting in the superior border moving posteroinferiorly and the inferior angle moving anterosuperiorly

## Synergy with the muscles of the glenohumeral joint

The shoulder joint and shoulder girdle work together in carrying out upper-extremity activities. It is critical to understand that movement of the shoulder girdle is not dependent on the shoulder joint and its muscles. However, the muscles of the shoulder girdle are essential in providing a scapula-stabilizing effect, so that the muscles of the shoulder joint will have a stable base from which to exert force for powerful movement involving the humerus. Consequently, the shoulder girdle muscles contract to maintain the scapula in a relatively static position during many shoulder joint actions.

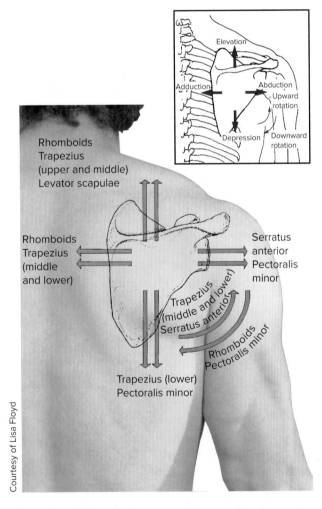

FIG. 4.6 • Actions of the scapular muscles. Posterior view with actions.

As the shoulder joint goes through more extreme ranges of motion, the scapular muscles contract to move the shoulder girdle so that its glenoid fossa will be in a more appropriate position from which the humerus can move. Without the accompanying movement of the scapula, we can raise our humerus only to around 90 to 120 degrees of total shoulder abduction and flexion. This works through the action of the appropriate muscles of both joints working in synergy to accomplish the desired action of the entire upper extremity. For example, if we want to raise our hand out to the side laterally as high as possible, the serratus anterior and trapezius (middle and lower fibers) muscles upwardly rotate the scapula as the supraspinatus and deltoid initiate glenohumeral abduction. This synergy between the scapula and shoulder joint muscles enhances movement of the entire upper extremity. Further discussion of the interaction and teamwork between these joints is provided at the beginning of Chapter 5, with Table 5.1 listing the shoulder girdle movements that usually accompany shoulder joint movements. Additional discussion of scapulohumeral rhythm is provided in Chapter 5 under Joint.

## Muscles

There are five muscles primarily involved in shoulder girdle movements, as shown in Fig. 4.7: the pectoralis minor, serratus anterior, trapezius, rhomboid, and levator scapulae. To avoid confusion, it is helpful to group muscles of the shoulder girdle separately from those of the shoulder joint. The subclavius muscle is also included in this group, but it is not regarded as a primary mover in any actions of the shoulder girdle. All five shoulder girdle muscles have their origin on the axial skeleton, with their insertion located on the scapula and/or the clavicle. Shoulder girdle muscles do not attach to the humerus, nor do they cause actions of the shoulder joint. The pectoralis minor and subclavius are located anteriorly in relation to the trunk. The serratus anterior is located anteriorly to the scapula but posteriorly and laterally to the trunk. Located posteriorly to the trunk and cervical spine are the trapezius, rhomboid, and levator scapulae.

The shoulder girdle muscles are essential in providing dynamic stability of the scapula so that it can serve as a relative base of support for shoulder joint activities such as throwing, batting, and blocking.

The scapula muscles also play a role in spinal posture. Typically, due to poor posture and the way we use our muscles through life, we tend to develop a forward shoulder posture that results in the scapula protractors and depressors becoming stronger and

tighter and the retractors becoming weaker. This leads to further depression and protraction, or a forward shoulder posture, which also contributes to increased kyphosis (increased posterior convexity of the thoracic spine) and a forward head with increased lordosis (increased posterior concavity of the cervical spine). See Chapter 12. This, in turn, places more stress on the posterior spinal muscles and also places the glenohumeral joint in a less functional and more compromised position. To avoid this, we should routinely practice good posture, beginning with an appropriate lumbar lordotic curve, and keep our shoulder girdle directly over our pelvis instead of forward. This will make it easier to maintain our head and cervical spine over the trunk in a correctly balanced position. An additional benefit of good scapular and spinal posture is easier inspiration due to less weight and mass over the rib cage and thoracic cavity.

Scapular winging is relatively rare but can affect normal functional activity of the upper extremity. Most commonly it affects the serratus anterior, leading to medial winging when pushing forward or raising the arm. The serratus anterior weakness or paralysis is typically due to injury of the long thoracic nerve, which can have a variety of causes. Much less commonly the trapezius and/or rhomboid may be affected, leading to lateral winging.

### Shoulder girdle muscles—location and action

Anterior
    Pectoralis minor—abduction, downward rotation, and depression
    Subclavius—depression and abduction
Posterior and laterally
    Serratus anterior—abduction and upward rotation
Posterior
    Trapezius
        Upper fibers—elevation and extension and rotation of the head at the neck
        Middle fibers—elevation, adduction, and upward rotation
        Lower fibers—adduction, depression, and upward rotation
    Rhomboid—adduction, downward rotation, and elevation
    Levator scapulae—elevation

It is important to understand that muscles may not necessarily be active throughout the absolute full range of motion for which they are noted as being agonists.

Table 4.2 provides a detailed breakdown of the muscles responsible for primary movements of the shoulder girdle.

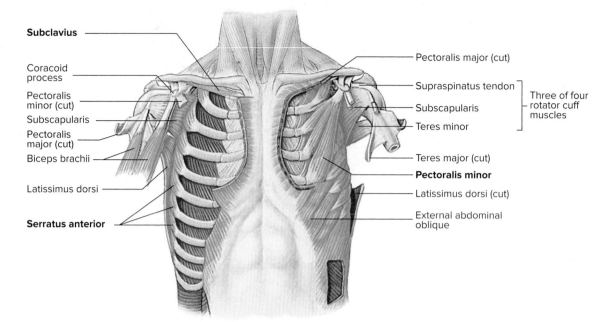

Subclavius

Coracoid process

Pectoralis minor (cut)

Subscapularis

Pectoralis major (cut)

Biceps brachii

Latissimus dorsi

**Serratus anterior**

Pectoralis major (cut)

Supraspinatus tendon

Subscapularis

Teres minor

} Three of four rotator cuff muscles

Teres major (cut)

**Pectoralis minor**

Latissimus dorsi (cut)

External abdominal oblique

**A**

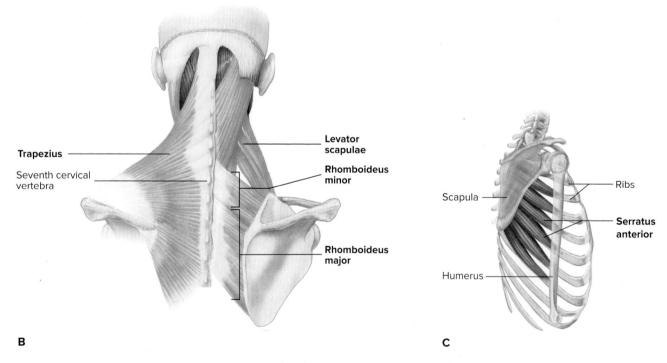

**Trapezius**

Seventh cervical vertebra

**Levator scapulae**

**Rhomboideus minor**

**Rhomboideus major**

Scapula

Humerus

Ribs

**Serratus anterior**

**B**

**C**

FIG. 4.7 • Muscles acting on the scapula. **A**, Anterior view: The pectoralis major is removed on both sides; **B**, Posterior view: The trapezius is removed on the right to reveal the deeper muscles; **C**, Lateral view: The serratus anterior.

## TABLE 4.2 · Agonist muscles of the shoulder girdle

| | Muscle | Origin | Insertion | Action | Plane of motion | Palpation | Innervation |
|---|---|---|---|---|---|---|---|
| **Anterior muscle** | Pectoralis minor | Anterior surfaces of the 3rd to 5th ribs | Coracoid process of the scapula | Abduction | Transverse | Difficult, but under the pectoralis major muscle and just inferior to the coracoid process during resisted depression; enhanced by placing the subject's hand behind the back and having the subject actively lift the hand away | Medial pectoral nerve (C8 and T1) |
| | | | | Downward rotation | Frontal | | |
| | | | | Depression | | | |
| **Posterior and lateral muscles** | Serratus anterior | Surface of the upper 9 ribs at the side of the chest | Anterior aspect of the whole length of the medial border of the scapula | Abduction | Transverse | Frontal and lateral side of the chest below the 5th and 6th ribs just proximal to the origin during abduction; best accomplished with the glenohumeral joint flexed 90 degrees; in the same position palpate the upper fibers between the lateral borders of the pectoralis major and latissimus dorsi in the axilla | Long thoracic nerve (C5–C7) |
| | | | | Upward rotation | Frontal | | |
| **Posterior muscles** | Trapezius upper fibers | Base of skull, occipital protuberance, and posterior ligaments of neck | Posterior aspect of the lateral 3rd of the clavicle | Elevation | Frontal | Between occipital protuberance and C6 and laterally to acromion, particularly during elevation and extension of the head at the neck | Spinal accessory nerve and branches of C3 and C4 |
| | | | | Upward rotation | | | |
| | | | | Extension of head at neck | Sagittal | | |
| | | | | Rotation of head at neck | Transverse | | |
| | Trapezius middle fibers | Spinous process of 7th cervical and upper 3 thoracic vertebrae | Medial border of the acromion process and superior border of the scapular spine | Elevation | Frontal | From C7 to T3 and laterally to acromion process and scapular spine, particularly during adduction | |
| | | | | Adduction | Transverse | | |
| | | | | Upward rotation | Frontal | | |
| | Trapezius lower fibers | Spinous process of 4th to 12th thoracic vertebrae | Triangular space at the base of the scapular spine | Adduction | Transverse | From T4 to T12 and medial aspect of scapular spine, particularly during depression and adduction | |
| | | | | Depression | | | |
| | | | | Upward rotation | Frontal | | |
| | Rhomboids | Spinous processes of the 7th cervical and first 5 thoracic vertebrae | Medial border of the scapula, inferior to the scapular spine | Adduction | Transverse | Difficult due to being deep to trapezius, but may be palpated through it during adduction; best accomplished with subject's ipsilateral hand behind the back to relax the trapezius and bring the rhomboid into action when the subject lifts the hand away from the back | Dorsal scapular nerve (C5) |
| | | | | Downward rotation | Frontal | | |
| | | | | Elevation | | | |
| | Levator scapulae | Transverse processes of the upper 4 cervical vertebrae | Medial border of the scapula from the superior angle to the scapular spine | Elevation | Frontal | Difficult to palpate due to being deep to trapezius; best palpated at insertion just medial to the superior angle of scapula, particularly during slight elevation | Dorsal scapular nerve C5 and branches of C3 and C4 |
| | | | | Weak downward rotation | | | |
| | | | | Weak adduction | Transverse | | |

**Note:** The subclavius is not listed because it is not a prime mover in shoulder girdle movements.

# Nerves

The muscles of the shoulder girdle are innervated primarily from the nerves of the cervical plexus and brachial plexus, as illustrated in Figs. 4.8 and 4.9. The trapezius is innervated by the spinal accessory nerve and from branches of C3 and C4. In addition to supplying the trapezius, C3 and C4 also innervate the levator scapula. The levator scapula receives further innervation from the dorsal scapular nerve originating from C5. The dorsal scapular nerve also innervates the rhomboid. The long thoracic nerve originates from C5, C6, and C7 and innervates the serratus anterior. The medial pectoral nerve arises from C8 and T1 to innervate the pectoralis minor. See Fig. 4.10 for a full listing of the shoulder girdle innervations.

Roots: C5, C6, C7, C8, T1
Trunks: upper, middle, lower
Anterior divisions
Posterior divisions
Cords: posterior, lateral, medial
Branches:  Axillary nerve
    Radial nerve
    Musculocutaneous nerve
    Median nerve
    Ulnar nerve

Roots (ventral rami)
Branches
Other nerves (not part of cervical plexus)

C1
C4

Hypoglossal nerve (XII)

Accessory nerve (XI)

Lesser occipital nerve

Nerve to sternocleidomastoid muscle

Greater auricular nerve

Superior root of ansa cervicalis

Transverse cervical nerve

Ansa cervicalis

Nerve to trapezius muscle

Inferior root of ansa cervicalis

Supraclavicular nerves

C1
C2
C3
C4
To brachial plexus
C5
Phrenic nerve

FIG. 4.8 • Cervical plexus, anterior view. The roots of the plexus are formed by the ventral rami of the spinal nerves C1–C4.

Dorsal scapular nerve

Suprascapular nerve

Subclavian nerve

Lateral cord

Posterior cord

Axillary nerve

Radial nerve

Musculo-cutaneous nerve

Medial and lateral pectoral nerves

Median nerve

Ulnar nerve

Medial brachial cutaneous nerve

C4
C5
Upper trunk
C6
Middle trunk
C7
Long thoracic nerve
C8
Lower trunk
T1
Medial cord

FIG. 4.9 • Brachial plexus, anterior view. The roots of the plexus are formed by the ventral rami of the spinal nerves C5–T1 and join to form an upper, middle, and lower trunk. Each trunk divides into anterior and posterior divisions. The divisions join together to form the posterior, lateral, and medial cords from which the major brachial plexus nerves arise.

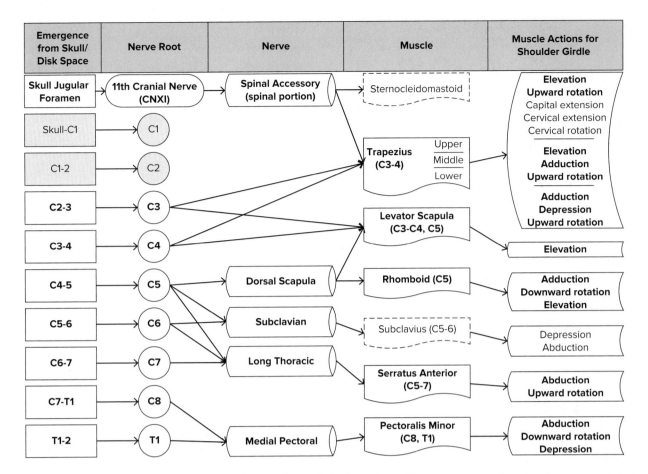

FIG. 4.10 • Innervation for shoulder girdle muscles and their actions. The neuromuscular structures associated with the primary movers for the scapula and their actions are bolded.

# Trapezius muscle   FIG. 4.11

(tra-pe ´zi-us)

## Origin

Upper fibers: base of skull, occipital protuberance, and posterior ligaments of neck

Middle fibers: spinous processes of seventh cervical and upper three thoracic vertebrae

Lower fibers: spinous processes of fourth through twelfth thoracic vertebrae

## Insertion

Upper fibers: posterior aspect of the lateral third of the clavicle

Middle fibers: medial border of the acromion process and upper border of the scapular spine

Lower fibers: triangular space at the base of the scapular spine

## Action

Upper fibers: elevation of the scapula, upward rotation, and extension and rotation of the head at the neck

Middle fibers: elevation, upward rotation, and adduction (retraction) of the scapula

Lower fibers: depression, adduction (retraction), and upward rotation of the scapula

## Palpation

Upper fibers: between occipital protuberance and C6 and laterally to acromion, particularly during elevation and extension of the head at the neck

Middle fibers: from C7 to T3 and laterally to acromion process and scapular spine, particularly during adduction

Lower fibers: from T4 to T12 and medial aspect of scapular spine, particularly during depression and adduction

Scapula elevation

Cervical extension

Cervical rotation unilaterally

Scapula upward rotation

Scapula adduction

Scapula depression

Upper fibers

Elevation (upper and middle fibers)

Middle fibers

Adduction (middle and lower fibers)

Upward rotation (middle and lower fibers)

Depression (lower fibers)

Lower fibers

O, Base of skull, occipital protuberance, posterior ligaments of neck, spinous processes of cervical (C7) and all thoracic vertebrae (T1–T12)

I, Posterior aspect of lateral third of clavicle, medial border of acromion process and upper border of scapular spine, triangular space at base of scapular spine

FIG. 4.11 • Trapezius muscle, posterior view. O, Origin; I, Insertion.

## Innervation

Spinal accessory nerve (cranial nerve XI) and
  branches of C3, C4

## Application, strengthening, and flexibility

The upper fibers are a thin and relatively weak part of the muscle. They provide some elevation of the clavicle. Due to their origin on the base of the skull, they assist in extension of the head.

The middle fibers are stronger and thicker and provide strong elevation, upward rotation, and adduction (retraction) of the scapula. Rarely is this portion of the muscle weak, because it is so active in positioning the shoulder for function and posture. As a result, it is often a source of tenderness and discomfort due to chronic tension.

The lower fibers depress the scapula, assist in adduction (retraction) and rotate the scapula upward. This portion is typically weak, particularly in individuals whose activities demand a significant amount of scapula abduction.

When all the parts of the trapezius are working together, they tend to pull upward and adduct at the same time. This may be seen in lifting the handles of a wheelbarrow. Typical action of the trapezius muscle is fixation of the scapula for deltoid action. Continuous action in upward rotation of the scapula permits the arms to be raised over the head. The muscle is always used in preventing the glenoid fossae from being pulled down during the lifting of objects with the arms. It is also typically seen in action during the holding of an object overhead. Holding the arm at the side horizontally shows typical fixation of the scapula by the trapezius muscle, while the deltoid muscle holds the arm in that position. The muscle is used strenuously when lifting with the hands, as in picking up a heavy wheelbarrow. The trapezius must prevent the scapula from being pulled downward. Carrying objects on the tip of the shoulder also calls this muscle into play. Strengthening of the upper and middle fibers can be accomplished through shoulder-shrugging exercises. The middle and lower fibers can be strengthened through bent-over rowing and shoulder joint horizontal abduction exercises from a prone position. The lower fibers can be emphasized with a chestproud shoulder retraction exercise attempting to place the elbows in the back pants pockets with depression. Parallel dips or body dips are also helpful for emphasizing the lower trapezius. See Appendix 2 for more commonly used exercises to address the trapezius and other muscles in this chapter.

To stretch the trapezius, each portion needs to be specifically addressed. The upper fibers may be stretched by using one hand to pull the head and neck forward into flexion or slight lateral flexion to the opposite side while the ipsilateral hand is hooked under a table edge to maintain the scapula in depression. The middle fibers are stretched to some extent with the procedure used for the upper fibers, but they may be stretched further by using a partner to passively pull the scapula into full protraction. The lower fibers are perhaps best stretched with the subject in a side-lying position while a partner grasps the lateral border and inferior angle of the scapula and moves it passively into maximal elevation and protraction.

# Levator scapulae muscle FIG. 4.12

(le-va´tor scap´u-lae)

## Origin

Transverse processes of the upper four cervical vertebrae

## Insertion

Medial border of the scapula from the superior angle to the scapular spine

## Action

Elevates the medial margin of the scapula
Weak downward rotation
Weak adduction

Scapula downward rotation

## Palpation

Difficult to palpate due to being deep to trapezius; best palpated at insertion just medial to the superior angle of the scapula, particularly during slight elevation

Scapula adduction

## Innervation

Dorsal scapular nerve C5 and branches of C3 and C4

## Application, strengthening, and flexibility

Shrugging the shoulders calls the levator scapulae muscle into play, along with the upper trapezius muscle. Fixation of the scapula by the pectoralis minor muscle allows the levator scapulae muscles on both sides to extend the neck or to flex laterally if used on one side only.

The levator scapulae is perhaps best stretched by rotating the head approximately 45 degrees contralaterally and flexing the cervical spine actively while maintaining the scapula in a relaxed, depressed position.

Like the trapezius, the levator scapulae is a very common site for tightness, tenderness, and discomfort secondary to chronic tension and from carrying items with straps over the shoulder.

Elevation

O, Transverse processes of upper four cervical vertebrae (C1–C4)

I, Medial border of scapula from superior angle to scapular spine

FIG. 4.12 • Levator scapulae muscle, posterior view. O, Origin; I, Insertion.

# Rhomboid muscles—major and minor  FIG. 4.13

(rom´boyd)

## Origin
Spinous processes of the seventh cervical and first five thoracic vertebrae

## Insertion
Medial border of the scapula, below the spine of the scapula

## Action
The rhomboid major and minor muscles work together. Adduction (retraction): draw the scapula toward the spinal column
Downward rotation: from the upward rotated position; draw the scapula into downward rotation
Elevation: slight upward movement accompanying adduction

## Palpation
Difficult to palpate due to being deep to the trapezius, but may be palpated through the relaxed trapezius during adduction. This is best accomplished by placing the subject's ipsilateral hand behind the back (glenohumeral internal rotation and scapula downward rotation), which relaxes the trapezius and brings the rhomboid into action when the subject lifts the hand away from the back.

## Innervation
Dorsal scapular nerve (C5)

## Application, strengthening, and flexibility
The rhomboid muscles fix the scapula in adduction (retraction) when the muscles of the shoulder joint adduct or extend the arm. These muscles are used powerfully in chinning. As one hangs from the horizontal bar, suspended by the hands, the scapula tends to be pulled away from the top of the chest. When the chinning movement begins, it is the rhomboid muscles that rotate the medial border of the scapula down and back toward the spinal column. Note their favorable position to do this. Related to this, the rhomboids work in a similar manner to prevent scapula winging.

The trapezius and rhomboid muscles working together produce adduction with slight elevation of the scapula. To prevent this elevation, the latissimus dorsi muscle is called into play.

Chin-ups, dips, and bent-over rowing are excellent exercises for developing strength in this muscle. The rhomboids may be stretched by passively moving the scapula into full protraction while maintaining depression. Upward rotation may assist in this stretch as well.

Scapula adduction

Scapula downward rotation

Scapula elevation

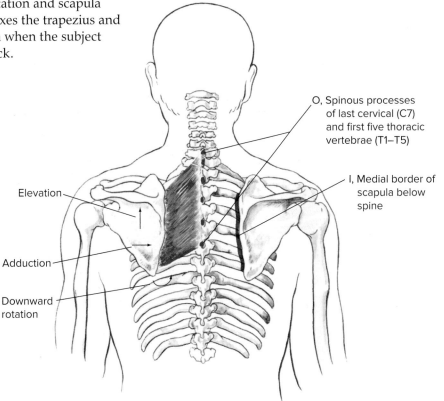

Elevation

Adduction

Downward rotation

O, Spinous processes of last cervical (C7) and first five thoracic vertebrae (T1–T5)

I, Medial border of scapula below spine

FIG. 4.13 • Rhomboid muscles (major and minor), posterior view. O, Origin; I, Insertion.

Scapula
abduction

# Serratus anterior muscle   FIG. 4.14

(ser-a´tus an-tir´e-or)

Chapter

4

## Origin

Surface of the upper nine ribs at the side of the chest

## Insertion

Anterior aspect of the whole length of the medial
border of the scapula

## Action

Abduction (protraction): draws the medial border of
the scapula away from the vertebrae
Upward rotation: longer, lower fibers tend to draw the
inferior angle of the scapula farther away from the
vertebrae, thus rotating the scapula upward slightly

## Palpation

Front and lateral side of the chest below the fifth
and sixth ribs just proximal to their origin during
abduction, which is best accomplished from a supine
position with the glenohumeral joint in 90 degrees
of flexion. The upper fibers may be palpated in the
same position between the lateral borders of the
pectoralis major and latissimus dorsi in the axilla.

Scapula
upward
rotation

## Innervation

Long thoracic nerve (C5–C7)

## Application, strengthening, and flexibility

The serratus anterior muscle is used commonly
in movements drawing the scapula forward with
slight upward rotation, such as throwing a baseball, punching in boxing, shooting and guarding in
basketball, and tackling in football. It works along
with the pectoralis major muscle in typical action,
such as throwing a baseball.

The serratus anterior muscle is used strongly
in doing push-ups, especially in the last 5 to
10 degrees of motion. The bench press and overhead press are good exercises for this muscle. A
winged scapula condition usually results from
weakness of the rhomboid and/or the serratus
anterior. Serratus anterior weakness may result
from an injury to the long thoracic nerve.

The serratus anterior can be stretched by standing, facing a corner and placing each hand at
shoulder level on the two walls. As you lean in and
attempt to place your nose in the corner, both scapulae are pushed into an adducted position, which
stretches the serratus anterior.

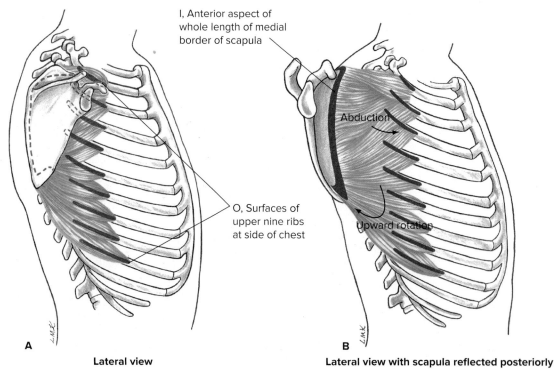

I, Anterior aspect of
whole length of medial
border of scapula

Abduction

O, Surfaces of
upper nine ribs
at side of chest

Upward rotation

**A**
**Lateral view**

**B**
**Lateral view with scapula reflected posteriorly
to reveal anterior surface**

FIG. 4.14 • Serratus anterior muscle, lateral view. O, Origin; I, Insertion.

# Pectoralis minor muscle   FIG. 4.15

(pek-to-ra´lis mi´nor)

## Origin

Anterior surfaces of the third to fifth ribs

## Insertion

Coracoid process of the scapula

## Action

Abduction (protraction): draws the scapula forward
  and tends to tilt the lower border away from the ribs
Downward rotation: as it abducts, it draws the scapula
  downward
Depression: when the scapula is rotated upward, it
  assists in depression

## Palpation

Difficult to palpate, but can be palpated under the
  pectoralis major muscle and just inferior to the coracoid process during resisted depression. This may
  be enhanced by placing the subject's hand behind
  the back and having him or her actively lift the hand
  away, which causes downward rotation.

## Innervation

Medial pectoral nerve (C8–T1)

## Application, strengthening, and flexibility

The pectoralis minor muscle is used, along with
the serratus anterior muscle, in true abduction
(protraction) without rotation. This is seen particularly in movements such as push-ups in which true
abduction of the scapula is necessary. Therefore,
the serratus anterior draws the scapula forward
with a tendency toward upward rotation, the pectoralis minor pulls forward with a tendency toward
downward rotation, and the two pulling together
give true abduction. These muscles will be seen
working together in most movements of pushing
with the hands.

The pectoralis minor is most used in depressing and rotating the scapula downward from an
upwardly rotated position, as in pushing the body
upward on dip bars or in body dips.

The pectoralis minor is often tight due to being
overused in activities involving abduction, which
may lead to forward and rounded shoulders. As a
result stretching may be indicated, which can be
accomplished with a wall push-up in the corner
as used for stretching the serratus anterior. Additionally, lying supine with a rolled towel directly
under the thoracic spine while a partner pushes
each scapula into retraction places this muscle on
stretch.

Scapula
abduction

Chapter
4

Scapula
downward
rotation

Scapula
depression

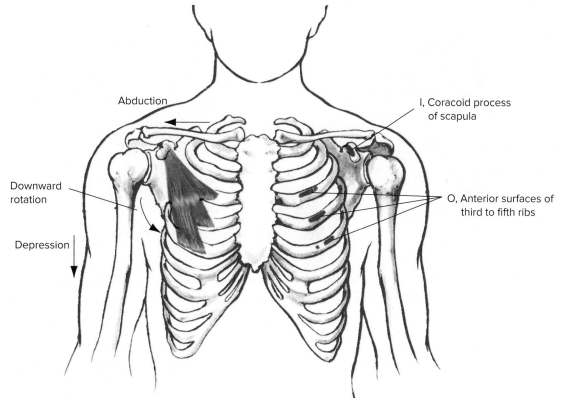

FIG. 4.15  •  Pectoralis minor muscle, anterior view. O, Origin; I, Insertion.

# Subclavius muscle   FIG. 4.16

(sub-klá ve-us)

## Origin

Superior aspect of first rib at its junction with its
costal cartilage

## Insertion

Inferior groove in the midportion of the clavicle

## Action

Stabilization and protection of the sternoclavicular
joint
Depression
Abduction (protraction)

## Palpation

Difficult to distinguish from the pectoralis major,
but may be palpated just inferior to the middle
third of the clavicle with the subject side-lying and
in a somewhat upwardly rotated position and the
humerus supported in a partially passively flexed
position. Slight active depression and abduction of
the scapula may enhance palpation.

## Innervation

Nerve fibers from C5 and C6

## Application, strengthening, and flexibility

The subclavius pulls the clavicle anteriorly and
inferiorly toward the sternum. In addition to
assisting in abducting and depressing the clavicle
and the shoulder girdle, it has a significant role
in protecting and stabilizing the sternoclavicular
joint during upper-extremity movements. It may
be strengthened during activities in which there
is active depression, such as dips, or active abduction, such as push-ups. Extreme elevation and
retraction of the shoulder girdle provide a stretch
to the subclavius.

Scapula
abduction

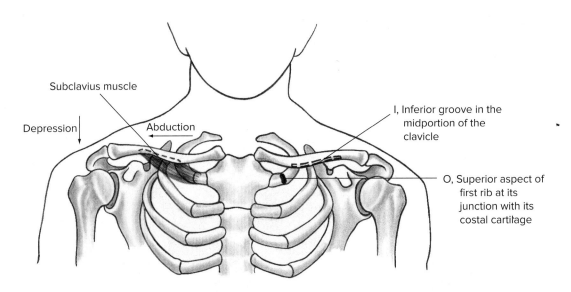

FIG. 4.16 • Subclavius muscle, anterior view. O, Origin; I, Insertion.

# REVIEW EXERCISES

1. List the planes in which each of the following shoulder girdle movements occurs. List the respective axis of rotation for each movement in each plane.
   a. Adduction
   b. Abduction
   c. Rotation upward
   d. Rotation downward
   e. Elevation
   f. Depression

2. **Muscle analysis chart** • Shoulder girdle

| Complete the chart by listing the muscles primarily involved in each movement. | |
|---|---|
| Abduction | Adduction |
| Elevation | Depression |
| Upward rotation | Downward rotation |

3. **Antagonistic muscle action chart** • Shoulder girdle

| Agonist | Antagonist |
|---|---|
| Serratus anterior | |
| Trapezius (upper fibers) | |
| Trapezius (middle fibers) | |
| Trapezius (lower fibers) | |
| Rhomboid | |
| Levator scapulae | |
| Pectoralis minor | |

Complete the chart by listing the muscle(s) or parts of muscles that are antagonist in their actions to the muscles in the left column.

# LABORATORY EXERCISES

1. Locate the following prominent skeletal features on a human skeleton and on a subject:
   a. Scapula
      1. Medial border
      2. Inferior angle
      3. Superior angle
      4. Coracoid process
      5. Spine of scapula
      6. Glenoid cavity
      7. Acromion process
      8. Supraspinatus fossa
      9. Infraspinatus fossa
   b. Clavicle
      1. Sternal end
      2. Acromial end
   c. Joints
      1. Sternoclavicular joint
      2. Acromioclavicular joint
2. Describe how and where you palpate the following muscles on a human subject:
   a. Serratus anterior
   b. Trapezius

   c. Rhomboid major and minor
   d. Levator scapulae
   e. Pectoralis minor
      *Note:* "How" means palpating during active contraction and possibly resisting a primary movement of the muscle. Some muscles have several primary movements, such as the trapezius with rotation upward and adduction. "Where" refers to the location on the body where the muscle can be felt.
3. Palpate the sternoclavicular and acromioclavicular joint movements and the muscles primarily involved while demonstrating the following shoulder girdle movements:
   a. Adduction
   b. Abduction
   c. Rotation upward
   d. Rotation downward
   e. Elevation
   f. Depression

4. Shoulder girdle movement analysis chart

After analyzing each exercise in the chart, break each into two primary movement phases, such as a lifting phase and a lowering phase. For each phase, determine the shoulder girdle movements occurring, and then list the shoulder girdle muscles primarily responsible for causing/controlling those movements. Beside each muscle in each movement, indicate the type of contraction as follows: I—isometric; C—concentric; E—eccentric.

| Exercise | Initial movement (lifting) phase | | Secondary movement (lowering) phase | |
|---|---|---|---|---|
| | Movement(s) | Agonist(s)—(contraction type) | Movement(s) | Agonist(s)—(contraction type) |
| Push-up | | | | |
| Chin-up | | | | |
| Bench press | | | | |
| Dip | | | | |
| Lat pull | | | | |
| Overhead press | | | | |
| Prone row | | | | |
| Barbell shrugs | | | | |

## 5. Shoulder girdle sport skill analysis chart

Analyze each skill in the chart and list the movements of the right and left shoulder girdle in each phase of the skill. You may prefer to list the initial position the shoulder girdle is in for the stance phase. After each movement, list the shoulder girdle muscle(s) primarily responsible for causing/controlling that movement. Beside each muscle in each movement, indicate the type of contraction as follows: I—isometric; C—concentric; E—eccentric. It may be desirable to review the concepts for analysis in Chapter 8 for the various phases.

| Exercise | | Stance phase | Preparatory phase | Movement phase | Follow-through phase |
|---|---|---|---|---|---|
| Baseball pitch | (R) | | | | |
| | (L) | | | | |
| Volleyball serve | (R) | | | | |
| | (L) | | | | |
| Tennis serve | (R) | | | | |
| | (L) | | | | |
| Softball pitch | (R) | | | | |
| | (L) | | | | |
| Tennis backhand | (R) | | | | |
| | (L) | | | | |
| Batting | (R) | | | | |
| | (L) | | | | |
| Bowling | (R) | | | | |
| | (L) | | | | |
| Basketball free throw | (R) | | | | |
| | (L) | | | | |

# References

Andrews JR, Zarins B, Wilk KE: *Injuries in baseball,* Philadelphia, 1998, Lippincott-Raven.

Dail NW, Agnew TA, Floyd RT. *Kinesiology for manual therapies,* McGraw-Hill, 2010.

DePalma MJ, Johnson EW: Detecting and treating shoulder impingement syndrome: the role of scapulothoracic dyskinesis, *The Physician and Sportsmedicine* 31(7), 2003.

Field D: *Anatomy: palpation and surface markings,* ed 3, Oxford, 2001, Butterworth-Heinemann.

Hislop HJ, Montgomery J: *Daniels and Worthingham's muscle testing: techniques of manual examination,* ed 8, Philadelphia, 2007, Saunders.

Houglum PA, Bertoti DB: *Brunstom's clinical kinesiology,* ed 6, Philadelphia, 2012, FA Davis.

Johnson RJ: Acromioclavicular joint injuries: identifying and treating "separated shoulder" and other conditions, Harmon K, Rubin A, eds: *The physician and sportsmedicine* 29(11), 2001.

Loftice JW, Fleisig GS, Wilk KE, Reinold MM, Chmielewski T, Escamilla RF, Andrews JR, eds: *Conditioning program for baseball pitchers,* Birmingham, AL, 2004, American Sports Medicine Institute.

McMurtrie H, Rikel JK: *The coloring review guide to human anatomy,* New York, 1991, McGraw-Hill.

Muscolino JE: *The muscular system manual: the skeletal muscles of the human body,* ed 3, St. Louis, 2010, Elsevier Mosby.

Neumann DA: *Kinesiology of the musculoskeletal system: foundations for physical rehabilitation,* ed 2, St. Louis, 2010, Mosby.

Norkin CC, Levangie PK: *Joint structure and function—a comprehensive analysis,* ed 5, Philadelphia, 2011, Davis.

Rasch PJ: *Kinesiology and applied anatomy,* ed 7, Philadelphia, 1989, Lea & Febiger.

Seeley RR, Stephens TD, Tate P: *Anatomy & physiology,* ed 8, Dubuque, IA, 2008, McGraw-Hill.

Seeley RR, VanPutte CL, Regan JL, Russo AF: *Seeley's anatomy & physiology,* ed 9, New York, 2011, McGraw-Hall.

Sobush DC, et al: The Lennie test for measuring scapula position in healthy young adult females: a reliability and validity study, *Journal of Orthopedic and Sports Physical Therapy* 23:39, January 1996.

Soderburg GL: *Kinesiology—application to pathological motion,* Baltimore, 1986, Williams & Wilkins.

Van De Graaff KM: *Human anatomy,* ed 6, Dubuque, IA, 2002, McGraw-Hill.

Wilk KE, Reinold MM, Andrews JR, eds: *The athlete's shoulder,* ed 2, Philadelphia, 2009, Churchill Livingstone Elsevier.

Williams CC: Posterior sternoclavicular joint dislocation emergencies series, Howe WB, ed.: *The physician and sportsmedicine* 27(2), 1999.

# Worksheet Exercises

For in- or out-of-class assignments, or for testing, utilize this tear-out worksheet.

## Worksheet 1

Using crayons or colored markers, draw and label on the worksheet the following muscles. Indicate the origin and insertion of each muscle with an "O" and an "I," respectively, and draw in the origin and insertion on the contralateral side of the skeleton.

a. Trapezius
b. Rhomboid major and minor
c. Serratus anterior
d. Levator scapulae
e. Pectoralis minor

# Worksheet Exercises

For in- or out-of-class assignments, or for testing, utilize this tear-out worksheet.

## Worksheet 2

Label each of lines 1 through 6 on the drawing with the letter, from the following list, that corresponds to the movements of the shoulder girdle indicated by the arrow.

a.  Adduction (retraction)

b.  Abduction (protraction)

c.  Rotation upward

d.  Rotation downward

e.  Elevation

f.  Depression

On the lines below, which correspond to the numbers of the arrows above, list the muscle(s) or parts of muscles primarily responsible for causing each movement.

1. _____

2. _____

3. _____

4. _____

5. _____

6. _____

# CHAPTER 5

# THE SHOULDER JOINT

## Objectives

- To identify on a human skeleton or human subject selected bony structures of the shoulder joint

- To label on a skeletal chart selected bony structures of the shoulder joint

- To draw on a skeletal chart the muscles of the shoulder joint and indicate, using arrows, shoulder joint movements

- To demonstrate with a partner all the movements of the shoulder joints and list their respective planes and axes of rotation

- To learn and understand how movements of the scapula accompany movements of the humerus in achieving movement of the entire shoulder complex

- To determine, list, and palpate the muscles of the shoulder joint and their antagonists and appreciate the role of the ligaments in providing stability

- To organize and list the muscles that produce the movements of the shoulder girdle and the shoulder joint

- To learn and understand the innervation of the shoulder joint muscles

- To determine, through analysis, the shoulder joint movements and muscles involved in selected skills and exercises

The only attachment of the shoulder joint to the axial skeleton is via the scapula and its attachment through the clavicle at the sternoclavicular joint. Movements of the shoulder joint are many and varied. It is unusual to have movement of the humerus without scapula movement. When the humerus is flexed above shoulder level, the scapula is elevated, rotated upward, and abducted. With glenohumeral abduction above shoulder level, the scapula is elevated and rotated upward. Adduction of the humerus results in depression and rotation downward, whereas extension of the humerus results in depression, rotation downward, and adduction of the scapula. The scapula abducts with humeral internal rotation and horizontal adduction. The scapula adducts with external rotation and horizontal abduction of the humerus. For a summary of these movements and the muscles primarily responsible for them, refer to Table 5.1.

Because the shoulder joint has such a wide range of motion in so many different planes, it also has a significant amount of laxity, which often results in instability problems such as rotator cuff impingement, subluxations, and dislocations. The price of mobility is reduced stability. The concept that the more mobile a joint is, the less stable it is and that the more stable it is, the less mobile it is applies generally throughout the body, but particularly in the shoulder joint. See the section on stability and mobility of diarthrodial joints in Chapter 1.

TABLE 5.1 · **Pairing of shoulder girdle and shoulder joint movements. When the muscles of the shoulder joint (second column) perform the actions in the first column through any substantial range of motion, the muscles of the shoulder girdle (fourth column) work in concert by performing the actions in the third column.**

| Shoulder joint actions | Shoulder joint agonists | Shoulder girdle actions | Shoulder girdle agonists |
| --- | --- | --- | --- |
| Abduction | Supraspinatus, deltoid, upper pectoralis major | Upward rotation/elevation | Serratus anterior, middle and lower trapezius, levator scapulae, rhomboids |
| Adduction | Latissimus dorsi, teres major, lower pectoralis major | Downward rotation | Pectoralis minor, rhomboids |
| Flexion | Anterior deltoid, upper pectoralis major, coracobrachialis | Elevation/upward rotation | Levator scapulae, serratus anterior, upper and middle trapezius, rhomboids |
| Extension | Latissimus dorsi, teres major, lower pectoralis major, posterior deltoid | Depression/downward rotation | Pectoralis minor, lower trapezius |
| Internal rotation | Latissimus dorsi, teres major, pectoralis major, subscapularis | Abduction (protraction) | Serratus anterior, pectoralis minor |
| External rotation | Infraspinatus, teres minor | Adduction (retraction) | Middle and lower trapezius, rhomboids |
| Horizontal abduction | Middle and posterior deltoid, infraspinatus, teres minor | Adduction (retraction) | Middle and lower trapezius, rhomboids |
| Horizontal adduction | Pectoralis major, anterior deltoid, coracobrachialis | Abduction (protraction) | Serratus anterior, pectoralis minor |
| Diagonal abduction (overhand activities) | Posterior deltoid, infraspinatus, teres minor | Adduction (retraction)/upward rotation/elevation | Trapezius, rhomboids, serratus anterior, levator scapulae |
| Diagonal adduction (overhand activities) | Pectoralis major, anterior deltoid, coracobrachialis | Abduction (protraction)/depression/downward rotation | Serratus anterior, pectoralis minor |

## Bones

The scapula, clavicle, and humerus serve as attachments for most of the muscles of the shoulder joint. Learning the specific location and importance of certain bony landmarks is critical to understanding the functions of the shoulder complex. Some of these scapular landmarks are the supraspinous fossa, infraspinous fossa, subscapular fossa, spine of the scapula, glenoid cavity, coracoid process, acr omion process, and inferior angle. Humeral landmarks are the head, greater tubercle, lesser tubercle, intertubercular groove, and deltoid tuberosity (Figs. 5.1 and 5.2, and review Figs. 4.1 and 4.3).

## Joint

The shoulder joint, specifically known as the glenohumeral joint, is a multiaxial ball-and-socket joint classified as enarthrodial (see Fig. 5.1). As such, it moves in all planes and is the most movable joint in the body. It is similar to the hip in its joint classification; however, the socket provided by the glenoid fossa is much shallower and relatively small in comparison to the rather large humeral head. Its stability is enhanced slightly by the glenoid labrum (see Fig. 5.5), a cartilaginous ring that surrounds the glenoid fossa just inside its periphery. It is further stabilized by the glenohumeral ligaments, especially anteriorly and inferiorly. The anterior glenohumeral ligaments become taut as external rotation, extension, abduction, and horizontal abduction occur, whereas the very thin posterior capsular ligaments become taut in internal rotation, flexion, and horizontal adduction (See Table 5.2). In recent years the importance of the inferior glenohumeral ligament in providing both anterior and posterior stability has come to light (Figs. 5.3 and 5.4). It is, however, important to note that, due to the wide

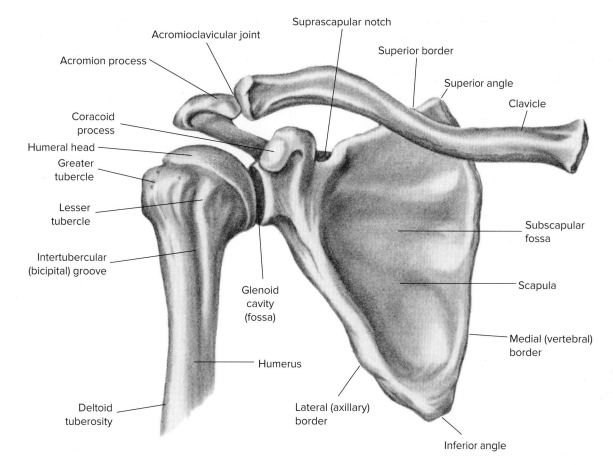

FIG. 5.1 • Right glenohumeral joint, anterior view.

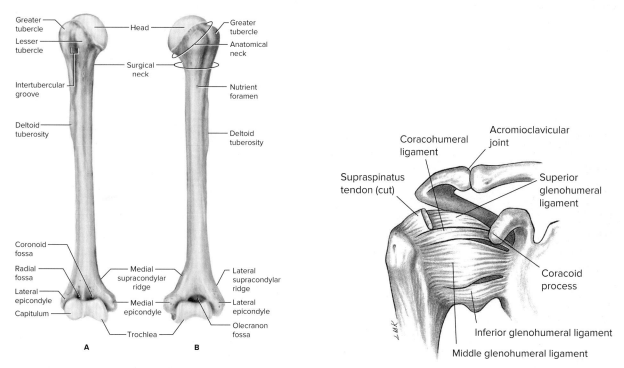

FIG. 5.2 • The right humerus. **A,** Anterior view; **B,** Posterior view.

FIG. 5.3 • Glenohumeral ligaments, anterior view.

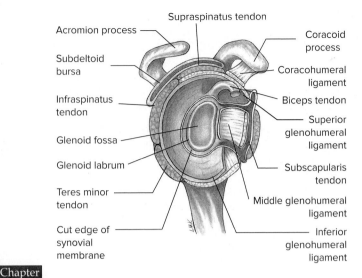

FIG. 5.4 • Right glenohumeral joint, lateral view with humerus removed.

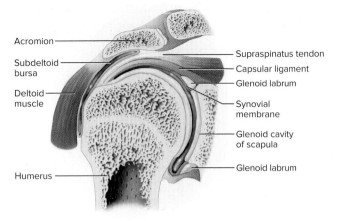

FIG. 5.5 • The right glenohumeral joint, frontal section.

range of motion involved in the glenohumeral joint, the ligaments are quite lax until the extreme ranges of motion are reached. This relative lack of static stability provided by the ligaments emphasizes the need for optimal dynamic stability to be provided by muscles such as the rotator cuff group. Stability is sacrificed to gain mobility.

Determining the exact range of each movement for the glenohumeral joint is difficult because of the accompanying shoulder girdle movement. However, if the shoulder girdle is prevented from moving, then the glenohumeral joint movements are generally thought to be in the following ranges: 90 to 100 degrees of abduction, 0 degrees adduction (prevented by the trunk) or 75 degrees anterior to the trunk, 40 to 60 degrees of extension, 90 to 100 degrees of flexion, 70 to 90 degrees of internal and external rotation, 45 degrees of horizontal abduction, and 135 degrees of horizontal adduction. If the shoulder girdle is free to move, then the total range of the combined joints is 170 to 180 degrees of abduction, 170 to 180 degrees of flexion, and 140 to 150 degrees of horizontal adduction.

As discussed in Chapter 4 and emphasized in Table 5.1, the glenohumeral joint is paired with the shoulder girdle to accomplish the total shoulder range of motion. As an example, the 170 to 180 degrees of total abduction includes approximately 60 degrees of scapula upward rotation, 25 degrees of scapula elevation, and 95 degrees of

glenohumeral abduction. These respective actions do not necessarily happen in a totally sequential fashion, but this synergistic relationship is often referred to as scapulohumeral rhythm. While the exact number of degrees in one segment compared to another may vary within and between individuals, the generally accepted ratio is 2 to 1; that is, for every 2 degrees of glenohumeral motion, there is 1 degree of scapula motion.

The shoulder joint is frequently injured because of its anatomical design. A number of factors contribute to its injury rate, including the shallowness of the glenoid fossa, the laxity of the ligamentous structures necessary to accommodate its wide range of motion, and the lack of strength and endurance in the muscles, which are essential in providing dynamic stability to the joint. As a result, anterior or anteroinferior glenohumeral subluxations and dislocations are quite common with physical activity. Although posterior dislocations are fairly rare, shoulder problems due to posterior instability are somewhat commonplace.

Contributing to glenohumeral stability is the glenoid labrum, a specialized cartilaginous structure that surrounds the periphery of the glenoid fossa. It enhances joint stability by deepening the concavity of the fossa and serving as a buttress to excessive humeral head translation. This structure may be injured with sudden overhead movements as in baseball and volleyball or traumatic

## TABLE 5.2 • Major ligaments of the glenohumeral joint

| Joint/ Structure | Ligament | | Proximal Attachment | Distal Attachments | Functional Application and Joint Stability |
|---|---|---|---|---|---|
| Glenohumeral joint | Coracoacromial ligament | | Lateral border of coracoid process | Medial border of acromion process of scapula | Passes superior to both the head of humerus and supraspinatus tendon; provides support to the head of humerus that prevents superior dislocation |
| | Coracohumeral ligament | | Coracoid process | Anterior band: lesser tubercle of humerus; posterior band: greater tubercle of humerus | Strengthens the upper aspect of the shoulder joint capsule; slight restriction to excessive external rotation and flexion, anterior and posterior translation |
| | Glenohumeral ligament | Superior | Apex of glenoid cavity close to root of coracoid process | Small depression above the lesser tubercle of humerus | Becomes taut in full adduction, limits inferior and anterior translation in anatomic position |
| | | Middle | Medial edge of glenoid cavity | Lower part of the lesser tubercle of humerus | Becomes taut in external rotation, limits anterior translation |
| | | Inferior | Inferior aspect of glenoid | Under anatomic neck of humerus | Anterior portion becomes taut in abduction and external rotation; Posterior portion becomes taut in abduction and internal rotation, Anterior portion limits anterior inferior translation; posterior portion limits posterior inferior translation |
| Humerus | Transverse humeral ligament | | Greater tubercle of the humerus | Lesser tubercle of the humerus | Spans the intertubercular sulcus of the humerus, acts to retain the tendon of the long head of the biceps brachii muscle in the intertubercular groove |

injuries such as glenohumeral subluxations or dislocations.

Another frequent injury is to the rotator cuff. The subscapularis, supraspinatus, infraspinatus, and teres minor muscles make up the rotator cuff. They are small muscles whose tendons cross the front, top, and rear of the head of the humerus to attach on the lesser and greater tubercles, respectively. Their point of insertion enables them to rotate the humerus, an essential movement in this freely movable joint. Most important, however, is the vital role that the rotator cuff muscles play in maintaining the humeral head in correct approximation within the glenoid fossa while the more powerful muscles of the joint move the humerus through its wide range of motion.

In recent years the phenomenon of glenohumeral internal rotation deficit, or GIRD, has received attention. GIRD represents a difference in internal rotation range of motion between an individual's throwing and nonthrowing shoulders. Studies have demonstrated that overhead athletes who had a GIRD of greater than 20% had a higher risk of injury than those who did not. Appropriate stretching exercises may be used to regain the amount of internal rotation necessary to improve performance and reduce the likelihood of injury.

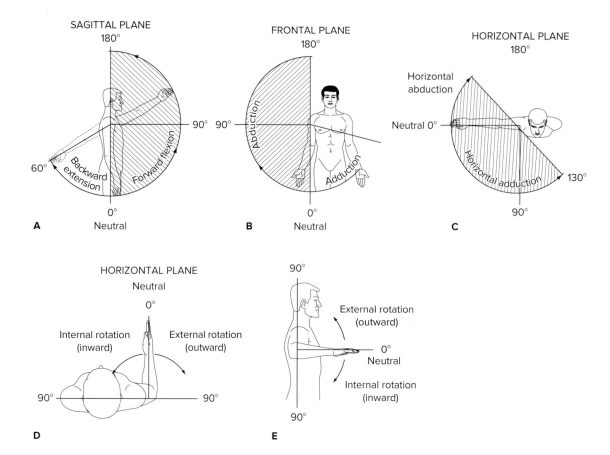

FIG. 5.6 • Range of motion of the shoulder. **A,** Flexion and extension; **B,** Abduction and adduction; **C,** Horizontal abduction and adduction; **D,** Internal and external rotation with the arm at the side of the body; **E,** Internal and external rotation with the arm abducted to 90 degrees.

## Movements FIGS. 5.6, 5.7

**Flexion:** Movement of the humerus straight anteriorly from any point in the sagittal plane

**Extension:** Movement of the humerus straight posteriorly from any point in the sagittal plane, sometimes referred to as hyperextension

**Abduction:** Upward lateral movement of the humerus in the frontal plane out to the side, away from the body

**Adduction:** Downward movement of the humerus in the frontal plane medially toward the body from abduction

**External rotation:** Movement of the humerus laterally in the transverse plane around its long axis away from the midline

**Internal rotation:** Movement of the humerus in the transverse plane medially around its long axis toward the midline

**Horizontal abduction (extension):** Movement of the humerus in a horizontal or transverse plane away from the chest

**Horizontal adduction (flexion):** Movement of the humerus in a horizontal or transverse plane toward and across the chest

**Diagonal abduction:** Movement of the humerus in a diagonal plane away from the midline of the body

**Diagonal adduction:** Movement of the humerus in a diagonal plane toward the midline of the body

Shoulder flexion

Shoulder extension

Shoulder abduction

Shoulder adduction

**Flexion**

**A**

**Extension**

**B**

Courtesy of Britt Jones (all)

**Abduction**

**C**

**Adduction**

**D**

FIG. 5.7 • Movements of the shoulder joint.

Shoulder
internal
rotation

Shoulder
external
rotation

Chapter
5

Shoulder
horizontal
abduction

Shoulder
horizontal
adduction

**Internal rotation**

E

**External rotation**

F

**Horizontal abduction**

G

**Horizontal adduction**

H

Courtesy of Britt Jones (all)

**Diagonal abduction**

I

**Diagonal adduction**

J

FIG. 5.7 (continued) • Movements of the shoulder joint.

# Muscles

In attempting to learn and understand the muscles of the glenohumeral joint, it may be helpful to group them according to their location and function. Muscles that originate on the scapula and clavicle and insert on the humerus may be thought of as muscles intrinsic to the glenohumeral joint, whereas muscles originating on the trunk and inserting on the humerus are considered extrinsic to the joint. The intrinsic muscles include the deltoid, the coracobrachialis, the teres major, and the rotator cuff group, which is composed of the subscapularis, the supraspinatus, the infraspinatus, and the teres minor. Extrinsic glenohumeral muscles are the latissimus dorsi and the pectoralis major. It may also be helpful to organize the muscles according to their general location. The pectoralis major, coracobrachialis, and subscapularis are anterior muscles. The deltoid and supraspinatus are located superiorly. The latissimus dorsi, teres major, infraspinatus, and teres minor are located posteriorly. Table 5.1 (p. 116) lists the glenohumeral joint movements and the muscles primarily responsible for them, and Table 5.3 provides the action of each muscle.

The biceps brachii and triceps brachii (long head) are also involved in glenohumeral movements. Primarily, the biceps brachii assists in flexing and horizontally adducting the shoulder, whereas the long head of the triceps brachii assists in extension, adduction and horizontal abduction. Further discussion of these muscles appears in Chapter 6.

Chapter

**5**

## TABLE 5.3 • Agonist muscles of the glenohumeral joint

| | Muscle | Origin | Insertion | Action | Plane of motion | Palpation | Innervation |
|---|---|---|---|---|---|---|---|
| Anterior muscles | Pectoralis major upper fibers | Medial half of anterior surface of clavicle | Flat tendon 2 or 3 inches wide to lateral lip of intertubercular groove of humerus | Internal rotation | Transverse | From medial end of the clavicle to the intertubercular groove of the humerus, during flexion and adduction from the anatomical position | Lateral pectoral nerve (C5, C6, C7) |
| | | | | Horizontal adduction | | | |
| | | | | Diagonal adduction | Diagonal | | |
| | | | | Flexion | Sagittal | | |
| | | | | Abduction | Frontal | | |
| | Pectoralis major lower fibers | Anterior surface of costal cartilages of first six ribs, and adjoining portion of sternum | Flat tendon 2 or 3 inches wide to lateral lip of intertubercular groove of humerus | Internal rotation | Transverse | From the lower ribs and sternum to the intertubercular groove of the humerus, during resisted extension from a flexed position | Medial pectoral nerve (C8, T1) |
| | | | | Horizontal adduction | | | |
| | | | | Diagonal adduction | Diagonal | | |
| | | | | Extension from flexed position | Sagittal | | |
| | | | | Adduction | Frontal | | |
| | Subscapularis | Entire anterior surface of subscapular fossa | Lesser tubercle of humerus | Internal rotation | Transverse | Mostly inaccessible, lateral portion may be palpated on supine subject (arm in slight flexion and adduction with elbow lying across abdomen); pull medial border laterally with one hand while palpating between the scapula and rib cage with other hand (subject actively internally rotates) | Upper and lower subscapular nerve (C5, C6) |
| | | | | Adduction | Frontal | | |
| | | | | Extension | Sagittal | | |
| | Coracobrachialis | Coracoid process of scapula | Middle of medial border of humeral shaft | Horizontal adduction | Transverse | The belly may be palpated high up on the medial arm just posterior to the short head of the biceps brachii and toward the coracoid process, particularly with resisted adduction | Musculotaneous nerve (C5, C6, C7) |
| | | | | Diagonal adduction | Diagonal | | |
| | | | | Flexion | Sagittal | | |
| | | | | Abduction | Frontal | | |

TABLE 5.3 (continued) • **Agonist muscles of the glenohumeral joint**

| | Muscle | Origin | Insertion | Action | Plane | Palpation | Nerve |
|---|---|---|---|---|---|---|---|
| **Superior muscles** | Deltoid anterior fibers | Anterior lateral third of clavicle | Deltoid tuberosity on lateral humerus | Abduction | Frontal | From the clavicle toward the anterior humerus during resisted flexion or horizontal adduction | Axillary nerve (C5, C6) |
| | | | | Flexion | Sagittal | | |
| | | | | Horizontal adduction | Transverse | | |
| | | | | Internal rotation | | | |
| | | | | Diagonal adduction | Diagonal | | |
| | Deltoid middle fibers | Lateral aspect of acromion | Deltoid tuberosity on lateral humerus | Abduction | Frontal | From the lateral border of the acromion down toward the deltoid tuberosity during resisted abduction | |
| | | | | Horizontal abduction | Transverse | | |
| | Deltoid posterior fibers | Inferior edge of spine of scapula | Deltoid tuberosity on lateral humerus | Abduction | Frontal | From the lower lip of the spine of the scapula toward the posterior humerus during resisted extension or horizontal abduction | |
| | | | | Extension | Sagittal | | |
| | | | | Horizontal abduction | Transverse | | |
| | | | | External rotation | | | |
| | | | | Diagonal abduction | Diagonal | | |
| | Supra-spinatus | Medial two-thirds of supraspinous fossa | Superiorly on greater tubercle of humerus | Abduction | Frontal | Above the spine of the scapula in supraspinous fossa during initial abduction in the scapula plane; tendon may be palpated just off acromion on greater tubercle | Suprascapula nerve (C5) |
| **Posterior muscles** | Latissimus dorsi | Posterior crest of ilium, back of sacrum and spinous processes of lumbar and lower six thoracic vertebrae, slips from lower three ribs | Medial side of intertubercular groove of humerus, just anterior to the insertion of the teres major | Extension | Sagittal | Tendon may be palpated as it passes under the teres major at the posterior axillary wall, particularly during resisted extension and internal rotation. The muscle can be palpated in the upper lumbar/lower thoracic area during extension from a flexed position and throughout most of its length during resisted adduction from a slightly abducted position | Thoracodorsal (C6, C7, C8) |
| | | | | Adduction | Frontal | | |
| | | | | Internal rotation | Transverse | | |
| | | | | Horizontal abduction | | | |
| | Teres major | Posteriorly on inferior third of lateral border of scapula and just superior to inferior angle | Medial lip of intertubercular groove of humerus, just posterior to the insertion of the latissimus dorsi | Extension | Sagittal | Just above the latissimus dorsi and below the teres minor on the posterior scapula surface, moving diagonally upward and laterally from the inferior angle of the scapula during resisted internal rotation | Lower subscapular nerve (C5, C6) |
| | | | | Adduction | Frontal | | |
| | | | | Internal rotation | Transverse | | |
| | Infra-spinatus | Infraspinous fossa just below spine of scapula | Posteriorly on greater tubercle of humerus | External rotation | Transverse | Just below the spine of the scapula passing upward and laterally to the humerus during resisted external rotation | Suprascapula nerve (C5, C6) |
| | | | | Horizontal abduction | | | |
| | | | | Extension | Sagittal | | |
| | | | | Diagonal abduction | Diagonal | | |
| | Teres minor | Posteriorly on upper and middle aspect of lateral border of scapula | Posteriorly on greater tubercle of humerus | External rotation | Transverse | Just above the teres major on the posterior scapula surface, moving diagonally upward and laterally from the inferior angle of the scapula during resisted external rotation | Axillary nerve (C5, C6) |
| | | | | Horizontal abduction | | | |
| | | | | Extension | Sagittal | | |
| | | | | Diagonal abduction | Diagonal | | |

**Note:** The biceps brachii assists in flexion, horizontal adduction, and diagonal adduction, while the long head of the triceps brachii assists in extension, adduction, horizontal abduction, and diagonal abduction. Because they are covered in Chapter 6, neither is listed above.

Chapter 5

## Shoulder joint muscles—location

Anterior
  Pectoralis major
  Coracobrachialis
  Subscapularis
Superior
  Deltoid
  Supraspinatus
Posterior
  Latissimus dorsi
  Teres major
  Infraspinatus
  Teres minor

## Muscle identification

Figs. 5.8 and 5.9 identify the anterior and posterior muscles of the shoulder joint and shoulder girdle. Compare Fig. 5.8 with Fig. 5.10 and Fig. 5.9 with Fig. 5.11, and refer to Table 5.3 for a detailed breakdown of the agonist muscles for the glenohumeral joint.

Courtesy of Lisa Floyd (left, right)

FIG. 5.8 • Anterior shoulder joint and shoulder girdle muscles.

FIG. 5.9 • Posterior shoulder joint and shoulder girdle muscles.

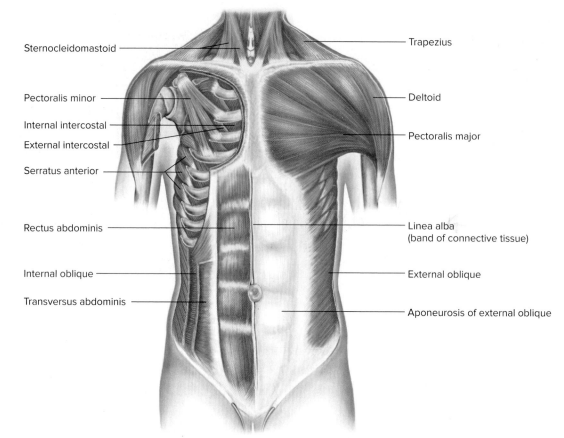

Sternocleidomastoid

Pectoralis minor

Internal intercostal

External intercostal

Serratus anterior

Rectus abdominis

Internal oblique

Transversus abdominis

Trapezius

Deltoid

Pectoralis major

Linea alba
(band of connective tissue)

External oblique

Aponeurosis of external oblique

**A**

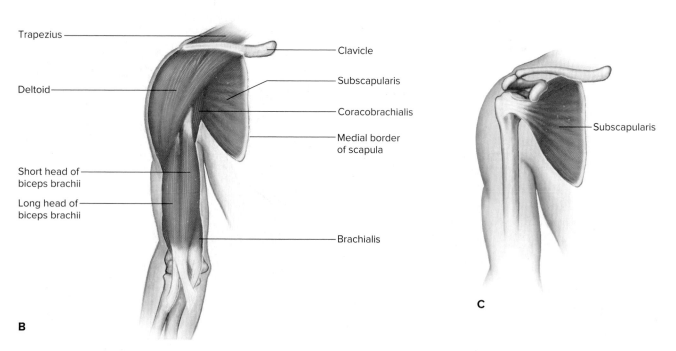

Trapezius

Deltoid

Short head of
biceps brachii

Long head of
biceps brachii

Clavicle

Subscapularis

Coracobrachialis

Medial border
of scapula

Brachialis

Subscapularis

**B**

**C**

FIG. 5.10 • Anterior muscles of the shoulder. **A,** The right pectoralis major is removed to show the pectoralis minor and serratus anterior; **B,** Muscles of the anterior right shoulder and arm, with the rib cage removed; **C,** Subscapularis.

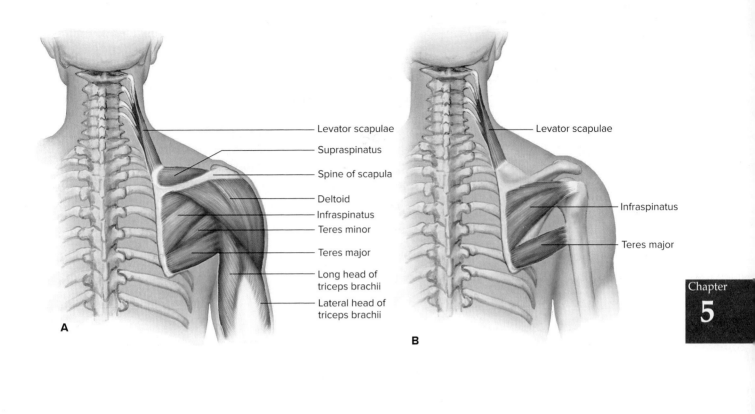

Levator scapulae

Supraspinatus

Spine of scapula

Deltoid

Infraspinatus

Teres minor

Teres major

Long head of triceps brachii

Lateral head of triceps brachii

A

Levator scapulae

Infraspinatus

Teres major

B

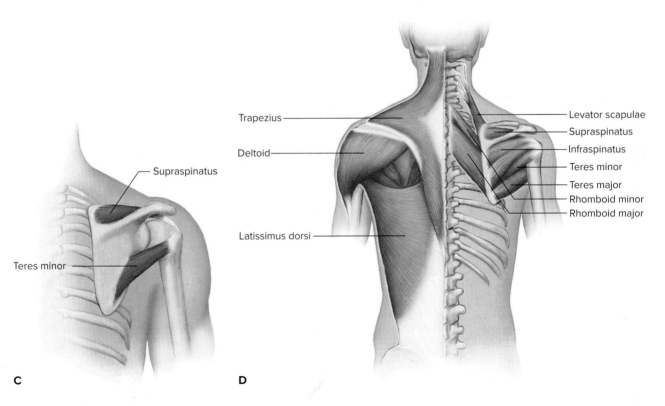

Supraspinatus

Teres minor

C

Trapezius

Deltoid

Latissimus dorsi

Levator scapulae

Supraspinatus

Infraspinatus

Teres minor

Teres major

Rhomboid minor

Rhomboid major

D

FIG. 5.11 • Posterior muscles of the shoulder. **A,** The right trapezius and deltoid are removed to show the underlying muscles; **B,** Levator scapulae, infraspinatus, and teres major; **C,** Supraspinatus and teres minor; **D,** Muscles of the posterior surface of the scapula and arm.

## Nerves   FIGS. 5.12, 5.13

The muscles of the shoulder joint are all innervated by the nerves of the brachial plexus. The pectoralis major is innervated by the pectoral nerves. Specifically, the lateral pectoral nerve arising from C5, C6, and C7 innervates the clavicular head, while the medial pectoral nerve arising from C8 and T1 innervates the sternal head. The thoracodorsal nerve, arising from C6, C7, and C8, supplies the latissimus dorsi. The axillary nerve (Fig. 5.12), branching from C5 and C6, innervates the deltoid and teres minor. A lateral patch of skin over the deltoid region of the arm is provided sensation by the axillary nerve. Both the upper and lower subscapular nerves arising from C5 and C6 innervate the subscapularis, while only the lower subscapular nerve supplies the teres major. The supraspinatus and infraspinatus are innervated by the suprascapular nerve, which originates from C5 and C6. The musculocutaneous nerve, as seen in Fig. 5.13, branches from C5, C6, and C7 and innervates the coracobrachialis. It supplies sensation to the radial aspect of the forearm. See Fig. 5.14 for a full listing of the glenohumeral innervations.

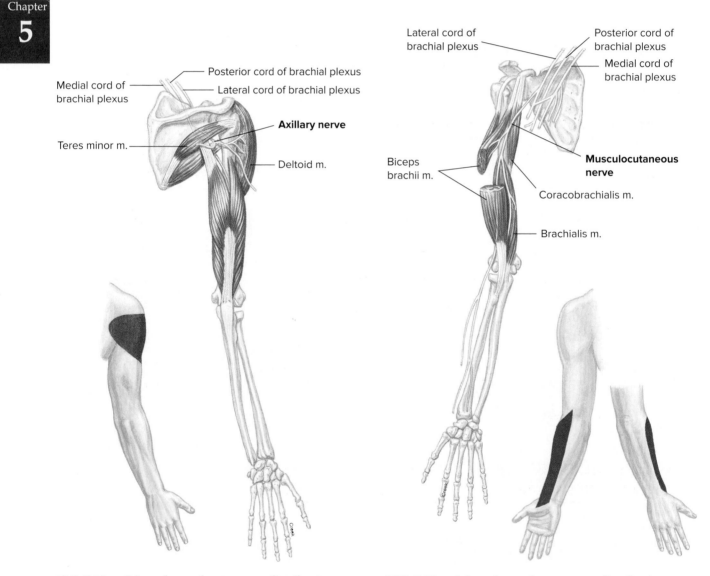

FIG. 5.12 • Muscular and cutaneous distribution of the axillary nerve.

FIG. 5.13 • Muscular and cutaneous distribution of the musculocutaneous nerve.

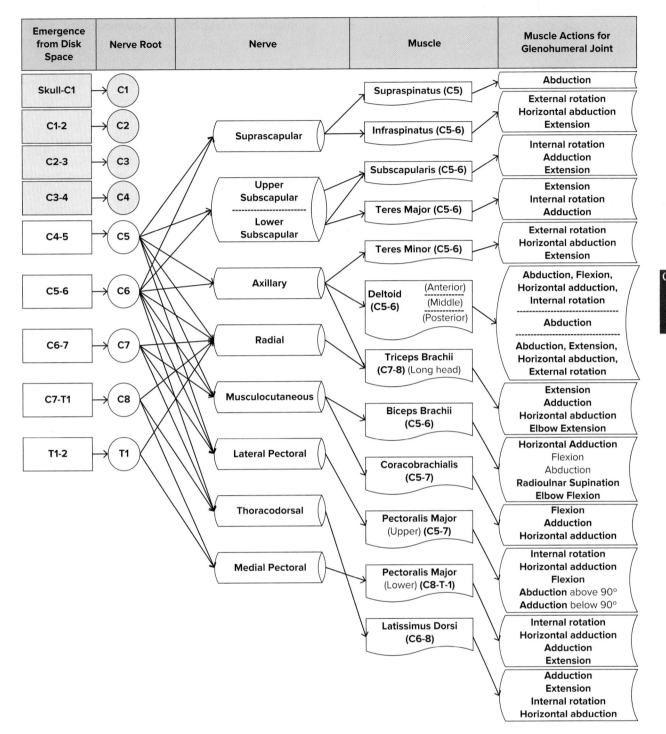

FIG. 5.14 • Innervation for glenohumeral muscles and their actions. The neuromuscular structures associated with the primary movers for the shoulder joint and their actions are bolded.

Shoulder
abduction

# Deltoid muscle    FIG. 5.15

(del-toyd´)

## Origin

Anterior fibers: anterior lateral third of the clavicle
Middle fibers: lateral aspect of the acromion
Posterior fibers: inferior edge of the spine of the scapula

Shoulder
flexion

## Insertion

Deltoid tuberosity on the lateral humerus

## Action

Anterior fibers: abduction, flexion, horizontal adduction, and internal rotation of the glenohumeral joint
Middle fibers: abduction and horizontal abduction of the glenohumeral joint
Posterior fibers: abduction, extension, horizontal abduction, and external rotation of the glenohumeral joint

Shoulder
horizontal
adduction

## Palpation

Anterior fibers: from the clavicle toward the anterior humerus during resisted flexion or horizontal adduction
Middle fibers: from the lateral border of the acromion down toward the deltoid tuberosity during resisted abduction
Posterior fibers: from the lower lip of the spine of the scapula toward the posterior humerus during resisted extension or horizontal abduction

Shoulder
internal
rotation

## Innervation

Axillary nerve (C5, C6)

Shoulder
extension

Shoulder
horizontal
abduction

Shoulder
external
rotation

## Application, strengthening, and flexibility

The deltoid muscle is used in any lifting movement. The trapezius muscle stabilizes the scapula as the deltoid pulls on the humerus. The anterior fibers of the deltoid muscle flex and internally rotate the humerus. The posterior fibers extend and externally rotate the humerus. The anterior fibers also horizontally adduct the humerus, while the posterior fibers horizontally abduct it. Any movement of the humerus on the scapula will involve part or all of the deltoid muscle.

Lifting the humerus from the side to the position of abduction is a typical action of the deltoid. Sidearm dumbbell raises are excellent for strengthening the deltoid, especially the middle fibers. By abducting the arm in a slightly horizontally adducted (30 degrees) position, the anterior deltoid fibers can be emphasized. The posterior fibers can be strengthened better by abducting the arm in a slightly horizontally abducted (30 degrees) position. See Appendix 2 for more commonly used exercises for the deltoid and other muscles in this chapter.

Stretching the deltoid requires varying positions, depending on the fibers to be stretched. The anterior deltoid is stretched by taking the humerus into extreme horizontal abduction or by extreme extension and adduction. The middle deltoid is stretched by taking the humerus into extreme adduction behind the back. Extreme horizontal adduction stretches the posterior deltoid.

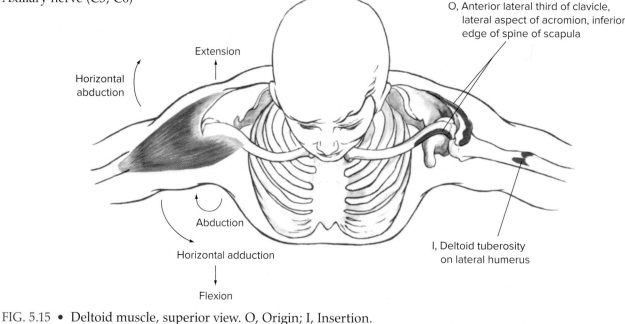

FIG. 5.15 • Deltoid muscle, superior view. O, Origin; I, Insertion.

# Pectoralis major muscle  FIG. 5.16
(pek-to-ra´lis ma´jor)

## Origin

Upper fibers (clavicular head): medial half of the anterior surface of the clavicle

Lower fibers (sternal head): anterior surface of the costal cartilages of the first six ribs, and adjacent portion of the sternum

## Insertion

Flat tendon 2 or 3 inches wide to the lateral lip of the intertubercular groove of the humerus

## Action

Upper fibers (clavicular head): internal rotation, horizontal adduction, flexion up to about 60 degrees, abduction (once the arm is abducted 90 degrees, the upper fibers assist in further abduction), and adduction (with the arm below 90 degrees of abduction) of the glenohumeral joint

Lower fibers (sternal head): internal rotation, horizontal adduction, and adduction and extension of the glenohumeral joint from a flexed position to the anatomical position

## Palpation

Upper fibers: from the medial end of the clavicle to the intertubercular groove of the humerus, during flexion and adduction from the anatomical position

Lower fibers: from the ribs and sternum to the intertubercular groove of the humerus, during resisted extension from a flexed position and resisted adduction from the anatomical position

## Innervation

Upper fibers: lateral pectoral nerve (C5–C7)
Lower fibers: medial pectoral nerve (C8, T1)

Shoulder internal rotation

Shoulder horizontal adduction

Shoulder flexion

Shoulder abduction

Shoulder adduction

Shoulder extension

Horizontal adduction

Adduction

Internal rotation

O, Medial half of anterior surface of clavicle, anterior surface of costal cartilages of first six ribs, adjoining portion of sternum

I, Flat tendon 2 or 3 inches wide to lateral lip of intertubercular groove of humerus

FIG. 5.16 • Pectoralis major muscle, anterior view. O, Origin; I, Insertion.

### Application, strengthening, and flexibility

The anterior axillary fold is formed primarily by the pectoralis major (Fig. 5.17). It aids the serratus anterior muscle in drawing the scapula forward as it moves the humerus in flexion and internal rotation. Even though the pectoralis major is not attached to the scapula, it is effective in this scapula protraction because of its anterior pull on the humerus, which joins to the scapula at the glenohumeral joint. Typical action is shown in throwing a baseball. As the glenohumeral joint is flexed, the humerus is internally rotated and the scapula is drawn forward with upward rotation. It also works as a helper of the latissimus dorsi muscle when extending and adducting the humerus from a raised position.

The pectoralis major and the anterior deltoid work closely together. The pectoralis major is used powerfully in push-ups, pull-ups, throwing, and tennis serves. With a barbell, the subject takes a supine position on a bench with the arms at the side and moves the arms to a horizontally adducted position. This exercise, known as bench pressing, is widely used for pectoralis major development.

Due to the popularity of bench pressing and other weight-lifting exercises that emphasize the pectoralis major and its use in most sporting activities, it is often overdeveloped in comparison to its antagonists. As a result, stretching is often needed and can be done by passive external rotation. It is also stretched when the shoulder is horizontally abducted. Extending the shoulder fully provides stretching to the upper pectoralis major, while full abduction stretches the lower pectoralis major.

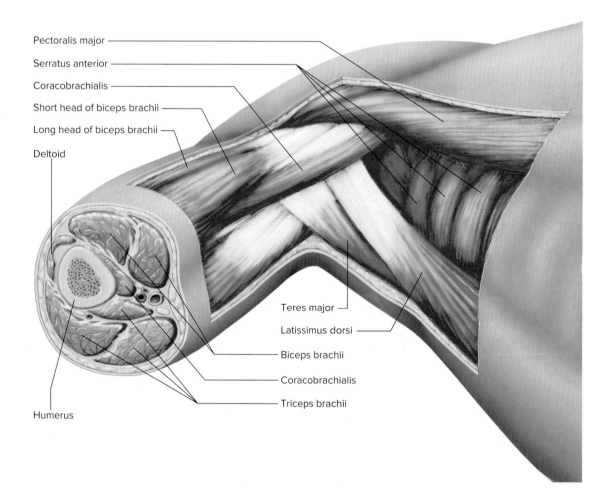

Pectoralis major
Serratus anterior
Coracobrachialis
Short head of biceps brachii
Long head of biceps brachii
Deltoid
Teres major
Latissimus dorsi
Biceps brachii
Coracobrachialis
Triceps brachii
Humerus

FIG. 5.17 • Cross section of right arm and relationship of glenohumeral muscles in axilla.

Chapter 5

# Latissimus dorsi muscle   FIG. 5.18
(lat-is´i-mus dor´si)

## Origin
Posterior crest of the ilium, back of the sacrum and spinous processes of the lumbar and lower six thoracic vertebrae (T6–T12); slips from the lower three ribs

## Insertion
Medial lip of the intertubercular groove of the humerus, just anterior to the insertion of the teres major

Shoulder adduction

Shoulder extension

Shoulder internal rotation

Shoulder horizontal abduction

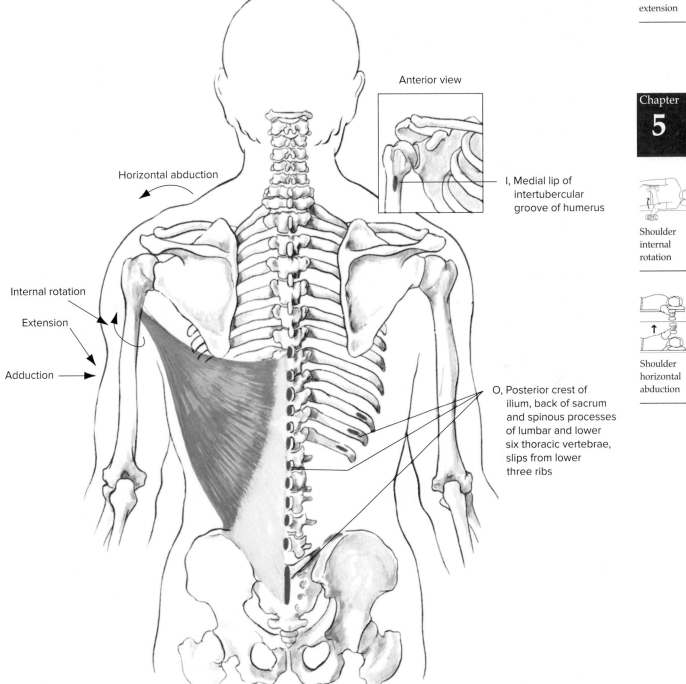

Anterior view

Horizontal abduction

Internal rotation

Extension

Adduction

I, Medial lip of intertubercular groove of humerus

O, Posterior crest of ilium, back of sacrum and spinous processes of lumbar and lower six thoracic vertebrae, slips from lower three ribs

FIG. 5.18 • Latissimus dorsi muscle, posterior view. O, Origin; I, Insertion.

## Action

Adduction of the glenohumeral joint
Extension of the glenohumeral joint
Internal rotation of the glenohumeral joint
Horizontal abduction of the glenohumeral joint

## Palpation

The tendon may be palpated as it passes under the teres major at the posterior wall of the axilla, particularly during resisted extension and internal rotation. The muscle can be palpated in the upper lumbar/ lower thoracic area during extension from a flexed position. The muscle may be palpated throughout most of its length during resisted adduction from a slightly abducted position.

## Innervation

Thoracodorsal nerve (C6–C8)

## Application, strengthening, and flexibility

*Latissimus dorsi* means broadest muscle of the back. This muscle, along with the teres major, forms the posterior axillary fold (see Fig. 5.17). It has a strong action in adduction, extension, and internal rotation of the humerus. Due to the upward rotation of the scapula that accompanies glenohumeral abduction, the latissimus effectively downwardly rotates the scapula by way of its action in pulling the entire shoulder girdle downward in active glenohumeral adduction. It is one of the most important extensor muscles of the humerus and contracts powerfully in chinning. The latissimus dorsi is assisted in all its actions by the teres major and is sometimes referred to as the swimmer's muscle because of its function in pulling the body forward in the water during internal rotation, adduction, and extension. Development of this muscle contributes significantly to what is known as a "swimmer's build."

Exercises in which the arms are pulled down bring the latissimus dorsi muscle into powerful contraction. Chinning, rope climbing, and other uprise movements on the horizontal bar are good examples. In barbell exercises, the basic rowing and pullover exercises are good for developing the "lats." Pulling the bar of an overhead pulley system down toward the shoulders, known as "lat pulls," is a common exercise for this muscle.

The latissimus dorsi is stretched with the teres major when the shoulder is externally rotated while in a 90-degree abducted position. This stretch may be accentuated further by abducting the shoulder fully while maintaining external rotation and then laterally flexing and rotating the trunk to the opposite side.

# Teres major muscle   FIG. 5.19

(te´rez ma´jor)

## Origin

Posteriorly on the inferior third of the lateral border of the scapula and just superior to the inferior angle

## Insertion

Medial lip of the intertubercular groove of the humerus just posterior to the insertion of the latissimus dorsi

## Action

Extension of the glenohumeral joint, particularly from the flexed position to the posteriorly extended position

Internal rotation of the glenohumeral joint

Adduction of the glenohumeral joint, particularly from the abducted position down to the side and toward the midline of the body

## Palpation

Just above the latissimus dorsi and below the teres minor on the posterior scapula surface, moving diagonally upward and laterally from the inferior angle of the scapula during resisted internal rotation

## Innervation

Lower subscapular nerve (C5, C6)

## Application, strengthening, and flexibility

The teres major muscle is effective only when the rhomboid muscles stabilize the scapula or move the scapula in downward rotation. Otherwise, the scapula would move forward to meet the arm.

This muscle works effectively with the latissimus dorsi. It assists the latissimus dorsi, pectoralis major, and subscapularis in adducting, internally rotating, and extending the humerus. It is said to be the latissimus dorsi's "little helper." It may be strengthened by lat pulls, rope climbing, and internal rotation exercises against resistance.

Externally rotating the shoulder in a 90-degree abducted position stretches the teres major.

Shoulder extension

Shoulder internal rotation

Chapter 5

Shoulder adduction

Anterior view with teres major insertion

Extension

Internal rotation

Adduction

I, Medial lip of the intertubercular groove of the humerus

O, Inferior third of lateral border of scapula

FIG. 5.19 • Teres major muscle, posterior view. O, Origin; I, Insertion.

# Coracobrachialis muscle  FIG. 5.20

(kor-a-ko-bra ´ki-a ´lis)

## Origin

Coracoid process of the scapula

## Insertion

Middle of the medial border of the humeral shaft

## Action

Flexion of the glenohumeral joint
Adduction of the glenohumeral joint
Horizontal adduction of the glenohumeral joint

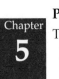

## Palpation

The belly may be palpated high up on the medial arm just posterior to the short head of the biceps brachii and toward the coracoid process, particularly with resisted adduction.

## Innervation

Musculocutaneous nerve (C5–C7)

## Application, strengthening, and flexibility

The coracobrachialis is not a powerful muscle, but it does assist in flexion and adduction and is most functional in moving the arm horizontally toward and across the chest. It is best strengthened by horizontally adducting the arm against resistance, as in bench pressing. It may also be strengthened by performing lat pulls (defined on p. 134).

The coracobrachialis is best stretched in extreme horizontal abduction, although extreme extension also stretches this muscle.

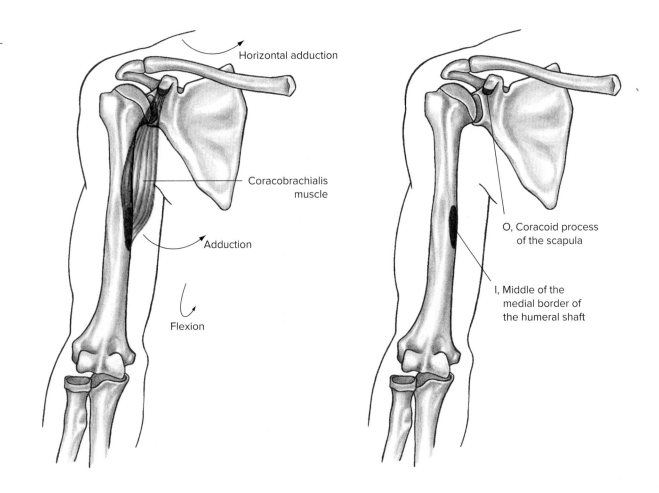

FIG. 5.20 • Coracobrachialis muscle, anterior view. O, Origin; I, Insertion.

## Rotator cuff muscles

Figs. 5.21 and 5.22 illustrate the rotator cuff muscle group, which, as previously mentioned, is most important in maintaining the humeral head in its proper location within the glenoid cavity. The acronym **SITS** may be used in learning the names of the supraspinatus, infraspinatus, teres minor, and subscapularis. These muscles, which are not very large in comparison with the deltoid and pectoralis major, must possess not only adequate strength but also a significant amount of muscular endurance to ensure their proper functioning, particularly in repetitive overhead activities such as throwing, swimming, and pitching. Quite often when these types of activities are conducted with poor technique, muscle fatigue, or inadequate warm-up and conditioning, the rotator cuff muscle group—particularly the supraspinatus—fails to dynamically stabilize the humeral head in the glenoid cavity, leading to further rotator cuff problems such as tendinitis and rotator cuff impingement within the subacromial space.

Rotator cuff **impingement syndrome** occurs when the tendons of these muscles, particularly the supraspinatus and infraspinatus, become irritated and inflamed as they pass through the subacromial space between the acromion process of the scapula and the head of the humerus. Pain, weakness, and loss of movement at the shoulder can result. Loss of function of the rotator cuff muscles, due to injury or loss of strength and endurance, may cause the humerus to move superiorly, resulting in this impingement.

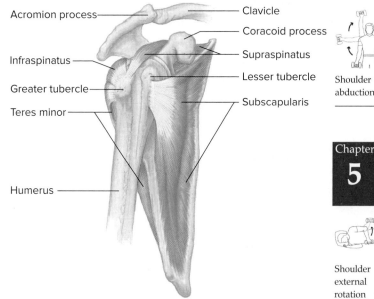

FIG. 5.21 • Rotator cuff muscles, anterolateral view, right shoulder.

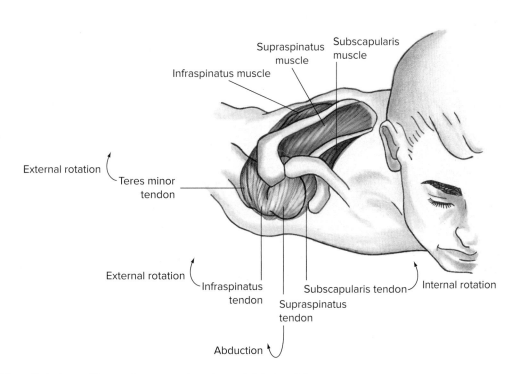

FIG. 5.22 • Rotator cuff muscles, superior view, right shoulder.

Shoulder internal rotation

Shoulder abduction

Chapter 5

Shoulder external rotation

Shoulder extension

Shoulder adduction

Shoulder horizontal abduction

# Subscapularis muscle  FIG. 5.23
(sub-skap-u-la´ris)

Shoulder internal rotation

### Origin
Entire anterior surface of the subscapular fossa

### Insertion
Lesser tubercle of the humerus

### Action
Internal rotation of the glenohumeral joint
Adduction of the glenohumeral joint
Extension of the glenohumeral joint
Stabilization of the humeral head in the glenoid fossa

Shoulder adduction

Shoulder extension

### Palpation
The subscapularis, latissimus dorsi, and teres major, in conjunction, form the posterior axillary fold. Most of the subscapularis is inaccessible on the anterior scapula behind the rib cage. The lateral portion may be palpated with the subject supine and arm in slight flexion and adduction so that the elbow is lying across the abdomen. Use one hand posteriorly to grasp the medial border and pull it laterally, while palpating between the scapula and rib cage with

FIG. 5.23 • Subscapularis muscle, anterior view. O, Origin; I, Insertion.

the other hand with the subject actively internally rotating by pressing the forearm against the chest.

### Innervation
Upper and lower subscapular nerve (C5, C6)

### Application, strengthening, and flexibility
The subscapularis muscle, another rotator cuff muscle, holds the head of the humerus in the glenoid fossa from in front and below. It acts with the latissimus dorsi and teres major muscles in its typical movement but is less powerful in its action because of its proximity to the joint. The muscle also requires the help of the rhomboid in stabilizing the scapula to make it effective in the movements described. The subscapularis is relatively hidden behind the rib cage in its location on the anterior aspect of the scapula in the subscapular fossa. It may be strengthened with exercises similar to those used for the latissimus dorsi and teres major, such as rope climbing and lat pulls. A specific exercise for its development is done by internally rotating the arm against resistance in the beside-the-body position at 0 degrees of glenohumeral abduction.

External rotation with the arm adducted by the side stretches the subscapularis.

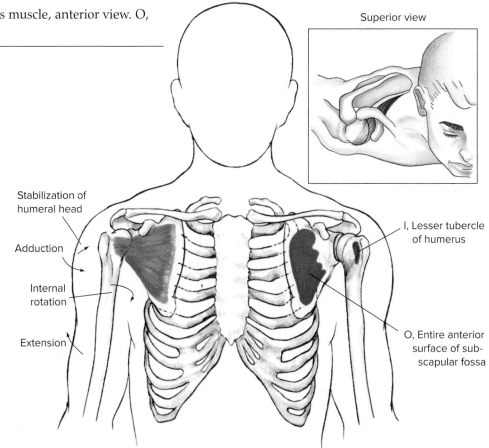

Superior view

Stabilization of humeral head

Adduction

Internal rotation

Extension

I, Lesser tubercle of humerus

O, Entire anterior surface of sub-scapular fossa

# Supraspinatus muscle FIG. 5.24

(su´pra-spi-na´tus)

## Origin
Medial two-thirds of the supraspinous fossa

## Insertion
Superiorly on the greater tubercle of the humerus

## Action
Abduction of the glenohumeral joint
Stabilization of the humeral head in the glenoid fossa

## Palpation
Anterior and superior to the spine of the scapula in the supraspinous fossa during initial abduction in the scapula plane. Also, the tendon may be palpated in a seated position just off the acromion on the greater tubercle.

## Innervation
Suprascapular nerve (C5)

## Application, strengthening, and flexibility

The supraspinatus muscle holds the head of the humerus in the glenoid fossa. In throwing movements, it provides important dynamic stability by maintaining the proper relationship between the humeral head and the glenoid fossa. In the cocking

Shoulder abduction

phase of throwing, there is a tendency for the humeral head to subluxate anteriorly. In the follow-through phase, the humeral head tends to move posteriorly.

The supraspinatus, along with the other rotator cuff muscles, must have excellent strength and endurance to prevent abnormal and excessive movement of the humeral head in the fossa.

The supraspinatus is the most often injured rotator cuff muscle. Acute severe injuries may occur with trauma to the shoulder. However, mild to moderate strains or tears often occur with athletic activity, particularly if the activity involves repetitious overhead movements, such as throwing or swimming.

Injury or weakness in the supraspinatus may be detected when the athlete attempts to substitute the scapula elevators and upward rotators to obtain humeral abduction. An inability to smoothly abduct the arm against resistance is indicative of possible rotator cuff injury.

The supraspinatus muscle may be called into play whenever the middle fibers of the deltoid muscle are used. A "full-can exercise" may be used to emphasize supraspinatus action. This is performed by placing the arm in thumbs-up position, followed by abducting the arm to 90 degrees in a 30- to 45-degree horizontally adducted position (scaption), as if one were holding a full can.

Adducting the arm behind the back with the shoulder internally rotated and extended stretches the supraspinatus.

Chapter

5

FIG. 5.24 • Supraspinatus muscle, posterior view. O, Origin; I, Insertion.

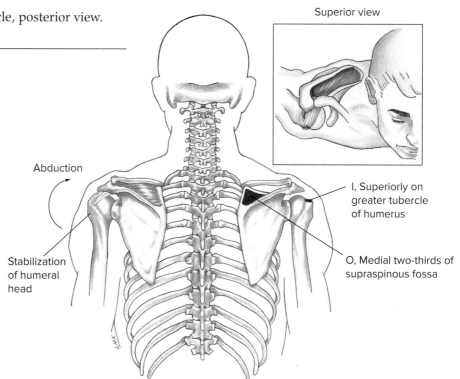

Superior view

Abduction

Stabilization of humeral head

I, Superiorly on greater tubercle of humerus

O, Medial two-thirds of supraspinous fossa

# Infraspinatus muscle  FIG. 5.25

(inˊfra-spi-naˊtus)

Shoulder external rotation

## Origin

Posterior surface of scapula below spine

## Insertion

Greater tubercle on posterior side of the humerus

Shoulder horizontal abduction

## Action

External rotation of the glenohumeral joint
Horizontal abduction of the glenohumeral joint
Extension of the glenohumeral joint
Stabilization of the humeral head in the glenoid fossa

Chapter 5

## Palpation

Just below the spine of the scapula passing upward and laterally to the humerus during resisted external rotation

## Innervation

Suprascapular nerve (C5, C6)

Shoulder extension

## Application, strengthening, and flexibility

The infraspinatus and teres minor muscles are effective when the rhomboid muscles stabilize the scapula. When the humerus is rotated outward, the rhomboid muscles flatten the scapula to the back and fixate it so that the humerus may be rotated.

An appropriate amount of strength and endurance is critical in both the infraspinatus and teres minor as they are called upon eccentrically to slow down the arm from high velocity internal rotation activities such as baseball pitching and serving in tennis. The infraspinatus is vital to maintaining the posterior stability of the glenohumeral joint. It is the most powerful of the external rotators and is the second most commonly injured rotator cuff muscle.

Both the infraspinatus and the teres minor can best be strengthened by externally rotating the arm against resistance in the 15- to 20-degree abducted position and the 90-degree abducted position.

Stretching of the infraspinatus is accomplished with internal rotation and extreme horizontal adduction.

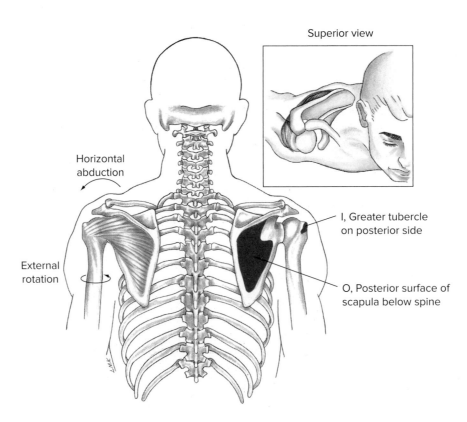

FIG. 5.25 • Infraspinatus muscle, posterior view. O, Origin; I, Insertion.

# Teres minor muscle   FIG. 5.26

(te´rez mi´nor)

## Origin

Posteriorly on the upper and middle aspect of the lateral border of the scapula

## Insertion

Posteriorly on the greater tubercle of the humerus

## Action

External rotation of the glenohumeral joint
Horizontal abduction of the glenohumeral joint
Extension of the glenohumeral joint
Stabilization of the humeral head in the glenoid fossa

## Palpation

Just above the teres major on the posterior scapula surface, moving diagonally upward and laterally from the inferior angle of the scapula during resisted external rotation

## Innervation

Axillary nerve (C5, C6)

## Application, strengthening, and flexibility

The teres minor functions very similarly to the infraspinatus in providing dynamic posterior stability to the glenohumeral joint. Both of these muscles perform the same actions together. The teres minor is strengthened with the same exercises that are used in strengthening the infraspinatus.

The teres minor is stretched similarly to the infraspinatus by internally rotating the shoulder while moving into extreme horizontal adduction.

Shoulder external rotation

Shoulder horizontal abduction

Chapter 5

Shoulder extension

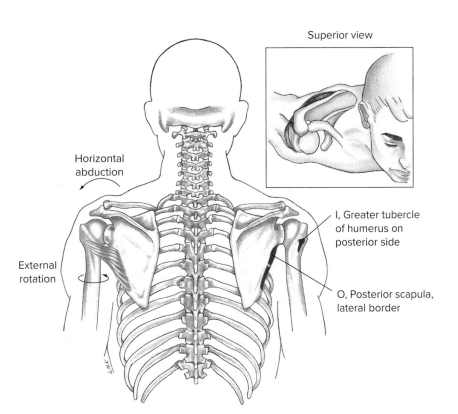

FIG. 5.26 • Teres minor muscle, posterior view. O, Origin; I, Insertion.

# REVIEW EXERCISES

1. List the planes in which each of the following glenohumeral joint movements occurs. List the respective axis of rotation for each movement in each plane.
   a. Abduction
   b. Adduction
   c. Flexion
   d. Extension
   e. Horizontal adduction
   f. Horizontal abduction
   g. External rotation
   h. Internal rotation

2. Why is it essential that both anterior and posterior muscles of the shoulder joint be properly developed? What are some activities or sports that would cause unequal or equal development?

3. What practical application do the activities or sports in Question 2 support if each of the following is true?

   a. The rotator cuff muscles are not functioning properly due to fatigue or lack of appropriate strength and endurance.
   b. The scapula stabilizers are not functioning properly due to fatigue or lack of strength and endurance.

4. The movements of the scapula in relation to the humerus can be explained by discussing the movement of the shoulder complex in its entirety. How does the position of the scapula affect shoulder joint abduction? How does the position of the scapula affect shoulder joint flexion?

5. Describe the bony articulations and movements specific to shoulder joint rotation during the acceleration phase of the throwing motion and how an athlete can work toward increasing the velocity of the throw. What factors affect the velocity of the throw?

6. Using the information from this chapter and other resources, how would you strengthen the four rotator cuff muscles? Give several examples of how these muscles are used in everyday activities.

7. **Muscle analysis chart** · Shoulder girdle and shoulder joint

| Complete the chart by listing the muscles primarily involved in each movement. | |
|---|---|
| **Shoulder girdle** | **Shoulder joint** |
| Upward rotation | Abduction |
| Downward rotation | Adduction |
| Depression | Extension |
| Elevation | Flexion |
| Abduction | Horizontal adduction |
| | Internal rotation |
| Adduction | Horizontal abduction |
| | External rotation |

## 8. **Antagonistic muscle action chart** · Shoulder joint

Complete the chart by listing the muscle(s) or parts of muscles that are antagonist in their actions to the muscles in the left column.

| Agonist | Antagonist |
|---|---|
| Deltoid (anterior fibers) | |
| Deltoid (middle fibers) | |
| Deltoid (posterior fibers) | |
| Supraspinatus | |
| Subscapularis | |
| Teres major | |
| Infraspinatus/Teres minor | |
| Latissimus dorsi | |
| Pectoralis major (upper fibers) | |
| Pectoralis major (lower fibers) | |
| Coracobrachialis | |

Chapter

5

# LABORATORY EXERCISES

1. Locate the following parts of the humerus and scapula on a human skeleton and on a subject:
   a. Greater tubercle
   b. Lesser tubercle
   c. Neck
   d. Shaft
   e. Intertubercular groove
   f. Medial epicondyle
   g. Lateral epicondyle
   h. Trochlea
   i. Capitulum
   j. Supraspinous fossa
   k. Infraspinous fossa
   l. Spine of the scapula

2. How and where can the following muscles be palpated on a human subject?
   a. Deltoid
   b. Teres major
   c. Infraspinatus
   d. Teres minor
   e. Latissimus dorsi
   f. Pectoralis major (upper and lower)

   *Note:* Using the pectoralis major muscle, indicate how various actions allow muscle palpation.

3. Demonstrate and locate on a human subject the muscles that are primarily used in the following shoulder joint movements:
   a. Abduction
   b. Adduction
   c. Flexion
   d. Extension
   e. Horizontal adduction
   f. Horizontal abduction
   g. External rotation
   h. Internal rotation

4. Using an articulated skeleton, compare the relationship of the greater tubercle to the undersurface of the acromion in each of the following situations:
   a. Flexion with the humerus internally versus externally rotated
   b. Abduction with the humerus internally versus externally rotated
   c. Horizontal adduction with the humerus internally versus externally rotated

5. Pair up with a partner with the back exposed. Use your hand to grasp your partner's right scapula along the lateral border to prevent scapula movement. Have your partner slowly abduct the glenohumeral joint as much as possible. Note the difference in total abduction possible normally versus when you restrict movement of the scapula. Repeat the same exercise, except hold the inferior angle of the scapula tightly against the chest wall while you have your partner internally rotate the humerus. Note the difference in total internal rotation possible normally versus when you restrict movement of the scapula.

6. Shoulder joint movement analysis chart

After analyzing each of the exercises in the chart, break each into two primary movement phases, such as a lifting phase and a lowering phase. For each phase, determine the shoulder joint movements occurring, and then list the shoulder joint muscles primarily responsible for causing/controlling those movements. Beside each muscle in each movement, indicate the type of contraction as follows: I—isometric; C—concentric; E—eccentric.

| Exercise | Initial movement (lifting) phase | | Secondary movement (lowering) phase | |
|---|---|---|---|---|
| | Movement(s) | Agonist(s)—(contraction type) | Movement(s) | Agonist(s)—(contraction type) |
| Push-up | | | | |
| Chin-up | | | | |
| Bench press | | | | |
| Dip | | | | |
| Lat pull | | | | |
| Overhead press | | | | |
| Prone row | | | | |
| Barbell shrugs | | | | |

## 7. Shoulder joint sport skill analysis chart

Analyze each skill in the chart and list the movements of the right and left shoulder joint in each phase of the skill. You may prefer to list the initial position the shoulder joint is in for the stance phase. After each movement, list the shoulder joint muscle(s) primarily responsible for causing/controlling that movement. Beside each muscle in each movement, indicate the type of contraction as follows: I—isometric; C—concentric; E—eccentric. It may be desirable to review the concepts for analysis in Chapter 8 for the various phases.

| Exercise | | Stance phase | Preparatory phase | Movement phase | Follow-through phase |
|---|---|---|---|---|---|
| Baseball pitch | (R) | | | | |
| | (L) | | | | |
| Volleyball serve | (R) | | | | |
| | (L) | | | | |
| Tennis serve | (R) | | | | |
| | (L) | | | | |
| Softball pitch | (R) | | | | |
| | (L) | | | | |
| Tennis backhand | (R) | | | | |
| | (L) | | | | |
| Batting | (R) | | | | |
| | (L) | | | | |
| Bowling | (R) | | | | |
| | (L) | | | | |
| Basketball free throw | (R) | | | | |
| | (L) | | | | |

# References

Andrews JR, Zarins B, Wilk KE: *Injuries in baseball,* Philadelphia, 1988, Lippincott-Raven.

Bach HG, Goldberg BA: Posterior capsular contracture of the shoulder, *Journal of the American Academy of Orthopaedic Surgery* 14(5): 265–277, 2006.

Dail NW, Agnew TA, Floyd RT: *Kinesiology for manual therapies,* 2010, McGraw-Hill.

Field D: *Anatomy: palpation and surface markings,* ed 3, Oxford, 2001, Butterworth-Heinemann.

Fongemie AE, Buss DD, Rolnick SJ: Management of shoulder impingement syndrome and rotator cuff tears, *American Family Physician,* 57(4):667–674, 680–682, Feburary 1998.

Garth WP, et al: Occult anterior subluxations of the shoulder in noncontact sports, *American Journal of Sports Medicine* 15:579, November–December 1987.

Hislop HJ, Montgomery J: *Daniels and Worthingham's muscle testing: techniques of manual examination,* ed 8, Philadelphia, 2007, Saunders.

Houglum PA, Bertoti DB: *Brunstom's clinical kinesiology,* ed 6, Philadelphia, 2012, FA Davis.

Loftice JW, Fleisig GS, Wilk KE, Reinold MM, Chmielewski T, Escamilla RF, Andrews JR, eds: *Conditioning program for baseball pitchers,* Birmingham, 2004, American Sports Medicine Institute.

Muscolino JE: *The muscular system manual: the skeletal muscles of the human body,* ed 3, St. Louis, 2010, Elsevier Mosby.

Myers JB, et al: Glenohumeral range of motion deficits and posterior shoulder tightness in throwers with pathologic internal impingement, *American Journal of Sports Medicine* 34:385–391, 2006.

Neumann DA: *Kinesiology of the musculoskeletal system: foundations for physical rehabilitation,* ed 2, St. Louis, 2010, Mosby.

Oatis CA: *Kinesiology: the mechanics and pathomechanics of human movement,* ed 2, Philadelphia, 2008, Lippincott Williams & Wilkins.

Perry JF, Rohe DA, Garcia AO: *The kinesiology workbook,* Philadelphia, 1992, Davis.

Rasch PJ: *Kinesiology and applied anatomy,* ed 7, Philadelphia, 1989, Lea & Febiger.

Reinold MM, Macrina LC, Wilk KE, Fleisig GS, Dun S, Barrentine SW, Ellerbusch MT, Andrews JR: Electromyographic analysis of the supraspinatus and deltoid muscles during 3 common rehabilitation exercises, *Journal of Athletic Training* 42(4): 464–469, 2007.

Seeley RR, VanPutte CL, Regan JL, Russo AF: *Seeley's anatomy & physiology,* ed 9, New York, 2011, McGraw-Hill.

Sieg KW, Adams SP: *Illustrated essentials of musculoskeletal anatomy,* ed 4, Gainesville, FL, 2002, Megabooks.

Spigelman T: Identifying and assessing glenohumeral internal-rotation deficit, *Athletic Therapy Today* 6:29–31, 2006.

Stacey E: Pitching injuries to the shoulder, *Athletic Journal* 65:44, January 1984.

Wilk KE, Reinold MM, Andrews JR, eds: *The athlete's shoulder,* ed 2, Philadelphia, 2009, Churchill Livingstone Elsevier.

Chapter
**5**

# Worksheet Exercises

For in- or out-of-class assignments, or for testing, utilize this tear-out worksheet.

## Worksheet 1

Using crayons or colored markers, draw and label on the worksheet the following muscles. Indicate the origin and insertion of each muscle with an "O" and an "I," respectively, and draw in the origin and insertion on the contralateral side of the skeleton.

a. Deltoid

b. Supraspinatus

c. Subscapularis

d. Teres major

e. Infraspinatus

f. Teres minor

g. Latissimus dorsi

h. Pectoralis major

i. Coracobrachialis

# Worksheet Exercises

For in- or out-of-class assignments, or for testing, utilize this tear-out worksheet.

## Worksheet 2

Label and indicate with arrows the following movements of the shoulder joint. For each motion, complete the sentence by supplying the plane in which it occurs and the axis of rotation.

a. Abduction occurs in the _____ plane about the _____ axis.

b. Adduction occurs in the _____ plane about the _____ axis.

c. Flexion occurs in the _____ plane about the _____ axis.

d. Extension occurs in the _____ plane about the _____ axis.

e. Horizontal adduction occurs in the _____ plane about the _____ axis.

f. Horizontal abduction occurs in the _____ plane about the _____ axis.

# THE ELBOW AND RADIOULNAR JOINTS

## Objectives

■ To identify on a human skeleton selected bony features of the elbow and radioulnar joints

■ To label selected bony features on a skeletal chart

■ To draw and label the muscles on a skeletal chart

■ To determine, list, and palpate the muscles of the elbow and radioulnar joints and their antagonists and appreciate the role of the ligaments in providing stability

■ To palpate the muscles on a human subject and list their antagonists

■ To list the planes of motion and their respective axes of rotation

■ To organize and list the muscles that produce the primary movements of the elbow joint and the radioulnar joint

■ To learn and understand the innervation of the elbow and radioulnar joint muscles

■ To determine, through analysis, the elbow and radioulnar joint movements and muscles involved in selected skills and exercises

Almost any movement of the upper extremity will involve the elbow and radioulnar joints. Quite often, these joints are grouped together because of their close anatomical relationship. The elbow joint is intimately associated with the radioulnar joint in that both bones of the radioulnar joint, the radius and ulna, share an articulation with the humerus to form the elbow joint. For this reason, some may confuse motions of the elbow with those of the radioulnar joint. In addition, radioulnar joint motion may be incorrectly attributed to the wrist joint because it appears to occur there. However, with close inspection, movements of the elbow joint can be clearly distinguished from those of the radioulnar joints, just as the radioulnar movements can be distinguished from those of the wrist. Even though the radius and ulna are both part of the articulation with the wrist, the relationship between them is not nearly as intimate as that of the elbow and radioulnar joints.

## Bones

The ulna is much larger proximally than the radius (Fig. 6.1), but distally the radius is much larger than the ulna (see Fig. 7.1 on p. 170). The scapula and humerus serve as the proximal attachments for the muscles that flex and extend the elbow. The ulna and radius serve as the distal attachments for the same muscles. The scapula, humerus, and ulna serve as proximal attachments for the muscles that pronate and supinate the radioulnar joints. The distal attachments of the radioulnar joint muscles are located on the radius.

The medial supracondylar ridge, olecranon process, coronoid process, and radial tuberosity are important bony landmarks for these muscles. Additionally, the medial epicondyle, lateral epicondyle, and lateral supracondylar ridge are key bony landmarks for the muscles of the wrist and hand, discussed in Chapter 7.

## Joints

The elbow joint is classified as a ginglymus or hinge-type joint that allows only flexion and extension (Fig. 6.1). The elbow may actually be thought of as two interrelated joints: the humeroulnar and the radiohumeral joints (Fig. 6.2). Elbow motions primarily involve movement between the articular surfaces of the humerus and ulna—specifically, the humeral trochlear fitting into the trochlear notch of the ulna. The head of the radius has a relatively small amount of contact with the capitulum of the humerus at the radiocapitellar joint. As the elbow reaches full extension, the olecranon process of the ulna is received by the olecranon fossa of the humerus. This arrangement provides increased joint stability when the elbow is fully extended. Similarly, when the elbow is forced into full flexion, the coranoid fossa receives the coranoid process unless the soft tissue mass on the anterior aspect of the arm and forearm prevents full flexion.

**A  Anterior view**

FIG. 6.1 • Right elbow joint. **A,** Anterior view; **B,** Lateral view; **C,** Medial view.

As the elbow flexes approximately 20 degrees or more, its bony stability is somewhat unlocked, allowing for more side-to-side laxity. The stability of the elbow in flexion is more dependent on the collateral ligaments, such as the lateral or radial collateral ligament and especially the medial or ulnar collateral ligament (Fig. 6.3). The ulnar collateral ligament is critical to providing medial support to prevent the elbow from abducting (not a normal movement of the elbow) when stressed in physical activity. Many contact sports, particularly sports with throwing activities, place stress on the medial aspect of the joint, resulting in injury. Often this injury involves either acute or chronic stress to

the anterior band of the ulnar collateral ligament, or UCL, resulting in partial to complete tears to it. The UCL is particularly crucial to those high-velocity sporting activities, such as baseball pitching, that require optimal stability of the medial elbow. Even moderate injury to this structure can seriously impact an athlete's ability to throw at the highest levels. Compromise of this structure often requires surgery using a tendon graft such as the palmaris longus tendon to reconstruct this ligament. This surgery, often referred to as the "Tommy John procedure," is particularly common among high school, collegiate, and professional pitchers. The radial collateral and lateral ulnar collateral ligaments on

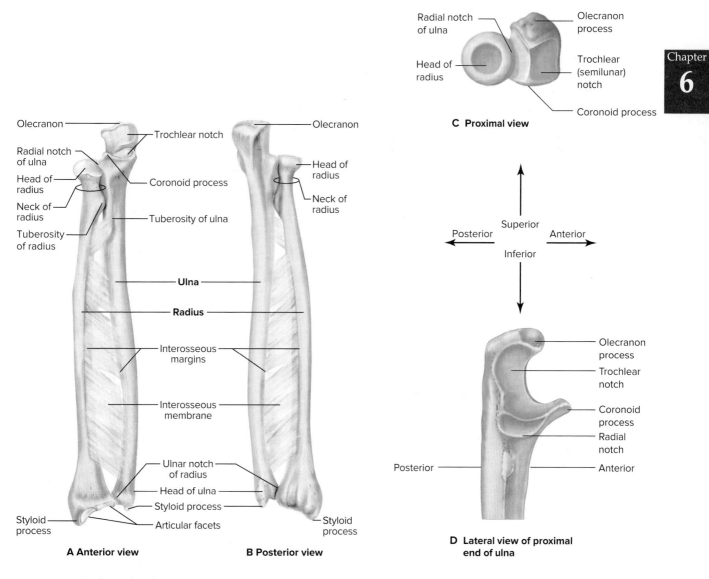

FIG. 6.2 • Right radioulnar joint in supination. **A,** Anterior view; **B,** Posterior view; **C,** Proximal view of radioulnar joint; **D,** Lateral view of proximal end of ulna.

the opposite side provide lateral stability and are rarely injured. Additionally, the annular ligament is located laterally, providing a sling effect around the radial head to secure its stability. See Table 6.1 for further details on the ligaments supporting the elbow and radioulnar joints.

In the anatomical position, it is common for the forearm to deviate laterally from the arm between 5 to 15 degrees. This is referred to as the **carrying angle** and permits the forearms to clear the hips in the swinging movements during walking. It is also important when carrying objects. Typically,

the angle is slightly greater in the dominant limb than in the nondominant limb. It is also common for females to have a slightly greater carrying angle than men (Fig. 6.4).

The elbow is capable of moving from 0 degrees of extension to approximately 145 to 150 degrees of flexion, as detailed in Fig. 6.5. Some people, more commonly females, may hyperextend the elbow up to approximately 15 degrees.

The radioulnar joint is classified as a trochoid or pivot-type joint. The radial head rotates around in its location at the proximal ulna. This rotary

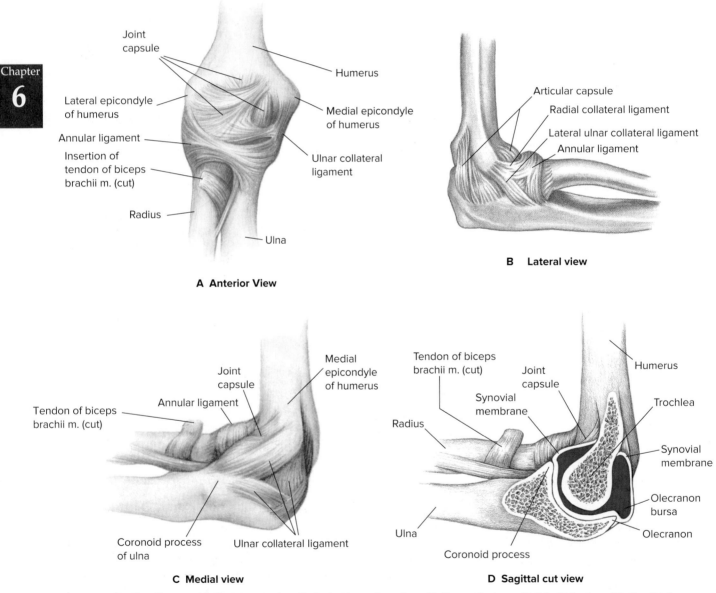

FIG. 6.3 • Right elbow with ligaments detailed. **A,** Anterior view; **B,** Lateral view; **C,** Medial view; **D,** Sagittal cut view.

**TABLE 6.1 · Major ligaments of the elbow and radioulnar joints**

| Joint/structure | Ligament | | Proximal attachments | Distal attachments | Functional application and joint stability |
|---|---|---|---|---|---|
| Elbow | Ulnar collateral ligament | Anterior bundle | Anteroinferior surface of medial epicondyle | Sublime tubercle at medial margin of coranoid process of ulna | Primary contributor to valgus stability, becomes taut in full extension and with valgus stress |
| | | Posterior bundle | Posterior aspect of medial epicondyle | Medial margin of coranoid process of ulna | Some contribution to valgus stability in flexion, reinforces medial side of elbow articular capsule, becomes taut in full flexion |
| | | Oblique (transverse) bundle | Medial aspect of olecranon process | Posterior aspect of coranoid process | Essentially no function in joint stability, helps deepen the trochlear notch |
| | Radial collateral ligament | | Anterior lateral epicondyle of humerus | Annular ligament of radius | Spans lateral side of elbow joint; connects lateral epicondyle of humerus with radius and annular ligament, reinforces lateral side of elbow articular capsule |
| | Lateral ulnar collateral ligament | | Lateral epicondyle of humerus | Supinator crest of lateral ulna | Primary contributor to varus stability, particularly as elbow flexes, provides posterolateral stability |
| Radioulnar joint (proximal) | Quadrate ligament (ligament of Denuce) | | Inferior border of radial notch of proximal ulna | Neck of proximal radius | Reinforces the inferior part of the elbow joint capsule, secures proximal radius against the radial notch by restricting excessive supination and pronation to a lesser degree |
| | Annular ligament of radius | | Anterior margin of radial (lesser sigmoid) notch of ulna | Posterior margin of radial (lesser sigmoid) notch of ulna | Forms 4/5 of a circle, other 1/5 is radial notch of ulna; encircles head of radius and holds it in radial notch of ulna, holds radial head against ulna and cups it distal to the head to provide restraint against distal dislocation of radius |
| Radioulnar joint | Interosseous membrane | | Interosseous border on shaft of ulna | Interosseous border on shaft of radius | Fibers are oriented obliquely downward from radius toward ulna; syndesmosis; proximally directed forces from the hand pass through radius and are transferred to ulna through this membrane |
| Radioulnar joint (distal) | Dorsal radioulnar ligament | | Posterior margin of ulnar notch of distal radius | Posterior aspect of head of ulna | Assists in maintaining ulna in place against radius, reinforces joint capsule, provides dorsal stability to distal radioulnar joint |
| | Volar radioulnar ligament | | Anterior margin of ulnar notch of distal radius | Anterior aspect of head of ulna | Assists in maintaining ulna in place against radius, reinforces joint capsule, provides volar stability to distal radioulnar joint |

movement is accompanied by rotation of the distal radius around the distal ulna. The radial head is maintained in its joint by the annular ligament. The radioulnar joint can supinate approximately 80 to 90 degrees from the neutral position. Pronation varies from 70 to 90 degrees (Fig. 6.6).

Due to the radius and ulna being held tightly together between the proximal and distal articulations by an interosseus membrane, the joint between the shafts of these bones is often referred to as a syndesmosis type of joint. This interosseus membrane is helpful in absorbing and transmitting forces received by the hand, particularly during upper extremity weight bearing. There is substantial rotary motion between the bones despite this classification.

Even though the elbow and radioulnar joints can and do function independently of each other, the muscles controlling each work together in synergy to perform actions at both to benefit the total function of the upper extremity. For this reason, dysfunction at one joint may affect normal function at the other.

FIG. 6.4 • Carrying angle of elbow.

FIG. 6.5 • ROM of the elbow: flexion, extension, and hyperextension. *Flexion:* zero to 150 degrees. *Extension:* 150 degrees to zero. *Hyperextension:* measured in degrees beyond the zero starting point. This motion is not present in all persons. When it is present, it may vary from 5 to 15 degrees.

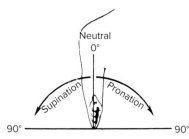

FIG. 6.6 • ROM of the forearm: pronation and supination. *Pronation:* zero to 80 or 90 degrees. *Supination:* zero to 80 or 90 degrees. *Total forearm motion:* 160 to 180 degrees. Persons may vary in the range of supination and pronation. Some may reach the 90-degree arc, and others may have only 70 degrees plus.

## Synergy among the glenohumeral, elbow, and radioulnar joint muscles

Just as there is synergy between the shoulder girdle and the shoulder joint in accomplishing upper extremity activities, there is also synergy between the glenohumeral joint and the elbow joint as well as the radioulnar joints.

As the radioulnar joint goes through its ranges of motion, the glenohumeral and elbow muscles contract to stabilize or assist in the effectiveness of movement at the radioulnar joints. For example, when attempting to fully tighten (with the right hand) a screw with a screwdriver that involves radioulnar supination, we tend to externally rotate and flex the glenohumeral and elbow joints, respectively. Conversely, when attempting to loosen a tight screw with pronation, we tend to internally rotate and extend the elbow and glenohumeral joints, respectively. In either case, we depend on both the agonists and the antagonists in the surrounding joints to provide an appropriate amount of stabilization and assistance with the required task.

## Movements FIGS. 6.5, 6.6, 6.7, 6.8, 6.9

### Elbow movements

**Flexion:** Movement of the forearm to the shoulder by bending the elbow to decrease its angle

**Extension:** Movement of the forearm away from the shoulder by straightening the elbow to increase its angle

### Radioulnar joint movements

**Pronation:** Internal rotary movement of the radius on the ulna that results in the hand moving from the palm-up to the palm-down position

**Supination:** External rotary movement of the radius on the ulna that results in the hand moving from the palm-down to the palm-up position

Elbow flexion

Elbow extension

Chapter

**6**

Radioulnar pronation

Radioulnar supination

**Flexion**

**A**

**Extension**

**B**

Courtesy of Britt Jones (all)

**Pronation**

**C**

**Supination**

**D**

FIG. 6.7 • Movements of the elbow and radioulnar joint. **A,** Elbow flexion; **B,** Elbow extension; **C,** Radioulnar pronation; **D,** Radioulnar supination.

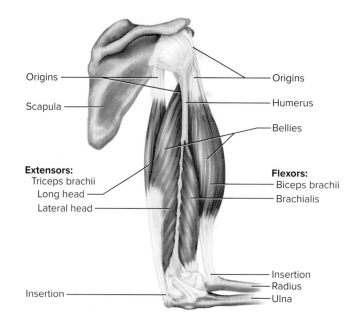

FIG. 6.8 • Right elbow, lateral view of flexors and extensors.

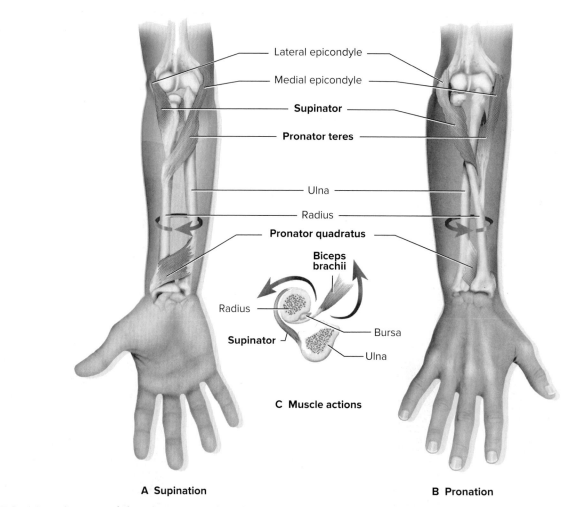

**A Supination**

**B Pronation**

**C Muscle actions**

FIG. 6.9 • Actions of the rotator muscles of the forearm. **A,** Supination; **B,** Pronation, **C,** Cross section just distal to the elbow, showing how the biceps brachii aids the supinator.

# Muscles

The muscles of the elbow and radioulnar joints may be more clearly understood when separated by function. The elbow flexors, located anteriorly, are the biceps brachii, the brachialis, and the brachioradialis, with some weak assistance from the pronator teres (Figs. 6.10, 6.11, and 6.14).

The triceps brachii, located posteriorly, is the primary elbow extensor, with assistance provided by the anconeus (Figs. 6.12, 6.13, and 6.14). The pronator group, located anteriorly, consists of the pronator teres, the pronator quadratus, and the brachioradialis. The brachioradialis also assists with supination, which is controlled mainly by the supinator muscle and the biceps brachii. The supinator muscle is located posteriorly. See Table 6.2.

A common problem associated with the muscles of the elbow is "tennis elbow," which usually involves the extensor digitorum muscle near its origin on the lateral epicondyle. This condition, known technically as **lateral epicondylitis**, is quite frequently associated with gripping and lifting activities. More recently, depending upon the specific pathology, this condition may be termed lateral epicondylagia or lateral epicondylosis. **Medial epicondylitis**, a somewhat less common problem frequently referred to as "golfer's elbow," is associated with the wrist flexor and pronator group near their origin on the medial epicondyle. Both of these conditions involve muscles that cross the elbow but act primarily on the wrist and hand. These muscles will be addressed in Chapter 7.

## Elbow and radioulnar joint muscles—location

Anterior
> Primarily flexion and pronation
>> Biceps brachii
>> Brachialis
>> Brachioradialis
>> Pronator teres
>> Pronator quadratus

Posterior
> Primarily extension and supination
>> Triceps brachii
>> Anconeus
>> Supinator

FIG. 6.10 • Anterior upper extremity muscles.

Courtesy of Lisa Floyd

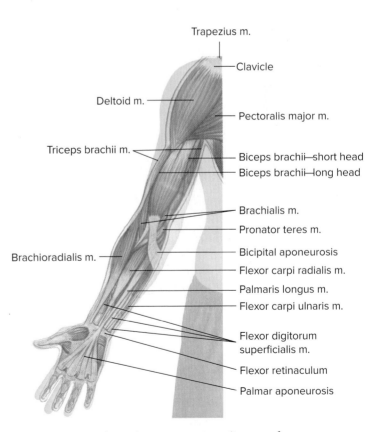

- Trapezius m.
- Clavicle
- Deltoid m.
- Pectoralis major m.
- Triceps brachii m.
- Biceps brachii—short head
- Biceps brachii—long head
- Brachialis m.
- Pronator teres m.
- Brachioradialis m.
- Bicipital aponeurosis
- Flexor carpi radialis m.
- Palmaris longus m.
- Flexor carpi ulnaris m.
- Flexor digitorum superficialis m.
- Flexor retinaculum
- Palmar aponeurosis

FIG. 6.11 • Anterior upper extremity muscles.

Courtesy of Lisa Floyd

FIG. 6.12 • Posterior upper extremity muscles.

Trapezius m. — Acromion process of scapula

Deltoid m.

Triceps brachii m.

Ulnar nerve

Anconeus

Flexor carpi ulnaris m.

Extensor carpi ulnaris m.

Extensor retinaculum

Brachioradialis m.

Extensor carpi radialis longus m.

Extensor carpi radialis brevis m.

Extensor digitorum m.

Extensor digiti minimi m.

FIG. 6.13 • Posterior upper extremity muscles.

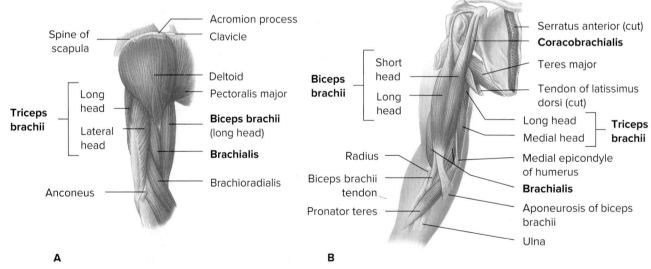

Spine of scapula

Acromion process

Clavicle

Deltoid

Pectoralis major

**Triceps brachii**

Long head

Lateral head

**Biceps brachii (long head)**

**Brachialis**

Anconeus

Brachioradialis

**A**

**Biceps brachii**

Short head

Long head

Radius

Biceps brachii tendon

Pronator teres

Serratus anterior (cut)

**Coracobrachialis**

Teres major

Tendon of latissimus dorsi (cut)

Long head

Medial head

**Triceps brachii**

Medial epicondyle of humerus

**Brachialis**

Aponeurosis of biceps brachii

Ulna

**B**

FIG. 6.14 • Muscles of the arm. **A,** Lateral view of the right shoulder and arm; **B,** Anterior view of the right shoulder and arm (deep). Deltoid, pectoralis major, and pectoralis minor muscles removed to reveal deeper structures.

## TABLE 6.2 · Agonist muscles of the elbow and radioulnar joints

| | Muscle | Origin | Insertion | Action | Plane of motion | Palpation | Innervation |
|---|---|---|---|---|---|---|---|
| Anterior muscles (primarily flexors and pronators) | Biceps brachii long head | Supraglenoid tubercle above the superior lip of the glenoid fossa | Tuberosity of the radius and bicipital aponeurosis (lacertus fibrosis) | Supination of the forearm | Transverse | Easily palpated on the anterior humerus; the long head and short head tendons may be palpated in the intertubercular groove and just inferomedial to the coracoid process, respectively; distally, the biceps tendon is palpated just anteromedial to the elbow joint during supination and flexion | Musculocutaneous nerve (C5, C6) |
| | | | | Horizontal adduction of the shoulder joint | | | |
| | | | | Flexion of the elbow | Sagittal | | |
| | Biceps brachii short head | Coracoid process of the scapula and upper lip of the glenoid fossa in conjunction with the proximal attachment of the coracobrachialis | | Weak flexion of the shoulder joint | | | |
| | | | | Weak abduction of the shoulder joint | Frontal | | |
| | Brachialis | Distal half of the anterior shaft of the humerus | Coronoid process of the ulna | Flexion of the elbow | Sagittal | Deep on either side of the biceps tendon during flexion/extension with forearm in partial pronation; lateral margin may be palpated between biceps brachii and triceps brachii; belly may be palpated through biceps brachii when forearm is in pronation during light flexion | Musculocutaneous nerve (C5, C6) |
| | Brachioradialis | Distal 2/3 of the lateral supracondylar ridge of the humerus | Lateral surface of the distal end of the radius at the styloid process | Flexion of the elbow | Sagittal | Anterolaterally on the proximal forearm during resisted elbow flexion with the radioulnar joint positioned in neutral | Radial nerve (C5, C6) |
| | | | | Pronation from supination to neutral | Transverse | | |
| | | | | Supination from pronation to neutral | | | |
| | Pronator teres | Distal part of the medial supracondylar ridge of the humerus and medial side of the proximal ulna | Middle 1/3 of the lateral surface of the radius | Pronation of the forearm | Transverse | Anteromedial surface of the proximal forearm during resisted mid- to full pronation | Median nerve (C6, C7) |
| | | | | Weak flexion of the elbow | Sagittal | | |
| | Pronator quadratus | Distal 1/4 of the anterior side of the ulna | Distal 1/4 of the anterior side of the radius | Pronation of the forearm | Transverse | Very deep and difficult to palpate, but with the forearm in supination, palpate immediately on either side of the radial pulse with resisted pronation | Anterior interosseous nerve from median nerve (C6, C7) |

TABLE 6.2 (continued) • **Agonist muscles of the elbow and radioulnar joints**

| Muscle | | Origin | Insertion | Action | Plane of motion | Palpation | Innervation |
|---|---|---|---|---|---|---|---|
| Posterior muscles (primarily extensors and supinators) | Triceps brachii long head | Infraglenoid tubercle below inferior lip of glenoid fossa of the scapula | Olecranon process of the ulna | Extension of the elbow joint | Sagittal | Proximally as a tendon on the posteromedial arm to underneath the posterior deltoid during resisted shoulder extension/ adduction | Radial nerve (C7, C8) |
| | | | | Extension of the shoulder joint | Sagittal | | |
| | | | | Adduction of the shoulder joint | Frontal | | |
| | | | | Horizontal abduction of the shoulder joint | Transverse | | |
| | Triceps brachii lateral head | Upper half of the posterior surface of the humerus | | Extension of the elbow joint | Sagittal | Easily palpated on the proximal 2/3 of the posterior humerus during resisted extension | |
| | Triceps brachii medial head | Distal 2/3 of the posterior surface of the humerus | | Extension of the elbow joint | | Deep head: medially and laterally just proximal to the medial and lateral epicondyles | |
| | Supinator | Lateral epicondyle of the humerus and neighboring posterior part of the ulna | Lateral surface of the proximal radius just below the head | Supination of the forearm | Transverse | Position elbow and forearm in relaxed flexion and pronation, respectively; palpate deep to the brachioradialis, extensor carpi radialis longus, extensor carpi radialis brevis on the lateral aspect of the proximal radius with slight resistance to supination | Posterior interosseous nerve from radial nerve (C6) |
| | Anconeus | Posterior surface of the lateral condyle of the humerus | Posterior surface of the lateral olecranon process and proximal 1/4 of the ulna | Extension of the elbow | Sagittal | Posterolateral aspect of the proximal ulna to the olecranon process during resisted extension of the elbow with the wrist in flexion | Radial nerve (C7, C8) |

**Note:** The flexor carpi radialis, palmaris longus, flexor carpi ulnaris, flexor digitorum superficialis, extensor carpi radialis brevis, and extensor carpi radialis longus assist in weak flexion of the elbow, while the extensor carpi ulnaris, extensor digitorum, and extensor digiti minimi assist in weak extension of the elbow. The flexor carpi radialis and extensor carpi radialis longus assist in weak pronation and the extensor indicis, extensor pollicis longus and abductor pollicis longus assist in weak supination. Because they are covered in Chapter 7, they are not listed above.

## Nerves FIGS. 5.13, 6.15, 6.16

The muscles of the elbow and radioulnar joints are all innervated from the median, musculocutaneous, and radial nerves of the brachial plexus. The radial nerve, originating from C5, C6, C7, C8, and T1, provides innervation for the triceps brachii, brachioradialis, supinator, and anconeus (Fig. 6.15). More specifically, the posterior interosseous nerve, derived from the radial nerve, supplies the supinator. The radial nerve also provides sensation to the posterolateral arm, forearm, and hand. The median nerve, illustrated in Fig. 6.16, innervates the pronator teres and further branches to become the anterior interosseus nerve, which supplies the pronator quadratus. The median nerve's most important related derivations are from C6 to C7. It provides sensation to the palmar aspect of the hand and first three phalanges. The palmar aspect of the radial side of the fourth finger is also provided sensation, along with the dorsal aspect of the index and long fingers. The musculocutaneous nerve, shown in Fig. 5.13, branches from C5 to C6 and supplies the biceps brachii and brachialis. Because there is no innervation to the muscles discussed in this chapter from the ulnar nerve, it is not addressed here, but it is often injured at the elbow in various ways. See Chapter 7 for some discussion related to this nerve and injuries to it.

Chapter
**6**

FIG. 6.15 • Muscular and cutaneous distribution of the radial nerve.

FIG. 6.16 • Muscular and cutaneous distribution of the median nerve.

FIG. 6.17 • Innervation for elbow and radioulnar joint muscles and their actions. The neuromuscular structures associated with the primary movers for the elbow and radioulnar joint and their actions are bolded.

# Biceps brachii muscle  FIG. 6.18

(bi´seps bra´ki-i)

## Origin

Long head: supraglenoid tubercle above the superior lip of the glenoid fossa

Short head: coracoid process of the scapula and upper lip of the glenoid fossa in conjunction with the proximal attachment of the coracobrachialis

## Insertion

Tuberosity of the radius and bicipital aponeurosis (lacertus fibrosis)

## Action

Flexion of the elbow

Supination of the forearm

Horizontal adduction of the shoulder joint

Weak flexion of the shoulder joint

Weak abduction of the shoulder joint when the shoulder joint is in external rotation

## Palpation

Easily palpated on the anterior humerus. The long head and short head tendons may be palpated in the intertubercular groove and just inferior to the coracoid process, respectively. Distally, the biceps tendon is palpated just anteromedial to the elbow joint during supination and flexion.

## Innervation

Musculocutaneous nerve (C5, C6)

## Application, strengthening, and flexibility

The biceps brachii, often referred to as the biceps is commonly known as a two-joint (shoulder and elbow), or biarticular, muscle. However, technically it should be considered a three-joint (multiarticular) muscle—shoulder, elbow, and radioulnar. It is weak in actions of the shoulder joint, although it does assist in providing dynamic anterior stability to maintain the humeral head in the glenoid fossa which in part contributes to pathology involving biceps tendinitis of the long head. It is more powerful in flexing the elbow when the radioulnar joint is supinated. It is also a strong supinator, particularly if the elbow is flexed. Palms away from the face (pronation) decrease the effectiveness of the biceps, partly as a result of the disadvantageous pull of the muscle as the radius rotates. The same muscles are used in elbow joint flexion, regardless of forearm pronation or supination.

Flexion of the forearm with a barbell in the hands, known as "curling," is an excellent exercise to develop the biceps brachii. This movement can be performed one arm at a time with dumbbells or both arms simultaneously with a barbell. Other activities in which there is powerful flexion of the forearm are chinning and rope climbing. See Appendix 2 for more commonly used exercises for the biceps brachii and other muscles in this chapter.

Due to the multiarticular orientation of the biceps brachii, all three joints must be positioned appropriately to achieve optimal stretching. The elbow must be extended maximally with the shoulder in full extension. The biceps may also be stretched by beginning with full elbow extension and progressing into full horizontal abduction at approximately 70 to 110 degrees of shoulder abduction. In all cases, the forearm should be fully pronated to achieve maximal lengthening of the biceps brachii.

Elbow flexion

Radioulnar supination

Chapter

**6**

Shoulder Horizontal abduction

Shoulder flexion

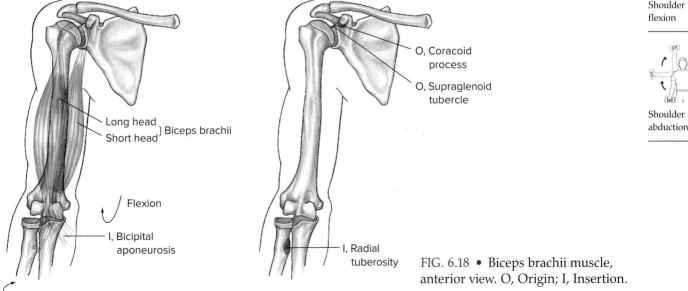

Long head

Short head

} Biceps brachii

Flexion

I, Bicipital aponeurosis

Supination

O, Coracoid process

O, Supraglenoid tubercle

I, Radial tuberosity

FIG. 6.18 • Biceps brachii muscle, anterior view. O, Origin; I, Insertion.

Shoulder abduction

Elbow
flexion

# Brachialis muscle  FIG. 6.19

(bra ´ki-a ´lis)

### Origin
Distal half of the anterior shaft of the humerus

### Insertion
Coronoid process of the ulna

### Action
True flexion of the elbow

### Palpation
Deep on either side of the biceps tendon during flexion/extension with forearm in partial pronation. The lateral margin may be palpated between the biceps brachii and the triceps brachii, and the belly may be palpated through the biceps brachii when the forearm is in pronation during light flexion.

### Innervation
Musculocutaneous nerve and sometimes branches from radial and median nerves (C5, C6)

## Application, strengthening, and flexibility

The brachialis muscle is used along with other flexor muscles, regardless of pronation or supination. It pulls on the ulna, which does not rotate, thus making this muscle the only pure flexor of this joint.

The brachialis muscle is called into action whenever the elbow flexes. It is exercised along with elbow curling exercises, as described for the biceps brachii, pronator teres, and brachioradialis muscles. Elbow flexion activities with the forearm pronated isolate the brachialis to some extent by reducing the effectiveness of the biceps brachii. Since the brachialis is a pure flexor of the elbow, it can be stretched maximally only by extending the elbow with the shoulder relaxed and flexed. Forearm positioning should not affect the stretch on the brachialis unless the forearm musculature itself limits elbow extension, in which case the forearm is probably best positioned in neutral.

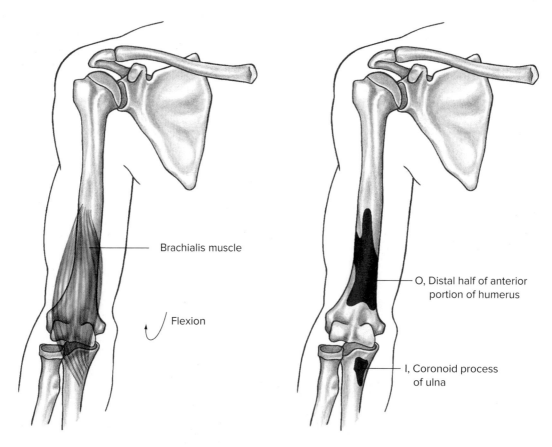

Brachialis muscle

Flexion

O, Distal half of anterior portion of humerus

I, Coronoid process of ulna

FIG. 6.19 • Brachialis muscle, anterior view. O, Origin; I, Insertion.

# Brachioradialis muscle FIG. 6.20

(bra´ki-o-ra´di-a´lis)

## Origin

Distal two-thirds of the lateral supracondylar ridge of the humerus

## Insertion

Lateral surface of the distal end of the radius at the styloid process

## Action

Flexion of the elbow

Pronation of the forearm from supinated position to neutral

Supination of the forearm from pronated position to neutral

## Palpation

Anterolaterally on the proximal forearm during resisted elbow flexion with the radioulnar joint positioned in neutral

## Innervation

Radial nerve (C5, C6)

## Application, strengthening, and flexibility

The brachioradialis is one of three muscles, sometimes known as the mobile wad of three, on the lateral forearm. The other two muscles are the extensor carpi radialis brevis and extensor carpi radialis longus, to which it lies directly anterior. The brachioradialis muscle acts as a flexor best in a midposition or neutral position between pronation and supination. In a supinated position of the forearm, it tends to pronate as it flexes. In a pronated position, it tends to supinate as it flexes. This muscle is favored in its action of flexion when the neutral position between pronation and supination is assumed, as previously suggested. Its insertion at the end of the radius makes it a strong elbow flexor. Its ability as a supinator decreases as the radioulnar joint moves toward neutral. Similarly, its ability to pronate decreases as the forearm reaches neutral. Because of its action of rotating the forearm to a neutral thumb-up position, it is referred to as the hitchhiker muscle, although it has no action at the thumb. As you will see in Chapter 7, nearly all the muscles originating off the lateral epicondyle have some action as weak elbow extensors. This is not the case with the brachioradialis, due to its line of pull being anterior to the elbow's axis of rotation.

The brachioradialis may be strengthened by performing elbow curls against resistance, particularly with the radioulnar joint in the neutral position. In addition, the brachioradialis may be developed by performing pronation and supination movements through the full range of motion against resistance.

The brachioradialis is stretched by maximally extending the elbow with the shoulder in flexion and the forearm in either maximal pronation or maximal supination.

Elbow flexion

Radioulnar pronation

Radioulnar supination

Brachioradialis m.

Supination    Pronation

O, Distal two-thirds of lateral condyloid (supracondylar) ridge of humerus

I, Lateral surface of distal end of radius at styloid process

FIG. 6.20 • Brachioradialis muscle, lateral view. O, Origin; I, Insertion.

# Triceps brachii muscle  FIG. 6.21

(tri´seps bra´ki-i)

## Origin

Long head: infraglenoid tubercle below inferior lip of glenoid fossa of the scapula

Lateral head: upper half of the posterior surface of the humerus

Medial head: distal two-thirds of the posterior surface of the humerus

## Insertion

Olecranon process of the ulna

## Action

All heads: extension of the elbow

Long head: extension, adduction, and horizonal abduction of the shoulder joint

## Palpation

Posterior arm during resisted extension from a flexed position and distally just proximal to its insertion on the olecranon process

Long head: proximally as a tendon on the posteromedial arm to underneath the posterior deltoid during

Lateral head: easily palpated on the proximal two-thirds of the posterior humerus

Medial head (deep head): medially and laterally just proximal to the medial and lateral epicondyles

## Innervation

Radial nerve (C7, C8)

## Application, strengthening, and flexibility

Typical action of the triceps brachii, often referred to as the triceps, is shown in push-ups when there is powerful extension of the elbow. It is used in hand balancing and in any pushing movement involving the upper extremity. The long head is an important extensor of the shoulder joint.

Two muscles extend the elbow—the triceps brachii and the anconeus. Push-ups demand strenuous contraction of these muscles. Dips on the parallel bars are more difficult to perform. Bench pressing a barbell or a dumbbell is an excellent exercise. Overhead presses and triceps curls (elbow extensions from an overhead position) emphasize the triceps.

The triceps brachii should be stretched with both the shoulder and the elbow in maximal flexion.

Long head

Lateral head

Extension

Medial head

O, Infraglenoid tubercle of scapula

O, Upper half of posterior surface of humerus

O, Distal two-thirds of posterior surface of humerus

I, Olecranon process of ulna

FIG. 6.21  •  Triceps brachii muscle, posterior view. O, Origin; I, Insertion.

# Anconeus muscle FIG. 6.22

(an-ko´ne-us)

## Origin
Posterior surface of the lateral condyle of the humerus

## Insertion
Posterior surface of the lateral olecranon process and proximal one-fourth of the lateral ulna

## Action
Extension of the elbow

## Palpation
Posterolateral aspect of the proximal ulna to the olecranon process during resisted extension of the elbow with the wrist in flexion

## Innervation
Radial nerve (C7, C8)

## Application, strengthening, and flexibility
The chief function of the anconeus muscle is to pull the synovial membrane of the elbow joint out of the way of the advancing olecranon process during extension of the elbow. It contracts along with the triceps brachii. It is strengthened with any elbow extension exercise against resistance. Maximal elbow flexion stretches the anconeus.

Elbow extension

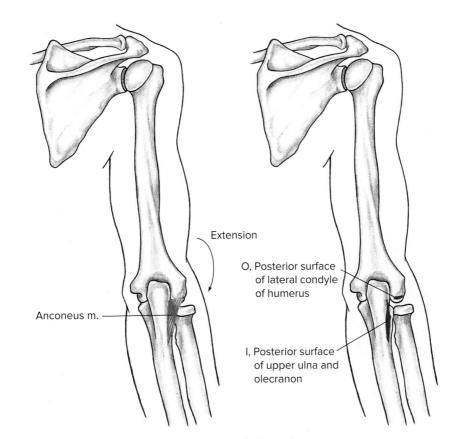

Extension

Anconeus m.

O, Posterior surface of lateral condyle of humerus

I, Posterior surface of upper ulna and olecranon

FIG. 6.22 • Anconeus muscle, posterior view. O, Origin; I, Insertion.

# Pronator teres muscle FIG. 6.23

(pro-na´tor te´rez)

## Origin

Distal part of the medial supracondylar ridge of the humerus and medial side of the proximal ulna

## Insertion

Middle third of the lateral surface of the radius

## Action

Pronation of the forearm
Weak flexion of the elbow

## Palpation

Anteromedial surface of the proximal forearm during resisted mid to full pronation

## Innervation

Median nerve (C6, C7)

Elbow flexion

Chapter
**6**

## Application, strengthening, and flexibility

Typical movement of the pronator teres muscle is with the forearm pronating as the elbow flexes. Movement is weaker in flexion with supination. The use of the pronator teres alone in movement tends to bring the back of the hand to the face as it contracts. Pronation of the forearm with a dumbbell in the hand localizes action and develops the pronator teres muscle. Strengthening this muscle begins with holding a hammer in the hand with the hammer head suspended from the ulnar side of the hand while the forearm is supported on a desk or table. The hammer should be hanging toward the floor, with the forearm pronated to the palm-down position.

The elbow must be fully extended while taking the forearm into full supination to stretch the pronator teres.

Pronator teres m.

Pronation

O, Distal part of medial condyloid ridge of humerus, medial side of proximal ulna

I, Middle third of lateral surface of radius

FIG. 6.23 • Pronator teres muscle, anterior view. O, Origin; I, Insertion.

# Pronator quadratus muscle  FIG. 6.24

(pro-na´tor kwad-ra´tus)

## Origin

Distal fourth of the anterior side of the ulna

## Insertion

Distal fourth of the anterior side of the radius

## Action

Pronation of the forearm

## Palpation

The pronator quadratus, because of its proximity and appearance in some anatomical drawings, is sometimes confused with the flexor retinaculum. See Fig. 6.16. It is very deep and difficult to palpate, but with the forearm in supination it may be palpated immediately on either side of the radial pulse with resisted pronation.

## Innervation

Anterior interosseous nerve from median nerve (C6, C7)

Radioulnar pronation

## Application, strengthening, and flexibility

The pronator quadratus muscle works in pronating the forearm in combination with the triceps brachii in extending the elbow. It is commonly used in turning a screwdriver, as in taking out a screw (with the right hand), when extension and pronation are needed. It is used also in throwing a screwball, when extension and pronation are needed. It may be developed with similar pronation exercises against resistance, as described for the pronator teres. The pronator quadratus is best stretched by using a partner to grasp the wrist and passively take the forearm into extreme supination.

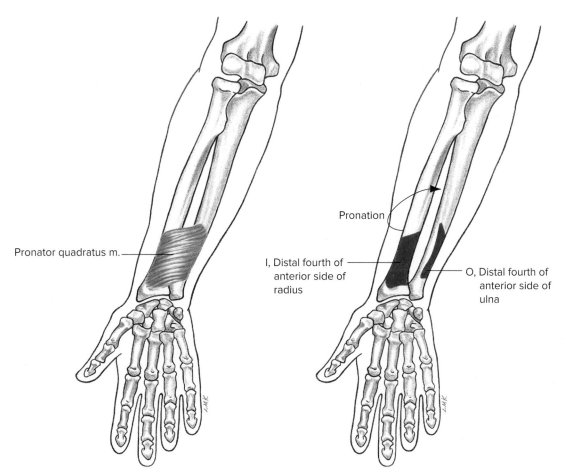

FIG. 6.24 • Pronator quadratus muscle, anterior view. O, Origin; I, Insertion.

# Supinator muscle   FIG. 6.25

(su´pi-na´tor)

## Origin

Lateral epicondyle of the humerus and neighboring
posterior part of the ulna

## Insertion

Lateral surface of the proximal radius just below the
head

## Action

Supination of the forearm

## Palpation

Position the elbow and forearm in relaxed flexion
and pronation, respectively, and palpate deep to the
brachioradialis, extensor carpi radialis longus, extensor carpi radialis brevis on the lateral aspect of the
proximal radius with slight resistance to supination

## Innervation

Posterior interosseous nerve from radial nerve (C6)

**Chapter 6**

## Application, strengthening, and flexibility

The supinator muscle is called into play when movements of extension and supination are required, as
when turning a screwdriver. The curve in throwing
a baseball calls this muscle into play as the elbow is
extended just before ball release. It is most isolated
in activities that require supination with elbow
extension, because the biceps brachii assist with
supination most when the elbow is flexed.

The hands should be grasped and the forearms
extended, in an attempt to supinate the forearms
against the grip of the hands. This localizes, to a
degree, the action of the supinator.

The hammer exercise used for the pronator teres
muscle may be modified to develop the supinator.
In the beginning, the forearm is supported and
the hand is free off the table edge. The hammer is
again held suspended out of the ulnar side of the
hand hanging toward the floor. The forearm is then
supinated to the palm-up position to strengthen
this muscle.

The supinator is stretched when the forearm is
maximally pronated.

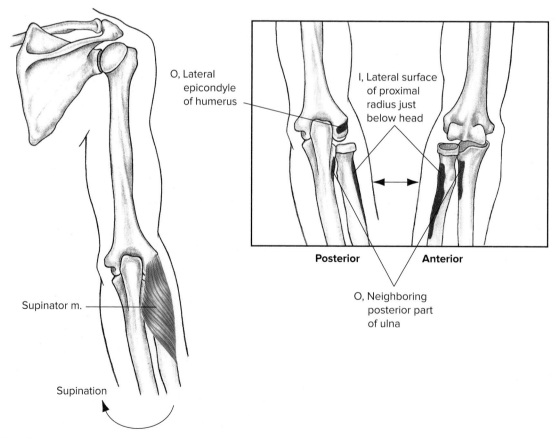

FIG. 6.25 • Supinator muscle, posterior view. O, Origin; I, Insertion.

# REVIEW EXERCISES

1. List the planes in which each of the following elbow and radioulnar joint movements occurs. List the respective axis of rotation for each movement in each plane.
   a. Flexion     c. Pronation
   b. Extension   d. Supination
2. Discuss the difference between chinning with the palms toward the face and chinning with the palms away from the face. Consider this muscularly and anatomically.
3. Analyze and list the differences in elbow and radioulnar joint muscle activity between turning a doorknob clockwise and pushing the door open, and turning the knob counterclockwise and pulling the door open.
4. The elbow joint is just proximal to the distal insertion of which biarticular muscles? Describe the motions each is involved in at the elbow and the superior joint of muscular origin.
5. Lifting a television set as you help your roommate move in requires appropriate lifting techniques and an effective angle of pull. Describe the angle of pull chosen at the elbow joint and why it is chosen instead of other angles.
6. List the muscles involved with "tennis elbow" and describe specifically how you would encourage someone to work on both the strength and the flexibility of these muscles.

## 7. Muscle analysis chart · Elbow and radioulnar joints

| Fill in the chart by listing the muscles primarily involved in each movement. | |
|---|---|
| Flexion | Extension |
| Pronation | Supination |

## 8. Antagonistic muscle action chart · Elbow and radioulnar joints

Fill in the chart by listing the muscle(s) or parts of muscles that are antagonist in their actions to the muscles in the left column.

| Agonist | Antagonist |
|---|---|
| Biceps brachii | |
| Brachioradialis | |
| Brachialis | |
| Pronator teres | |
| Supinator | |
| Triceps brachii | |
| Anconeus | |

# LABORATORY EXERCISES

1. Locate the following parts of the humerus, radius, and ulna on a human skeleton and on a subject:
   a. Skeleton
      1. Medial epicondyle
      2. Lateral epicondyle
      3. Lateral supracondylar ridge
      4. Trochlea
      5. Capitulum
      6. Olecranon fossa
      7. Olecranon process
      8. Coronoid process
      9. Coronoid fossa
      10. Tuberosity of the radius
      11. Ulnar tuberosity
   b. Subject
      1. Medial epicondyle
      2. Lateral epicondyle
      3. Lateral supracondylar ridge

      4. Proximal radioulnar joint
      5. Radiocapitellar joint
      6. Olecranon process
      7. Olecranon fossa
2. How and where can the following muscles be palpated on a human subject?
   a. Biceps brachii
   b. Brachioradialis
   c. Brachialis
   d. Pronator teres
   e. Supinator
   f. Triceps brachii
   g. Anconeus
3. Palpate and list the muscles primarily responsible for the following movements as you demonstrate each:
   a. Flexion
   b. Extension
   c. Pronation
   d. Supination

4. Elbow and radioulnar joint movement analysis chart

After analyzing each exercise in the chart, break each into two primary movement phases, such as a lifting phase and a lowering phase. For each phase, determine the elbow and radioulnar joint movements occurring, and then list the elbow and radioulnar joint muscles primarily responsible for causing/controlling those movements. Beside each muscle in each movement, indicate the type of contraction as follows: I—isometric; C—concentric; E—eccentric.

| Exercise | Initial movement (lifting) phase | | Secondary movement (lowering) phase | |
| --- | --- | --- | --- | --- |
| | Movement(s) | Agonist(s)—(contraction type) | Movement(s) | Agonist(s)—(contraction type) |
| Push-up | | | | |
| Chin-up | | | | |
| Bench press | | | | |
| Dip | | | | |
| Lat pull | | | | |
| Overhead press | | | | |
| Prone row | | | | |

5. Elbow and radioulnar joint sport skill analysis chart

Analyze each skill in the chart and list the movements of the right and left elbow and radioulnar joints in each phase of the skill. You may prefer to list the initial positions that the elbow and radioulnar joints are in for the stance phase. After each movement, list the elbow and radioulnar joint muscle(s) primarily responsible for causing/controlling that movement. Beside each muscle in each movement, indicate the type of contraction as follows: I—isometric; C—concentric; E—eccentric. It may be desirable to review the concepts for analysis in Chapter 8 for the various phases.

| Exercise | | Stance phase | Preparatory phase | Movement phase | Follow-through phase |
|---|---|---|---|---|---|
| Baseball pitch | (R) | | | | |
| | (L) | | | | |
| Volleyball serve | (R) | | | | |
| | (L) | | | | |
| Tennis serve | (R) | | | | |
| | (L) | | | | |
| Softball pitch | (R) | | | | |
| | (L) | | | | |
| Tennis backhand | (R) | | | | |
| | (L) | | | | |
| Batting | (R) | | | | |
| | (L) | | | | |
| Bowling | (R) | | | | |
| | (L) | | | | |
| Basketball free throw | (R) | | | | |
| | (L) | | | | |

Chapter

6

# References

Andrews JR, Zarins B, Wilk KE: *Injuries in baseball,* Philadelphia, 1998, Lippincott-Raven.

Back BR Jr, et al: Triceps rupture: a case report and literature review, *American Journal of Sports Medicine* 15:285, May–June 1987.

Bryce CD, Armstrong AD: Anatomy and biomechanics of the elbow, *Orthopedic Clinics of North America* 39:141–154, 2008.

Dail NW, Agnew TA, Floyd RT: *Kinesiology for manual therapies,* McGraw-Hill, 2010.

Gabbard CP, et al: Effects of grip and forearm position on flex arm hang performance, *Research Quarterly for Exercise and Sport,* July 1983.

Guarantors of Brain: *Aids to the examination of the peripheral nervous system,* ed 4, London, 2000, Saunders.

Herrick RT, Herrick S: Ruptured triceps in powerlifter presenting as cubital tunnel syndrome—a case report, *American Journal of Sports Medicine* 15:514, September–October 1987.

Hislop HJ, Montgomery J: *Daniels and Worthingham's muscle testing: techniques of manual examination,* ed 8, Philadelphia, 2007, Saunders.

Houglum PA, Bertoti DB: *Brunstom's clinical kinesiology,* ed 6, Philadelphia, 2012, FA Davis.

Loftice JW, Fleisig GS, Wilk KE, Reinold MM, Chmielewski T, Escamilla RF, Andrews JR (eds): *Conditioning program for baseball pitchers,* Birmingham, 2004, American Sports Medicine Institute.

Magee DJ: *Orthopedic physical assessment,* ed 6, St. Louis, 2014, Elsevier Saunders.

Muscolino JE: *The muscular system manual: the skeletal muscles of the human body,* ed 3, St. Louis, 2010, Elsevier Mosby.

Oatis CA: *Kinesiology: the mechanics and pathomechanics of human movement,* ed 2, Philadelphia, 2008, Lippincott Williams & Wilkins.

Rasch PJ: *Kinesiology and applied anatomy,* ed 7, Philadelphia, 1989, Lea & Febiger.

Seeley RR, VanPutte CL, Regan JL, Russo AF: *Seeley's anatomy & physiology,* ed 9, New York, 2011, McGraw-Hill.

Shier D, Butler J, Lewis R: *Hole's human anatomy & physiology,* ed 12, New York, 2010, McGraw-Hill.

Sieg KW, Adams SP: *Illustrated essentials of musculoskeletal anatomy,* ed 4, Gainesville, FL, 2002, Megabooks.

Sisto DJ, et al: An electromyographic analysis of the elbow in pitching, *American Journal of Sports Medicine* 15:260, May–June 1987.

Springer SI: Racquetball and elbow injuries, *National Racquetball* 16:7, March 1987.

Thomas BP, Sreekanth R: Distal radioulnar joint injuries, *Indian Journal of Orthopaedics,* 45(5):493–504, 2012.

Van De Graaff KM: *Human anatomy,* ed 6, Dubuque, IA, 2002, McGraw-Hill.

Van Roy P, Baeyens JP, Fauvart D, Lanssiers R, Clarijs JP: Arthrokinematics of the elbow: study of the carrying angle, *Ergonomics* 48(11–14):1645–1656, 2005.

Wilk KE, Reinold MM, Andrews JR (eds): *The athlete's shoulder,* ed 2, Philadelphia, 2009, Churchill Livingstone Elsevier.

Yilmaz E, Karakurt L, Belhan O, Bulut M, Serin E, Avci M: Variation of carrying angle with age, sex, and special reference to side, *Orthopedics* 28(11):1360–1363, 2005.

Chapter
**6**

# Worksheet Exercises

For in- or out-of-class assignments, or for testing, utilize this tear-out worksheet.

## Worksheet 1

Using crayons or colored markers, draw and label on the worksheet the following muscles. Indicate the origin and insertion of each muscle with an "O" and an "I," respectively.

a. Biceps brachii
b. Brachioradialis
c. Brachialis
d. Pronator teres

e. Supinator
f. Triceps brachii
g. Pronator quadratus
h. Anconeus

# Worksheet Exercises

For in- or out-of-class assignments, or for testing, utilize this tear-out worksheet.

## Worksheet 2

Label and indicate by arrows the following movements of the elbow and radioulnar joints. Then below, for each motion, list the agonist muscle(s), the plane in which the motion occurs, and its axis of rotation.

1. Elbow joints
   a. Flexion
   b. Extension

2. Radioulnar joints
   a. Pronation
   b. Supination

1. a. The _____ muscle(s) cause(s) _____ to occur in the

   _____ plane about the _____ axis.

   b. The _____ muscle(s) cause(s) _____ to occur in the

   _____ plane about the _____ axis.

2. a. The _____ muscle(s) cause(s) _____ to occur in the

   _____ plane about the _____ axis.

   b. The _____ muscle(s) cause(s) _____ to occur in the

   _____ plane about the _____ axis.

# THE WRIST AND HAND JOINTS

## Objectives

■ To identify on a human skeleton selected bony features of the wrist, hand, and fingers

■ To label selected bony features on a skeletal chart

■ To draw and label the muscles on a skeletal chart

■ To determine, list, and palpate the muscles of the wrist and hand joints and their antagonists and appreciate the role of the ligaments in providing stability

■ To palpate the muscles on a human subject while demonstrating their actions

■ To list the planes of motion and their respective axes of rotation

■ To organize and list the muscles that produce the primary movements of the wrist, hand, and fingers

■ To learn and understand the innervation of the wrist and hand joint muscles

■ To determine, through analysis, the wrist and hand movements and muscles involved in selected skills and exercises

The importance of the joints of the wrist, hand, and fingers is often overlooked in comparison with the larger joints needed for ambulation. Even though some sports and activities do not require the fine motor skills characteristic of this area, many activities in sports and life require precise functioning of the wrist and hand. Several sports, such as archery, bowling, golf, baseball, and tennis, require the combined use of all these joints. Beyond this, appropriate function in the joints and muscles of our hands is critical for daily activities throughout our life.

Because of the numerous muscles, bones, and ligaments, along with relatively small joint size, the functional anatomy of the wrist and hand is complex and overwhelming to some. This complexity may be simplified by relating the functional anatomy to the major actions of the joints: flexion, extension, abduction, and adduction of the wrist and hand separately, then together.

A large number of muscles are used in these movements. Anatomically and structurally, the human wrist and hand have highly developed, complex mechanisms capable of a variety of movements—a result of the arrangement of 29 bones, more than 25 joints, and more than 30 muscles, of which 18 are intrinsic muscles (both origin and insertion found in the hand).

For most who use this text, an extensive knowledge of these intrinsic muscles is not necessary. However, athletic trainers, physical therapists, occupational therapists, chiropractors, anatomists, physicians, and nurses require a more extensive knowledge. The intrinsic muscles are listed, illustrated, and discussed to a limited degree at the end of this chapter. References at the end of this chapter provide additional sources from which to gain further information.

Our discussion is limited to a review of the muscles, joints, nerves, and movements involved in gross motor activities. The muscles discussed are those of the forearm and the extrinsic muscles of the wrist, hand, and fingers. The larger, more important extrinsic muscles of each joint are included, providing a basic knowledge of this area. The prescription of exercises for strengthening these muscles will be somewhat redundant, as there are primarily only four movements accomplished by their combined actions. One exercise that will strengthen many of these muscles is fingertip push-ups.

## Bones

The wrist and hand contain 29 bones, including the radius and ulna (Fig. 7.1). Eight carpal bones in two rows of four bones form the wrist. The proximal row consists, from the radial (thumb) side to the ulnar (little finger) side, of the scaphoid (boat-shaped) or navicular as it is commonly called, the lunate (moon-shaped), the triquetrum (three-cornered), and the pisiform (pea-shaped) bones. The distal row, from the radial to the ulnar side, consists of the trapezium (greater multangular), trapezoid (lesser multangular), capitate (head-shaped), and hamate (hooked) bones. These bones form a three-sided arch that is concave on the palmar side. This bony arch is spanned by the transverse carpal and volar carpal ligaments creating the **carpal tunnel**, which is frequently a source of problems known as carpal tunnel syndrome (see Fig. 7.8). Of these carpal bones, the scaphoid is by far the most commonly fractured, usually by severe wrist hyperextension from falling on the outstretched hand. Unfortunately, this particular fracture is often dismissed as a sprain after initial injury, only to cause significant problems in the long term if not properly treated. Treatment often requires precise immobilization for periods longer than that of many fractures and/or surgery. Five metacarpal bones, numbered one to five from the thumb to the little finger, join the wrist bones. There are 14 phalanxes (digits), three for each phalange except the thumb, which has only two.

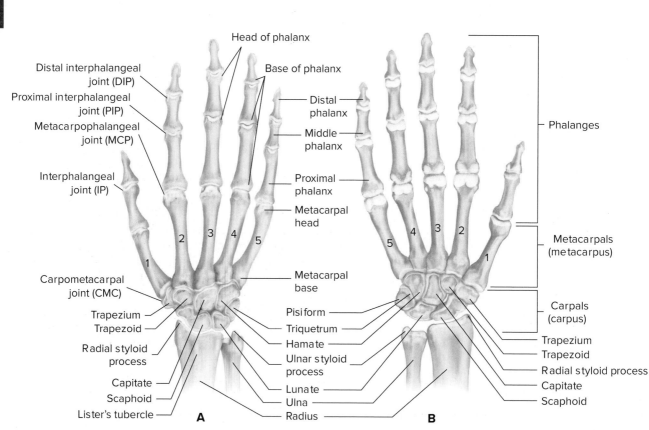

FIG. 7.1 • The right hand. **A,** Posterior (dorsal) view; **B,** Anterior (palmar) view.

They are indicated as proximal, middle, and distal from the metacarpals. Additionally, the thumb has a sesamoid bone within its flexor tendon, and other sesamoids may occur in the fingers.

The medial epicondyle, medial condyloid ridge, and coronoid process serve as a point of origin for many of the wrist and finger flexors, whereas the lateral epicondyle and lateral supracondylar ridge serve as the point of origin for many extensors of the wrist and fingers (Figs. 6.1 and 6.3). Distally, the key bony landmarks for the muscles involved in wrist motion are the base of the second, third, and fifth

metacarpals and the pisiform and hamate. The muscles of the fingers, which are also involved in wrist motion, insert on the base of the proximal, middle, and distal phalanxes (Figs. 7.1 and 7.2). The base of the first metacarpal and the proximal and distal phalanxes of the thumb serve as key insertion points for the muscles involved in thumb motion (Fig. 7.1). The hand consists of three distinct regions: the wrist, the palm, and the finger digits. The palm of the hand may be further separated into the thenar, hypothenar, and midpalmar or intermediate regions.

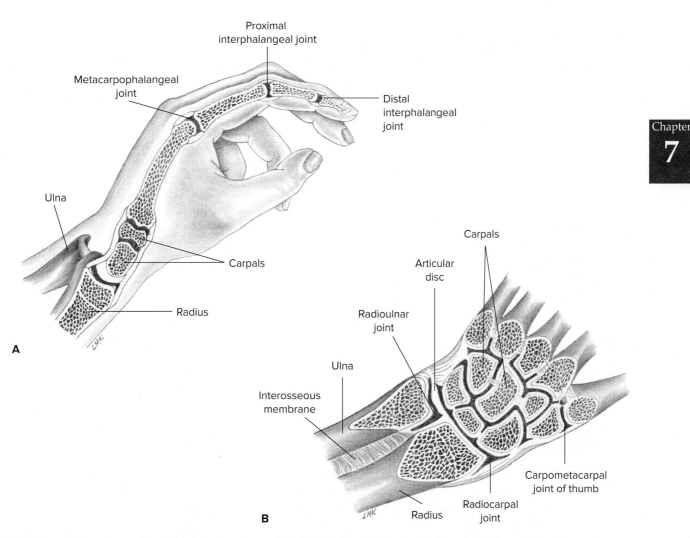

FIG. 7.2 • Left wrist and hand joint structures. **A,** Medial view; **B,** Posterior view with frontal section through wrist.

## Joints

The wrist joint is classified as a condyloid-type joint, allowing flexion, extension, abduction (radial deviation), and adduction (ulnar deviation) (Fig. 7.2). Wrist motion occurs primarily between the distal radius and the proximal carpal row, consisting of the scaphoid, lunate, and triquetrum. As a result, the wrist is often referred to as the radiocarpal joint. The joint allows 70 to 90 degrees of flexion and 65 to 85 degrees of extension. The wrist can abduct 15 to 25 degrees and adduct 25 to 40 degrees (Fig. 7.3).

Each finger has three joints. The metacarpophalangeal (MCP) joints are classified as condyloid. In these joints, 0 to 40 degrees of extension and 85 to 100 degrees of flexion are possible. The proximal interphalangeal (PIP) joints, classified as ginglymus, can move from full extension to approximately 90 to 120 degrees of flexion. The distal interphalangeal (DIP) joints, also classified as ginglymus, can flex 80 to 90 degrees from full extension (Fig. 7.4).

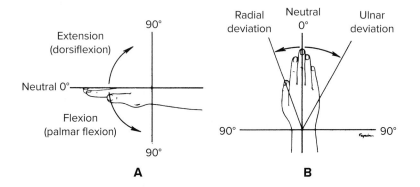

FIG. 7.3 • ROM of the wrist. **A,** Flexion and extension. *Flexion* (palmar flexion): 0 to ± 80 degrees. *Extension* (dorsiflexion): 0 to ± 70 degrees; **B,** Radial and ulnar deviation. *Radial deviation* 0 to 20 degrees. *Ulnar deviation:* 0 to 30 degrees.

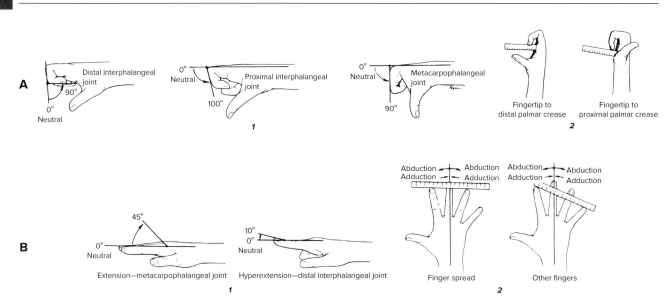

FIG. 7.4 • ROM of the fingers. **A,** Flexion. *1,* Motion can be estimated or measured in degrees. *2,* Motion can be estimated by a ruler as the distance from the tip of the finger to the distal palmar crease (*left*) (measures flexion of the middle and distal joints) and to the proximal palmar crease (*right*) (measures the distal, middle, and proximal joints of the fingers); **B,** Extension, abduction, and adduction. *1,* Extension and hyperextension. *2,* Abduction and adduction. These motions take place in the plane of the palm away from and to the long, or middle, finger of the hand. Abduction of the middle finger occurs when it moves laterally toward the thumb, and adduction occurs when it moves medially toward the little finger. The spread of fingers can be measured from the tip of the index finger to the tip of the little finger (*right*). Individual fingers spread from tip to tip of indicated fingers (*left*).

The thumb has only two joints, both of which are classified as ginglymus. The MCP joint moves from full extension into 40 to 90 degrees of flexion. The interphalangeal (IP) joint can flex 80 to 90 degrees. The carpometacarpal (CMC) joint of the thumb is a unique saddle-type joint having 50 to 70 degrees of abduction. It can flex approximately 15 to 45 degrees and extend 0 to 20 degrees (Fig. 7.5). Its structure also allows slight internal and external rotation.

Although there are far too many ligaments in the wrist and hand to allow a detailed discussion, injuries involving the ulnar and radial collateral ligaments of the metacarpophalangeal and proximal interphalangeal joints are very common because of the medial and lateral stresses they often encounter. One of the more common injuries involves a sprain of the ulnar collateral ligament of the thumb metacarpophalangeal joint. This is often referred to as a gamekeeper's or skier's thumb which may require surgery when severely sprained. The wrist, hand, and fingers depend heavily on the ligaments to provide support and static stability. Table 7.1 provides a listing of the major ligaments of the wrist and hand and some of the finger ligaments are detailed in Fig. 7.6.

FIG. 7.5 • ROM of the thumb. **A,** Abduction. *1,* Zero starting position: the extended thumb alongside the index finger, which is in line with the radius. *Abduction* is the angle created between the metacarpal bones of the thumb and the index finger. This motion may take place in two planes. *2,* Radial abduction or *extension* takes place parallel to the plane of the palm; **B,** Flexion. *1,* Zero starting position: the extended thumb. *2,* Flexion of the interphalangeal joint: 0 to ±80 degrees. *3,* Flexion of the metacarpophalangeal joint: 0 to ±50 degrees. *4,* Flexion of the carpometacarpal joint: 0 to ±15 degrees; **C,** Opposition. Zero starting position (*far left*): the thumb in line with the index fingers. *Opposition* is a composite motion consisting of three elements: (1) abduction, (2) rotation, and (3) flexion. Motion is usually considered complete when the tip of the thumb touches the tip of the fifth finger. Some consider the arc of opposition complete when the tip of the thumb touches the base of the fifth finger. Both methods are illustrated.

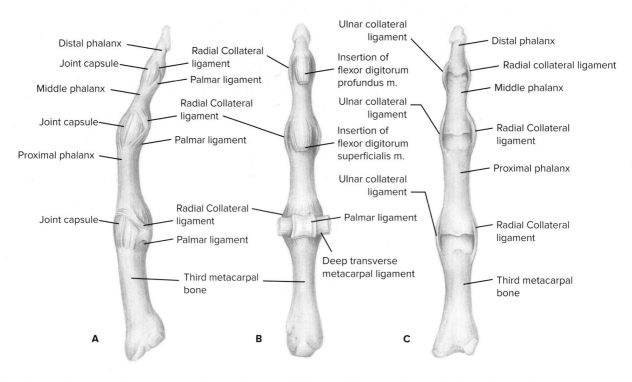

Distal phalanx
Joint capsule
Middle phalanx
Joint capsule
Proximal phalanx
Joint capsule

Radial Collateral ligament
Palmar ligament
Radial Collateral ligament
Palmar ligament
Radial Collateral ligament
Palmar ligament
Third metacarpal bone

**A**

Ulnar collateral ligament
Insertion of flexor digitorum profundus m.
Ulnar collateral ligament
Insertion of flexor digitorum superficialis m.
Ulnar collateral ligament
Palmar ligament
Deep transverse metacarpal ligament

**B**

Distal phalanx
Radial collateral ligament
Middle phalanx
Radial Collateral ligament
Proximal phalanx
Radial Collateral ligament
Third metacarpal bone

**C**

FIG. 7.6 • Metacarpophalangeal and interphalangeal joints of left long finger. **A,** Lateral view; **B,** Anterior (palmar) view; **C,** Posterior view.

Chapter

7

## TABLE 7.1 · Major ligaments of the wrist and hand joints

| Joint/structure | Ligament | Proximal attachments | Distal attachments | Functional application and joint stability |
|---|---|---|---|---|
| Radiocarpal joint | Radial carpal collateral ligament | Tip of styloid process of radius | Radial side of scaphoid with some fibers attaching to radial side of trapezium | Reinforces the articular capsule on the lateral side of the wrist, limits ulnar deviation of wrist |
| | Dorsal radiocarpal ligaments (radiotriquetral ligament, radiolunate ligament, radioscaphoid ligament) | Posterior border of the lower end of the distal radius | Dorsal surfaces of the scaphoid, lunate and triquetrum | Reinforces dorsal wrist fibrous capsule to assist in maintaining joint integrity, slight limitation to wrist flexion, limits excessive supination movement between joint surfaces |
| | Palmar radiocarpal ligament (radioscaphocapitate, radioscapholunate, long and short radiolunate) | Anterior rim of distal radius | Volar surface of the scaphoid, capitate and lunate | Reinforces volar wrist fibrous capsule to assist in maintaining joint integrity, slight limitation to wrist extension, limits excessive supination movement between joint surfaces |
| Ulnocarpal joint | Palmar ulnocarpal ligament (ulnocapitate, ulnolunate and ulnotriquetral) | Styloid process of ulna | Volar surfaces of capitate, lunate and triquetrum | Reinforces the ulnar side of the wrist, prevents dorsal migration of the distal ulna relative to the lunate and triquetrum |
| | Ulnar carpal collateral ligament | Styloid process of ulna | Two fasciculi, one of which is attached to the ulnar side of the triquetral bone, the other to the pisiform and flexor retinaculum | Connects the styloid process of the ulna to the pisiform and triquetrum, reinforces the articular capsule on the medial side of the wrist, limits radial deviation of wrist |
| Intercarpal joint | Transverse carpal ligament | Hamate and pisiform, medially | Scaphoid and trapezium, laterally | Reinforces fibers in antebrachial fascia, palmar surface of wrist, encloses the carpal tunnel on the palmar surface |
| Intercarpal joints (proximal row) | Scapholunate ligament (dorsal, volar, and proximal) and lunotriquetral ligament (dorsal, volar, and proximal) | Scaphoid and lunate | Lunate and triquetrum | Reinforces articular capsule of the intercarpal joints, prevents excessive movements in scapholunate and lunotriquetral joints |
| Intercarpal joints (midcarpal) | Scaphotrapeziotrapezoid ligament | Volar and dorsal aspect of distal pole of scaphoid | Volar and dorsal surface of trapezium and trapezoid | Reinforces articular capsule of the intercarpal joints, prevents excessive movements in between surface bones |
| | Scaphocapitate ligament | Scaphoid (volar surface) | Capitate (volar surface) | |
| | Triquetralcapitate ligament | Triquetrum (volar surface) | Capitate (volar surface) | |
| | Triquetralhamate ligament | Triquetrum (volar surface) | Hamate (volar surface) | |
| Intercarpal joints (distal row) | Capitohamate ligament (dorsal, deep, and palmar) | Dorsal, intraarticular and palmar surface of capitate | Dorsal, intraarticular and palmar surface of hamate | Reinforces articular capsule of the intercarpal joints, prevents excessive movements in between surface bones |
| | Trapeziotrapezoid ligament (dorsal) | Dorsal surface of trapezium | Dorsal surface of trapezoid | |
| | Capitotrapezoid ligament (dorsal, palmar) | Dorsal and palmar surface of capitate | Dorsal and palmar surface of trapezoid | |

Chapter

**7**

TABLE 7.1 (continued) • **Major ligaments of the wrist and hand joints**

| Joint/structure | Ligament | Proximal attachments | Distal attachments | Functional application and joint stability |
|---|---|---|---|---|
| Carpometacarpal joint | Dorsal carpometacarpal ligaments | Dorsal surface of hamate, capitate, trapezoid, and trapezium | Bases of 2nd, 3rd, 4th, and 5th metacarpal (dorsal surface) | Reinforces the carpometacarpal joints (on dorsal aspect), restricts dorsal subluxation of carpometacarpal joints |
| | Palmar carpometacarpal ligaments | Volar surface of hamate, capitate, trapezoid, and trapezium | Bases of 2nd, 3rd, 4th, and 5th metacarpal (volar surface) | Reinforces the carpometacarpal joints (on palmar aspect), restricts volar subluxation of carpometacarpal joints |
| | Pisometacarpal ligament | Pisiform (distal lateral aspect) | Bases of the 5th metacarpal | Forms the insertion for flexor carpi ulnaris tendon along with the pisohamate ligament, stabilizes the pisiform and metacarpal |
| Trapeziometacarpal (1st carpometacarpal) joint | Anterior oblique ligament | Trapezium (volar surface) | Base of 1st metacarpal (volar surface) | Taut in abduction, extension, and pronation, primary stabilizer of 1st CMC joint |
| Intermetacarpal joints | Dorsal metacarpal ligaments | Base of 3rd, 4th, and 5th metacarpals (dorsal surface) | Base of 2nd, 3rd, and 4th metacarpals (dorsal surface) | Reinforces metacarpal bones, prevents dorsal subluxation of intermetacarpal joints |
| | Interosseous metacarpal ligaments | Base of 3rd, 4th, and 5th metacarpals (lateral surface) | Base of 2nd, 3rd, and 4th metacarpals (medial surface) | Reinforces metacarpal bones, maintains intermetacarpal joint stability |
| | Palmar metacarpal ligaments | Base of 3rd, 4th, and 5th metacarpal (palmar surface) | Base of 2nd, 3rd, and 4th metacarpals (palmar surface) | Reinforces metacarpal bones, prevents palmar subluxation of intermetacarpal joints |
| Metacarpophalangeal joint | Radial collateral ligaments | Metacarpal heads (lateral aspect) | Proximal phalanx bases (lateral and palmar aspect) | Slack in extension, tightens during flexion, prevents excessive radial opening of joint |
| | Ulnar collateral ligaments | Metacarpal heads (medial aspect) | Proximal phalanx bases (medial and palmar aspect) | Slack in extension, tightens during flexion, prevents excessive ulnar opening of joint |
| | Volar ligament and volar plate | Metacarpal heads (palmar surface) | Proximal phalanx bases (palmar surface) | Slack in flexion, tightens in full extension, prevents hyperextension |
| Proximal interphalangeal joints | Radial collateral ligaments | Proximal phalanx heads (lateral aspect) | Middle phalanx bases (lateral and palmar aspect) | Slack in extension, tightens during flexion, prevents excessive radial opening of joint |
| | Ulnar collateral ligaments | Proximal phalanx heads (medial aspect) | Middle phalanx bases (medial and palmar aspect) | Slack in extension, tightens during flexion, prevents excessive ulnar opening of joint |
| | Volar ligament and volar plate | Proximal phalanx heads (palmar surface) | Middle phalanx bases (palmar surface) | Slack in flexion, tightens in full extension, prevents hyperextension |
| Distal interphalangeal joints | Radial collateral ligaments | Middle phalanx heads (lateral aspect) | Distal phalanx bases (lateral and palmar aspect) | Slack in extension, tightens during flexion, prevents excessive radial opening of joint |
| | Ulnar collateral ligaments | Middle phalanx heads (medial aspect) | Distal phalanx bases (medial and palmar aspect) | Slack in extension, tightens during flexion, prevents excessive ulnar opening of joint |
| | Volar ligament and volar plate | Middle phalanx heads (palmar surface) | Distal phalanx bases (palmar surface) | Slack in flexion, tightens in full extension, prevents hyperextension |

## Movements

The common actions of the wrist are flexion, extension, abduction, and adduction (Fig. 7.7, *A–D*). The fingers can only flex and extend (Fig. 7.7, *E* and *F*), except at the metacarpophalangeal joints, where abduction and adduction (Fig. 7.7, *G* and *H*) are controlled by the intrinsic hand muscles. In the hand, the middle phalange is regarded as the reference point by which to differentiate abduction and adduction. Abduction of the index and middle fingers occurs when they move laterally toward the radial side of the forearm. Abduction of the ring and little fingers occurs when they move medially toward the ulnar side of the hand. Movement medially of the index and middle fingers toward the ulnar side of the forearm is adduction. Ring and little finger adduction occurs when these fingers move laterally toward the radial side of the hand. The thumb is abducted when it moves away from the palm and is adducted when it moves toward the palmar aspect of the second metacarpal. These movements, together with pronation and supination of the forearm, make possible the many fine, coordinated movements of the forearm, wrist, and hand.

**Flexion (palmar flexion):** Movement of the palm of the hand and/or the phalanges toward the anterior or volar aspect of the forearm

**Extension (dorsiflexion):** Movement of the back of the hand and/or the phalanges toward the posterior or dorsal aspect of the forearm; sometimes referred to as hyperextension

**Abduction (radial deviation, radial flexion):** Movement of the thumb side of the hand toward the lateral aspect or radial side of the forearm; also, movement of the fingers away from the middle finger

**Adduction (ulnar deviation, ulnar flexion):** Movement of the little finger side of the hand toward the medial aspect or ulnar side of the forearm; also, movement of the fingers back together toward the middle finger

**Opposition:** Movement of the thumb across the palmar aspect to oppose any or all of the phalanges

**Reposition:** Movement of the thumb as it returns to the anatomical position from opposition with the hand and/or fingers

Wrist flexion

Wrist extension

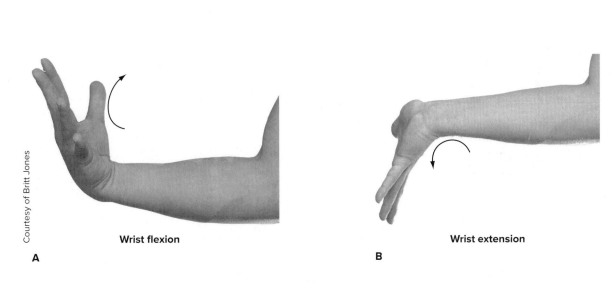

Courtesy of Britt Jones

**Wrist flexion**

**A**

**Wrist extension**

**B**

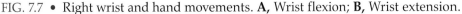

FIG. 7.7 • Right wrist and hand movements. **A,** Wrist flexion; **B,** Wrist extension.

Wrist
abduction

Wrist
adduction

2nd–5th
MCP, PIP,
and DIP
flexion

Thumb
CMC
flexion

2nd–5th
MCP, PIP,
and DIP
extension

Thumb
CMC
extension

**Wrist abduction
(radial deviation)**

**C**

**Wrist adduction
(ulnar deviation)**

**D**

**Flexion of fingers and thumb, opposition**

**E**

**Extension of fingers and thumb, reposition**

**F**

Courtesy of Britt Jones (all)

**Adduction of metacarpophalangeal
joints and the thumb**

**G**

**Abduction of metacarpophalangeal
joints and the thumb**

**H**

FIG. 7.7 (continued) • Right wrist and hand movements. **C,** Wrist abduction; **D,** Right wrist adduction; **E,**
Flexion of the fingers and thumb, opposition; **F,** Extension of the fingers and thumb, reposition; **G,** Adduction
of metacarpophalangeal joints and the thumb; **H,** Abduction of metacarpophalangeal joints and the thumb.

# Muscles TABLE 7.2

The extrinsic muscles of the wrist and hand can be grouped according to function and location (Table 7.2). There are six muscles that move the wrist but do not cross the hand to move the fingers and thumb. The three wrist flexors in this group are, from medial to lateral: the flexor carpi ulnaris, palmaris longus, and flexor carpi radialis,—all of which have their origin on the medial epicondyle of the humerus. The extensors of the wrist have their origins on the lateral epicondyle and include, from lateral to medial, the extensor carpi radialis longus, extensor carpi radialis brevis, and extensor carpi ulnaris (Figs. 6.11 and 6.13). These muscles are important in stabilizing the wrist in extension to enhance the effectiveness of the finger flexors in gripping.

Another nine muscles function primarily to move the phalanges but are also involved in wrist joint actions because they originate on the forearm and cross the wrist. These muscles generally are weaker in their actions on the wrist. The flexor digitorum superficialis and the flexor digitorum profundus are finger flexors; however, they also assist in wrist flexion along with the flexor pollicis longus, which is a thumb flexor. The extensor digiorum, the extensor indicis, and the extensor digiti minimi are finger extensors but also assist in wrist extension, along with the extensor pollicis longus and extensor pollicis brevis, which extend the thumb. The abductor pollicis longus abducts the thumb and assists in wrist abduction.

## TABLE 7.2 • Agonist muscles of the wrist and hand joints

| | Muscle | Origin | Insertion | Action | Plane of motion | Palpation | Innervation |
|---|---|---|---|---|---|---|---|
| Anterior muscles (wrist flexors) | Flexor carpi radialis | Medial epicondyle of humerus | Base of 2nd and 3rd metacarpals on palmar surface | Flexion of the wrist | Sagittal | Anterior surface of the distal forearm and wrist, slightly lateral, in line with the 2nd and 3rd metacarpals with resisted flexion and abduction | Median nerve (C6, C7) |
| | | | | Abduction of the wrist | Frontal | | |
| | | | | Weak flexion of the elbow | Sagittal | | |
| | | | | Weak pronation of the forearm | Transverse | | |
| | Palmaris longus | Medial epicondyle of humerus | Palmar aponeurosis of the 2nd, 3rd, 4th, and 5th metacarpals | Flexion of the wrist | Sagittal | Anteromedial and central aspect of the anterior forearm just proximal to the wrist, particularly with slight wrist flexion and opposition of thumb to the 5th finger | Median nerve (C6, C7) |
| | | | | Weak flexion of the elbow | | | |
| | Flexor carpi ulnaris | Medial epicondyle of humerus and posterior aspect of proximal ulna | Pisiform, hamate, and base of 5th metacarpal (palmar surface) | Flexion of the wrist | Sagittal | Anteromedial surface of the forearm, a few inches below the medial epicondyle of the humerus to just proximal to the wrist, with resisted flexion/adduction | Ulnar nerve (C8, T1) |
| | | | | Adduction of the wrist | Frontal | | |
| | | | | Weak flexion of the elbow | Sagittal | | |

## TABLE 7.2 (continued) • Agonist muscles of the wrist and hand joints

| | Muscle | Origin | Insertion | Action | Plane of motion | Palpation | Innervation |
|---|---|---|---|---|---|---|---|
| **Anterior muscles (wrist and phalangeal flexors)** | Flexor digitorum superficialis | Medial epicondyle of humerus<br><br>Ulnar head: medial coronoid process<br><br>Radial head: upper 2/3 of anterior border of the radius just distal to the radial tuberosity | Each tendon splits and attaches to the sides of the middle phalanx of the four fingers on the palmar surface | Flexion of the fingers at the metacarpo-phalangeal and proximal interphalangeal joints | Sagittal | In depressed area between palmaris longus and flexor carpi ulnaris tendons, particularly when making a fist but keeping the distal interphalangeals extended and with slightly resisted wrist flexion; also on anterior mid-forearm during same activity | Median nerve (C7, C8, T1) |
| | | | | Flexion of the wrist | | | |
| | | | | Weak flexion of the elbow | | | |
| | Flexor digitorum profundus | Proximal 3/4 of anterior and medial ulna | Base of distal phalanges of four fingers | Flexion of the four fingers at the metacar-pophalangeal, proximal inter-phalangeal, and distal interphalangeal joints | Sagittal | Deep to the flexor digitorum superficialis, but on anterior mid-forearm while flexing the distal interphalangeal joints and keeping the proximal interphalangeal joints extended; over the palmar surface of the 2nd, 3rd, 4th, and 5th metacarpo-phalangeal joints during finger flexion against resistance | Median nerve, anterior interosseous branch (C8, T1) to 2nd and 3rd fingers; ulnar nerve (C8, T1) to 4th and 5th fingers |
| | | | | Flexion of the wrist | | | |
| | Flexor pollicis longus | Middle anterior surface of the radius and ante-rior medial bor-der of the ulna just distal to the coronoid process; occasionally a small head is present attaching on the medial epicondyle of the humerus | Base of distal phalanx of thumb on palmar surface | Flexion of the thumb carpo-metacarpal, metacarpo-phalangeal, and interphalangeal joints | Sagittal | Anterior surface of the thumb on the proximal phalanx, and just lateral to the palmaris longus and medial to the flexor carpi radi-alis on the anterior distal forearm, especially during active flexion of the thumb interphalan-geal joint | Median nerve, anterior inter-osseous branch (C8, T1) |
| | | | | Flexion of the wrist | | | |
| | | | | Abduction of the wrist | Frontal | | |

TABLE 7.2 (continued) • **Agonist muscles of the wrist and hand joints**

| | Muscle | Origin | Insertion | Action | Plane of motion | Palpation | Innervation |
|---|---|---|---|---|---|---|---|
| Posterior muscles (wrist extensors) | Extensor carpi ulnaris | Lateral epicondyle of humerus and middle two-fourths of the posterior border of the ulna | Base of 5th metacarpal on dorsal surface | Extension of the wrist | Sagittal | Just lateral to the ulnar styloid process and crossing the posteromedial wrist, particularly with wrist extension/ adduction | Radial nerve, posterior interosseous branch (C6, C7, C8) |
| | | | | Adduction of the wrist | Frontal | | |
| | | | | Weak extension of the elbow | Sagittal | | |
| | Extensor carpi radialis brevis | Lateral epicondyle of humerus | Base of 3rd metacarpal on dorsal surface | Extension of the wrist | Sagittal | Just proximal to the dorsal aspect of the wrist and approximately 1 cm medial to the radial styloid process, the tendon may be felt during extension and traced to base of 3rd metacarpal, particularly when making a fist; proximally and posteriorly, just medial to the bulk of the brachioradialis | Radial nerve (C6, C7) |
| | | | | Abduction of the wrist | Frontal | | |
| | | | | Weak flexion of the elbow | Sagittal | | |
| | Extensor carpi radialis longus | Distal third of lateral supracondylar ridge of humerus and lateral epicondyle of the humerus | Base of 2nd metacarpal on dorsal surface | Extension of the wrist | Sagittal | Just proximal to the dorsal aspect of the wrist and approximately 1 cm medial to the radial styloid process, the tendon may be felt during extension and traced to base of 2nd metacarpal, particularly when making a fist; proximally and posteriorly, just medial to the bulk of the brachioradialis | Radial nerve (C6, C7) |
| | | | | Abduction of the wrist | Frontal | | |
| | | | | Weak flexion of the elbow | Sagittal | | |
| | | | | Weak pronation | Transverse | | |

Chapter

**7**

TABLE 7.2 (continued) • **Agonist muscles of the wrist and hand joints**

| | Muscle | Origin | Insertion | Action | Plane of motion | Palpation | Innervation |
|---|---|---|---|---|---|---|---|
| Posterior muscles (wrist and phalangeal extensors) | Extensor digitorum | Lateral epicondyle of humerus | Four tendons to bases of middle and distal phalanxes of four fingers on dorsal surface | Extension of the 2nd, 3rd, 4th, and 5th phalanges at the metacarpophalangeal joints | Sagittal | With all four fingers extended, on the posterior surface of the distal forearm immediately medial to extensor pollicis longus tendon and lateral to the extensor carpi ulnaris and extensor digiti minimi, then dividing into four separate tendons that are over the dorsal aspect of the hand and metacarpophalangeal joints | Radial nerve, posterior interosseous branch (C6, C7, C8) |
| | | | | Extension of the wrist | | | |
| | | | | Weak extension of the elbow | | | |
| | Extensor indicis | Middle to distal 1/3 of posterior ulna | Base of the middle and distal phalanxes of the 2nd phalange on dorsal surface | Extension of the index finger at the metacarpophalangeal joint | Sagittal | With forearm pronated on the posterior aspect of the distal forearm and dorsal surface of the hand just medial to the extensor digitorum tendon of the index finger with extension of the index fingers and flexion of the 3rd, 4th, and 5th fingers | Radial nerve, posterior interosseous branch (C6, C7, C8) |
| | | | | Weak wrist extension | | | |
| | | | | Weak supination | Transverse | | |
| | Extensor digiti minimi | Lateral epicondyle of humerus | Base of the middle and distal phalanxes of the 5th phalange on dorsal surface | Extension of the little finger at the metacarpophalangeal joint | Sagittal | Passing over the dorsal aspect of the distal radioulnar joint, particularly with relaxed flexion of other fingers and alternating 5th finger extension and relaxation; dorsal surface of forearm immediately medial to the extensor digitorum and lateral to the extensor carpi ulnaris | Radial nerve, posterior interosseous branch (C6, C7, C8) |
| | | | | Weak wrist extension | | | |
| | | | | Weak elbow extension | | | |
| | Extensor pollicis longus | Posterior lateral surface of the lower middle ulna | Base of distal phalanx of thumb on dorsal surface | Extension of the thumb at the carpometacarpal, metacarpophalangeal, and interphalangeal joints | Sagittal | Dorsal aspect of the hand to its insertion on the base of the distal phalanx; also on posterior surface of lower forearm between radius and ulnar just proximal to the extensor indicis and medial to the extensor pollicis brevis and abductor pollicis longus with forearm pronated and fingers in relaxed flexion while actively extending the thumb | Radial nerve, posterior interosseous branch (C6, C7, C8) |
| | | | | Extension of the wrist | | | |
| | | | | Abduction of the wrist | Frontal | | |
| | | | | Weak supination | Transverse | | |

TABLE 7.2 (continued) • **Agonist muscles of the wrist and hand joints**

| | Muscle | Origin | Insertion | Action | Plane of motion | Palpation | Innervation |
|---|---|---|---|---|---|---|---|
| Posterior muscles (wrist and phalangeal extensors) | Extensor pollicis brevis | Posterior surface of lower middle radius | Base of proximal phalanx of thumb on dorsal surface | Extension of the thumb at the carpometacarpal and metacarpophalangeal joints | Sagittal | Just lateral to the extensor pollicis longus tendon on the dorsal side of the hand to its insertion on the proximal phalanx with extension of the thumb carpometacarpal and metacarpophalangeal and flexion of the interphalangeal joints | Radial nerve, posterior interosseous branch (C6, C7) |
| | | | | Weak wrist extension | | | |
| | | | | Abduction of the wrist | Frontal | | |
| Posterior muscles | Abductor pollicis longus | Posterior aspect of radius and midshaft of the ulna | Base of 1st metacarpal on dorsal lateral surface | Abduction of the thumb at the carpometacarpal joint | Frontal | Lateral aspect of the wrist joint just proximal to the 1st metacarpal during active thumb and wrist abduction | Radial nerve, posterior interosseous branch (C6, C7) |
| | | | | Abduction of the wrist | | | |
| | | | | Extension of the thumb at the carpometacarpal joint | Sagittal | | |
| | | | | Weak wrist extension | | | |
| | | | | Weak supination | Transverse | | |

All of the wrist flexors generally have their origins on the anteromedial aspect of the proximal forearm and the medial epicondyle of the humerus, whereas their insertions are on the anterior aspect of the wrist and hand. All of the flexor tendons except for the flexor carpi ulnaris and palmaris longus pass through the carpal tunnel, along with the median nerve (Fig. 7.8). Conditions leading to swelling and inflammation in this area can result in increased pressure in the carpal tunnel, which interferes with normal function of the median nerve, leading to reduced motor and sensory function of its distribution. Known as **carpal tunnel syndrome**, this condition is particularly common with repetitive use of the hand and wrist in manual labor and clerical work such as typing and keyboarding. Often, slight modifications in work habits and the positions of the hand and wrist during these activities can be preventive. Additionally, flexibility exercises for the wrist and finger flexors may be helpful.

The wrist extensors generally have their origins on the posterolateral aspect of the proximal forearm and the lateral humeral epicondyle, whereas their insertions are located on the posterior aspect of the wrist and hand. The flexor and extensor tendons at the distal forearm immediately proximal to the wrist are held in place on the palmar and dorsal aspects by transverse bands of tissue. These bands, known respectively as the flexor and extensor retinaculum, prevent these tendons from bowstringing during flexion and extension.

The wrist abductors are the flexor carpi radialis, extensor carpi radialis longus, extensor carpi radialis brevis, abductor pollicis longus, extensor pollicis longus, and extensor pollicis brevis. These muscles generally cross the wrist joint anterolaterally and posterolaterally to insert on the radial side of the hand. The flexor carpi ulnaris and extensor carpi ulnaris adduct the wrist and cross the wrist joint anteromedially and posteromedially to insert on the ulnar side of the hand.

Chapter

**7**

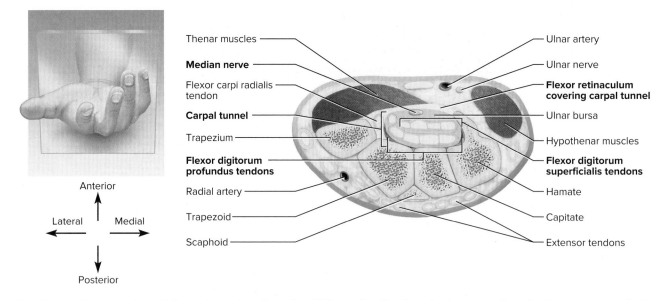

Thenar muscles

**Median nerve**

Flexor carpi radialis tendon

**Carpal tunnel**

Trapezium

**Flexor digitorum profundus tendons**

Radial artery

Trapezoid

Scaphoid

Ulnar artery

Ulnar nerve

**Flexor retinaculum covering carpal tunnel**

Ulnar bursa

Hypothenar muscles

**Flexor digitorum superficialis tendons**

Hamate

Capitate

Extensor tendons

Anterior

Lateral       Medial

Posterior

FIG. 7.8 • Cross section of the right wrist, viewed as if from the distal end of a person's right forearm extended toward you with the palm up. Note how the flexor tendons and median nerve are confined in a tight space between the carpal bones and the flexor retinaculum.

The intrinsic muscles of the hand (see Table 7.2 and Fig. 7.26) have their origins and insertions on the bones of the hand. Sorting the intrinsic muscles into three groups according to location is helpful in understanding and learning these muscles. On the radial side are four muscles of the thumb—the opponens pollicis, the abductor pollicis brevis, the flexor pollicis brevis, and the adductor pollicis. On the ulnar side are three muscles of the little finger—the opponens digiti minimi, the abductor digiti minimi, and the flexor digiti minimi brevis. In the remainder of the hand are 11 muscles, which can be further grouped as the 4 lumbricals, the 3 palmar interossei, and the 4 dorsal interossei.

### Wrist and hand muscles—location

**Anteromedial at the elbow and forearm and anterior at the hand (Fig. 7.9, A–C)**

Primarily wrist flexion
    Flexor carpi radialis
    Flexor carpi ulnaris
    Palmaris longus
Primarily wrist and phalangeal flexion
    Flexor digitorum superficialis
    Flexor digitorum profundus
    Flexor pollicis longus
**Posterolateral at the elbow and forearm and posterior at the hand (Fig. 7.9, D)**
Primarily wrist extension
    Extensor carpi radialis longus
    Extensor carpi radialis brevis
    Extensor carpi ulnaris
Primarily wrist and phalangeal extension
    Extensor digitorum
    Extensor indicis
    Extensor digiti minimi
    Extensor pollicis longus
    Extensor pollicis brevis
    Abductor pollicis longus

FIG. 7.9 • Muscles of the forearm. **A,** Anterior view of the right forearm (superficial). Brachioradialis muscle is removed; **B,** Anterior view of the right forearm (deeper than *A*). Pronator teres, flexor carpi radialis and ulnaris, and palmaris longus muscles are removed; **C,** Anterior view of the right forearm (deeper than *A* or *B*). Brachioradialis, pronator teres, flexor carpi radialis and ulnaris, palmaris longus, and flexor digitorum superficialis muscles are removed; **D,** Deep muscles of the right posterior forearm, with extensor digitorum, extensor digiti minimi, and extensor carpi ulnaris muscles cut to reveal deeper muscles.

# Nerves

The muscles of the wrist and hand are all innervated from the radial, median, and ulnar nerves of the brachial plexus, as illustrated in Figs. 6.15, 6.16, and 7.10. The radial nerve, originating from C6, C7, and C8, provides innervation for the extensor carpi radialis brevis and extensor carpi radialis longus. It then branches to become the posterior interosseous nerve, which supplies the extensor carpi ulnaris, extensor digitorum, extensor digiti minimi, abductor pollicis longus, extensor pollicis longus, extensor pollicis brevis, and extensor indicis. The median nerve, arising from C6, C7, C8, and T1, innervates the flexor carpi radialis, palmaris longus, and flexor digitorum superficialis. It then branches to become the anterior interosseous nerve, which innervates the flexor digitorum profundus for the index and long fingers as well as the flexor pollicis longus. Regarding the intrinsic muscles of the hand, the median nerve innervates the abductor pollicis brevis, flexor pollicis brevis (superficial head), opponens pollicis, and first and second lumbrical. The ulnar nerve, branching from C8 and T1, supplies the flexor digitorum profundus for the fourth and fifth fingers and the flexor carpi ulnaris. Additionally, it innervates the remaining intrinsic muscles of the hand (the deep head of the flexor pollicis brevis, adductor pollicis, palmar interossei, dorsal interossei, third and fourth lumbrical, opponens digiti minimi, abductor digiti minimi, and flexor digiti minimi brevis). Sensation to the ulnar side of the hand, the ulnar half of the ring finger, and the entire little finger is provided by the ulnar nerve. Of all the nerves in the upper extremity, the ulnar nerve is traumatized the most. Most people have hit their "funny bone" and experienced a painful tingling sensation into the ulnar side of their forearm and fourth and fifth fingers. This is actually a contusion to the ulnar nerve at the medial elbow. Usually it subsides fairly quickly, but chronic contusions or pressure over this area may lead to a hypersensitivity of this nerve, making it more easily irritated. Additionally, throwing athletes may experience a traction injury to this nerve with the medial stress placed on the elbow during pitching or in conjunction with an ulnar collateral ligament sprain. The ulnar nerve also may become inflamed from subluxing or slipping out of its groove, especially in chronic cases. See Fig. 7.11 for a full listing of the wrist and hand muscular innervations.

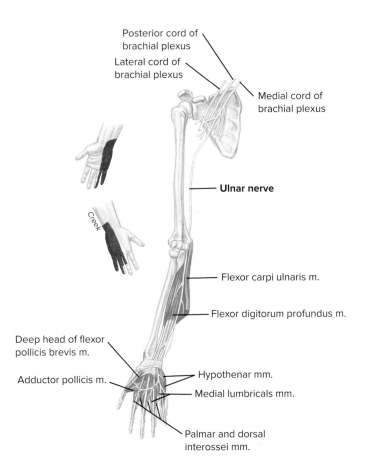

FIG. 7.10 • Muscular and cutaneous distribution of the ulnar nerve.

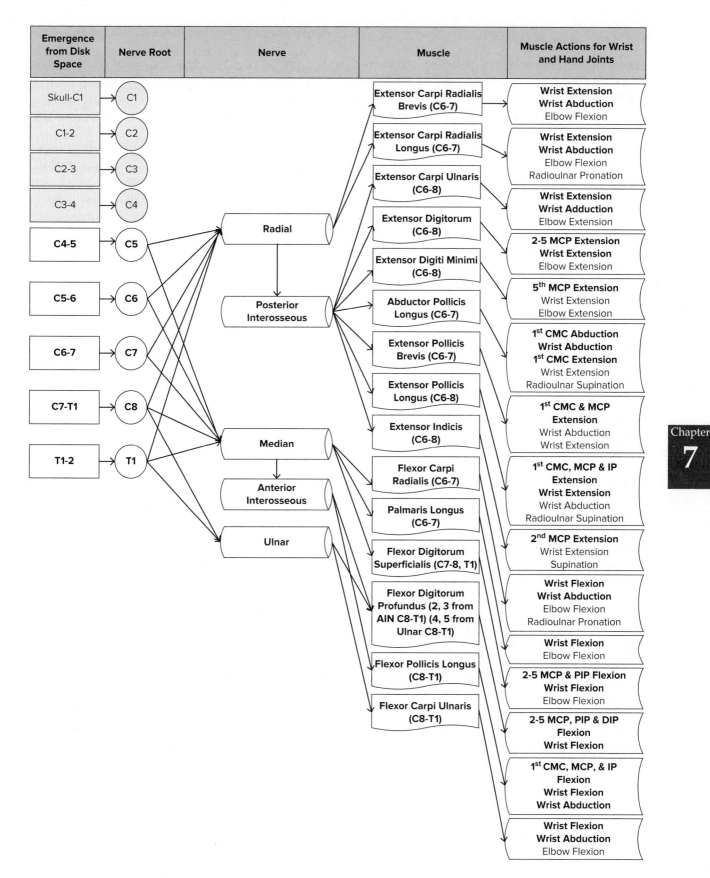

FIG. 7.11 • Innervation for wrist and hand joint muscles and their actions. The neuromuscular structures associated with the primary movers for the wrist and hand joints and their actions are bolded.

# Flexor carpi radialis muscle  FIG. 7.12

(fleks´or kar´pi ra´di-a´lis)

Wrist flexion

### Origin

Medial epicondyle of the humerus

### Insertion

Base of the second and third metacarpals, anterior (palmar surface)

Wrist abduction

### Action

Flexion of the wrist
Abduction of the wrist
Weak flexion of the elbow
Weak pronation of the forearm

### Palpation

Anterior surface of the distal forearm and wrist, slightly lateral, in line with the second and third metacarpals with resisted flexion and abduction

### Innervation

Median nerve (C6, C7)

## Application, strengthening, and flexibility

The flexor carpi radialis, flexor carpi ulnaris, and palmaris longus are the most powerful of the wrist flexors. They are brought into play during any activity that requires wrist curling or stabilization of the wrist against resistance, particularly if the forearm is supinated.

The flexor carpi radialis may be developed by performing wrist curls against a handheld resistance. This may be accomplished when the supinated forearm is supported by a table, with the hand and wrist hanging over the edge to allow full range of motion. The extended wrist is then flexed or curled up to strengthen this muscle. See Appendix 2 for more commonly used exercises for the flexor carpi radialis and other muscles in this chapter.

To stretch the flexor carpi radialis, the elbow must be fully extended with the forearm supinated while a partner passively extends and adducts the wrist.

Elbow flexion

Radioulnar pronation

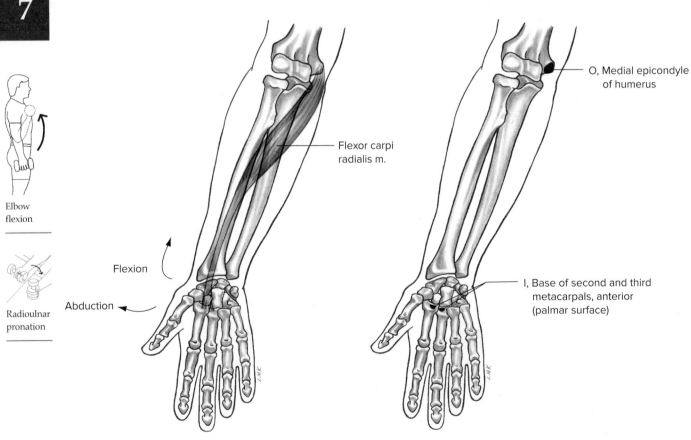

FIG. 7.12 • Flexor carpi radialis muscle, anterior view. O, Origin; I, Insertion.

# Palmaris longus muscle FIG. 7.13
### (pal-ma´ris lon´gus)

## Origin
Medial epicondyle of the humerus

## Insertion
Palmar aponeurosis of the second, third, fourth, and fifth metacarpals

## Action
Flexion of the wrist
Weak flexion of the elbow

## Palpation
The palmaris longus is absent in either one or both forearms in some people. Anteromedial and central aspect of the anterior forearm is just proximal to the wrist, particularly with slight wrist flexion and opposition of thumb to the fifth finger. This opposition makes the tendon, when present, easily viewable.

## Innervation
Median nerve (C6, C7)

## Application, strengthening, and flexibility

Unlike the flexor carpi radialis and flexor carpi ulnaris, which are not only wrist flexors but also abductors and adductors, respectively, the palmaris longus is involved only in wrist flexion from the anatomical position because of its central location on the anterior forearm and wrist. It can, however, assist in abducting the wrist from an extremely adducted position back to neutral, and assist in adducting the wrist from an extremely abducted position back to neutral. It may also assist slightly in forearm pronation because of its slightly lateral insertion in relation to its origin on the medial epicondyle. It may also be strengthened with any type of wrist-curling activity, such as those described for the flexor carpi radialis muscle.

Maximal elbow and wrist extension stretches the palmaris longus.

Wrist flexion

Elbow flexion

O, Medial epicondyle of humerus

Palmaris longus m.

Flexion

I, Palmaris aponeurosis of second, third, fourth, and fifth metacarpals

FIG. 7.13 • Palmaris longus muscle, anterior view. O, Origin; I, Insertion.

# Flexor carpi ulnaris muscle  FIG. 7.14

(fleks´or kar´pi ul-na´ris)

### Origin

Medial epicondyle of the humerus
Posterior aspect of the proximal ulna

**Wrist flexion**

### Insertion

Pisiform, hamate, and base of the fifth metacarpal
(palmar surface)

### Action

Flexion of the wrist
Adduction of the wrist, together with the extensor
carpi ulnaris muscle
Weak flexion of the elbow

**Wrist adduction**

### Palpation

Anteromedial surface of the forearm, a few inches
below the medial epicondyle of the humerus to
just proximal to the wrist, with resisted flexion/
adduction

### Innervation

Ulnar nerve (C8, T1)

### Application, strengthening, and flexibility

The flexor carpi ulnaris is very important in wrist
flexion or curling activities. In addition, it is one
of only two muscles involved in wrist adduction or
ulnar flexion. It may be strengthened with any type
of wrist-curling activity against resistance, such as
those described for the flexor carpi radialis muscle.

To stretch the flexor carpi ulnaris, the elbow
must be fully extended with the forearm supinated
while a partner passively extends and abducts the
wrist.

**Elbow flexion**

FIG. 7.14 • Flexor carpi ulnaris muscle, anterior view. O, Origin; I, Insertion.

# Extensor carpi ulnaris muscle  FIG. 7.15
(eks-ten´sor kar´pi ul-na´ris)

## Origin

Lateral epicondyle of the humerus
Middle two-fourths of the posterior border of the ulna

## Insertion

Base of the fifth metacarpal (dorsal surface)

## Action

Extension of the wrist
Adduction of the wrist together with the flexor carpi
  ulnaris muscle
Weak extension of the elbow

## Palpation

Just lateral to the ulnar styloid process and crossing the posteromedial wrist, particularly with wrist
  extension/adduction

## Innervation

Radial nerve, posterior interosseous branch (C6–C8)

## Application, strengthening, and flexibility

Besides being a powerful wrist extensor, the extensor carpi ulnaris muscle is the only muscle other
than the flexor carpi ulnaris involved in wrist
adduction or ulnar deviation. The extensor carpi
ulnaris, the extensor carpi radialis brevis, and the
extensor carpi radialis longus are the most powerful of the wrist extensors. These muscles act as
antagonists to wrist flexion to allow the finger flexors to function more effectively in gripping. Any
activity requiring wrist extension or stabilization of
the wrist against resistance, particularly if the forearm is pronated, depends greatly on the strength
of these muscles. They are often brought into play
with the backhand in racquet sports.

The extensor carpi ulnaris may be developed by
performing wrist extension against a handheld resistance. This may be accomplished with the pronated
forearm being supported by a table with the hand
hanging over the edge to allow full range of motion.
The wrist is then moved from the fully flexed position
to the fully extended position against the resistance.

Stretching the extensor carpi ulnaris requires
the elbow to be extended with the forearm pronated while the wrist is passively flexed and slightly
abducted.

Wrist
extension

Wrist
adduction

Elbow
extension

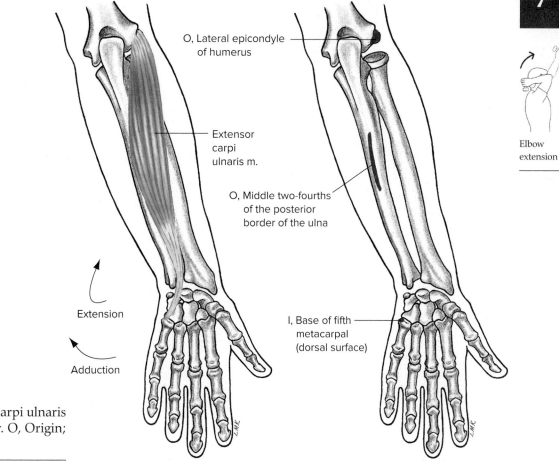

O, Lateral epicondyle
of humerus

Extensor
carpi
ulnaris m.

O, Middle two-fourths
of the posterior
border of the ulna

Extension

Adduction

I, Base of fifth
metacarpal
(dorsal surface)

FIG. 7.15 • Extensor carpi ulnaris
muscle, posterior view. O, Origin;
I, Insertion.

# Extensor carpi radialis brevis muscle FIG. 7.16

(eks-ten´sor kar´pi ra´di-a´lis bre´vis)

## Origin

Lateral epicondyle of the humerus

## Insertion

Base of the third metacarpal (dorsal surface)

## Action

Extension of the wrist
Abduction of the wrist
Weak flexion of the elbow

## Palpation

Dorsal side of the forearm, and difficult to distinguish from the extensor carpi radialis longus and the extensor digitorum

Just proximal to the dorsal aspect of the wrist and approximately 1 cm medial to the radial styloid process, the tendon may be felt during extension and traced to the base of the third metacarpal, particularly when making a fist; proximally and posteriorly, just medial to the bulk of the brachioradialis

## Innervation

Radial nerve (C6, C7)

## Application, strengthening, and flexibility

The extensor carpi radialis brevis is important in any sports activity that requires powerful wrist extension, such as golf or tennis. Wrist extension exercises, such as those described for the extensor carpi ulnaris, are appropriate for development of the muscle.

Stretching the extensor carpi radialis brevis and longus requires the elbow to be extended with the forearm pronated while the wrist is passively flexed and slightly adducted.

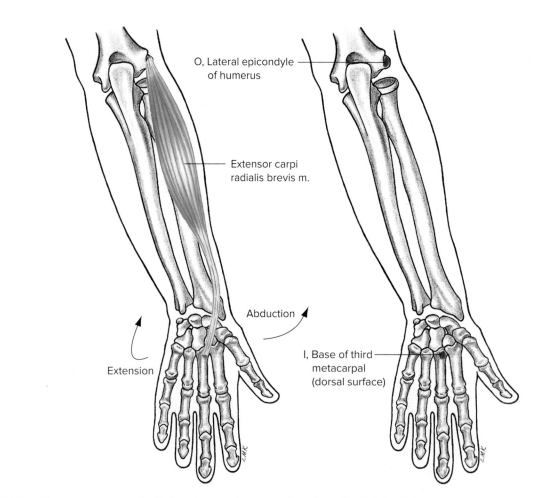

O, Lateral epicondyle of humerus

Extensor carpi radialis brevis m.

Abduction

Extension

I, Base of third metacarpal (dorsal surface)

FIG. 7.16 • Extensor carpi radialis brevis muscle, posterior view. O, Origin; I, Insertion.

# Extensor carpi radialis longus muscle  FIG. 7.17

(eks-ten´sor kar´pi ra´di-a´lis lon´gus)

## Origin

Distal third of lateral supracondylar ridge of the humerus and lateral epicondyle of the humerus

## Insertion

Base of the second metacarpal (dorsal surface)

## Action

Extension of the wrist
Abduction of the wrist
Weak flexion of the elbow
Weak pronation to neutral from a fully supinated position

## Palpation

Just proximal to the dorsal aspect of the wrist and approximately 1 cm medial to the radial styloid process, the tendon may be felt during extension and traced to the base of the second metacarpal, particularly when making a fist; proximally and posteriorly, just medial to the bulk of the brachioradialis

## Innervation

Radial nerve (C6, C7)

## Application, strengthening, and flexibility

The extensor carpi radialis longus, like the extensor carpi radialis brevis, is important in any sports activity that requires powerful wrist extension. In addition, both muscles are involved in abduction of the wrist. The extensor carpi radialis longus may be developed with the same wrist extension exercises as described for the extensor carpi ulnaris muscle.

The extensor carpi radialis longus is stretched in the same manner as the extensor carpi radialis brevis.

Wrist extension

Wrist abduction

Chapter 7

Elbow flexion

Radioulnar pronation

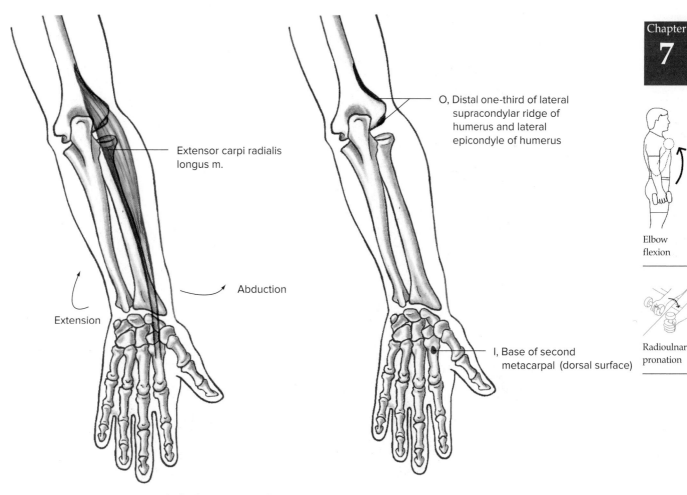

Extensor carpi radialis longus m.

Abduction

Extension

O, Distal one-third of lateral supracondylar ridge of humerus and lateral epicondyle of humerus

I, Base of second metacarpal (dorsal surface)

FIG. 7.17 • Extensor carpi radialis longus muscle, posterior view. O, Origin; I, Insertion.

# Flexor digitorum superficialis muscle FIG. 7.18

2nd–5th PIP flexion

(fleks´or dij-i-to´rum su´per-fish-e-al´is)

## Origin

Medial epicondyle of the humerus
Ulnar head: medial coronoid process
Radial head: upper two-thirds of anterior border of the radius just distal to the radial tuberosity

## Insertion

2nd–5th MCP flexion

Each tendon splits and attaches to the sides of the middle phalanx of the four fingers (palmar surface)

## Action

Flexion of the fingers at the metacarpophalangeal and proximal interphalangeal joints
Flexion of the wrist
Weak flexion of the elbow

## Palpation

In depressed area between palmaris longus and flexor carpi ulnaris tendons, particularly when making a fist but keeping the distal interphalangeal extended and slightly resisted wrist flexion; also on anterior mid-forearm during the same activity

## Innervation

Median nerve (C7, C8, T1)

## Application, strengthening, and flexibility

The flexor digitorum superficialis muscle, also known as the flexor digitorum sublimis, divides into four tendons on the palmar aspect of the wrist and hand to insert on each of the four fingers. The flexor digitorum superficialis and the flexor digitorum profundus are the only muscles involved in flexion of all four fingers. Both of these muscles are vital in any type of gripping activity.

Squeezing a sponge rubber ball in the palm of the hand, along with other gripping and squeezing activities, can be used to develop these muscles.

The flexor digitorum superficialis is stretched by passively extending the elbow, wrist, metacarpophalangeal, and proximal interphalangeal joints while maintaining the forearm in full supination.

2nd–5th MCP and PIP flexion

Wrist flexion

Elbow flexion

Flexor digitorum superficialis m.

Flexion of wrist

Flexion of fingers

O, Medial epicondyle of humerus. Ulnar head: medial coronoid process. Radial head: upper two-thirds of anterior border of radius

I, Split tendons attach to sides of middle phalanx of four fingers (palmar surface)

FIG. 7.18 • Flexor digitorum superficialis muscle, anterior view. O, Origin; I, Insertion.

# Flexor digitorum profundus muscle FIG. 7.19

(fleks´or dij-i-to´rum pro-fun´dus)

## Origin

Proximal three-fourths of the anterior and medial ulna

## Insertion

Base of the distal phalanxes of the four fingers

## Action

Flexion of the four fingers at the metacarpophalangeal, proximal interphalangeal, and distal interphalangeal joints

Flexion of the wrist

## Palpation

Difficult to distinguish, deep to the flexor digitorum superficialis, but on anterior mid-forearm while flexing the distal interphalangeal joints and keeping the proximal interphalangeal joints in extension; over the palmar surface of the second, third, fourth, and fifth metacarpophalangeal joints during finger flexion against resistance

## Innervation

Median nerve, anterior interosseous branch (C8, T1) to the second and third fingers. Ulnar nerve (C8, T1) to the fourth and fifth fingers

## Application, strengthening, and flexibility

Both the flexor digitorum profundus muscle and the flexor digitorum superficialis muscle assist in wrist flexion because of their palmar relationship to the wrist. The flexor digitorum profundus is used in any type of gripping, squeezing, or handclenching activity, such as gripping a racket or climbing a rope. It is the only muscle that flexes the finger DIP joints and disruption of the tendon at or near its insertion is known as a jersey or sweater finger and should be addressed surgically within approximately one week.

The flexor digitorum profundus muscle may be developed through these activities, in addition to the strengthening exercises described for the flexor digitorum superficialis muscle.

The flexor digitorum profundus is stretched similarly to the flexor digitorum superficialis, except that the distal interphalangeal joints must be passively extended in addition to the wrist, metacarpophalangeal, and proximal interphalangeal joints while maintaining the forearm in full supination.

2nd–5th DIP flexion

2nd–5th PIP flexion

2nd–5th MCP flexion

2nd–5th MCP and PIP flexion

Wrist flexion

Flexor digitorum profundus m.

Flexion of wrist

Flexion of fingers

O, Proximal three-fourths of anterior and medial ulna

I, Base of distal phalanxes of the four fingers

FIG. 7.19 • Flexor digitorum profundus muscle, anterior view. O, Origin; I, Insertion.

Thumb
CMC
flexion

# Flexor pollicis longus muscle  FIG. 7.20

(fleks´or pol´i-sis lon´gus)

Thumb
MCP
flexion

## Origin

Middle anterior surface of the radius and the anterior medial border of the ulna just distal to the coronoid process; occasionally a small head is present attaching on the medial epicondyle of the humerus

## Insertion

Base of the distal phalanx of the thumb (palmar surface)

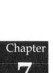

Thumb IP
flexion

## Action

Flexion of the thumb carpometacarpal, metacarpophalangeal, and interphalangeal joints
Flexion of the wrist
Abduction of the wrist

## Palpation

Anterior surface of the thumb on the proximal phalanx, and just lateral to the palmaris longus and medial to the flexor carpi radialis on the anterior distal forearm, especially during active flexion of the thumb interphalangeal joint

## Innervation

Median nerve, anterior interosseous branch (C8, T1)

## Application, strengthening, and flexibility

The primary function of the flexor pollicis longus muscle is flexion of the thumb, which is vital in gripping and grasping activities of the hand. Because of its palmar relationship to the wrist, it provides some assistance in wrist flexion.

It may be strengthened by pressing a sponge rubber ball into the hand with the thumb and by many other gripping or squeezing activities.

The flexor pollicis longus is stretched by passively extending the entire thumb while simultaneously maintaining maximal wrist extension.

Wrist
flexion

Wrist
abduction

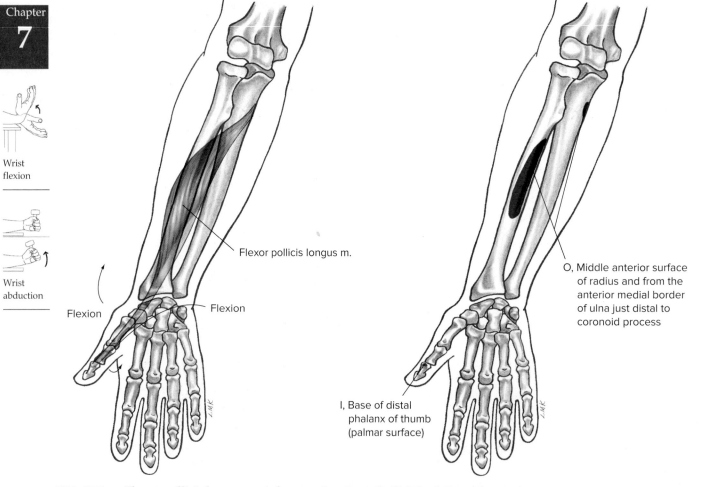

FIG. 7.20 • Flexor pollicis longus muscle, anterior view. O, Origin; I, Insertion.

# Extensor digitorum muscle   FIG. 7.21

(eks-ten´sor dij-i-to´rum)

## Origin

Lateral epicondyle of the humerus

## Insertion

Four tendons to bases of middle and distal phalanxes of the four fingers (dorsal surface)

## Action

Extension of the second, third, fourth, and fifth phalanges at the metacarpophalangeal joints
Extension of the wrist
Weak extension of the elbow

## Palpation

With all four fingers extended, on the posterior surface of the distal forearm immediately medial to the extensor pollicis longus tendon and lateral to the extensor carpi ulnaris and extensor digiti minimi, then dividing into four separate tendons that are over the dorsal aspect of the hand and metacarpophalangeal joints

## Innervation

Radial nerve, posterior interosseous branch (C6–C8)

## Application, strengthening, and flexibility

The extensor digitorum, also known as the extensor digitorum communis, is the only muscle involved in extension of all four fingers. This muscle divides into four tendons on the dorsum of the wrist to insert on each of the fingers. It also assists with wrist extension movements. It may be developed by applying manual resistance to the dorsal aspect of the flexed fingers and then extending the fingers fully. When performed with the wrist in flexion, this exercise increases the workload on the extensor digitorum.

To stretch the extensor digitorum, the fingers must be maximally flexed at the metacarpophalangeal, proximal interphalangeal, and distal interphalangeal joints while the wrist is fully flexed.

2nd–5th MCP extension

2nd–5th MCP, PIP, and DIP extension

Wrist extension

Elbow extension

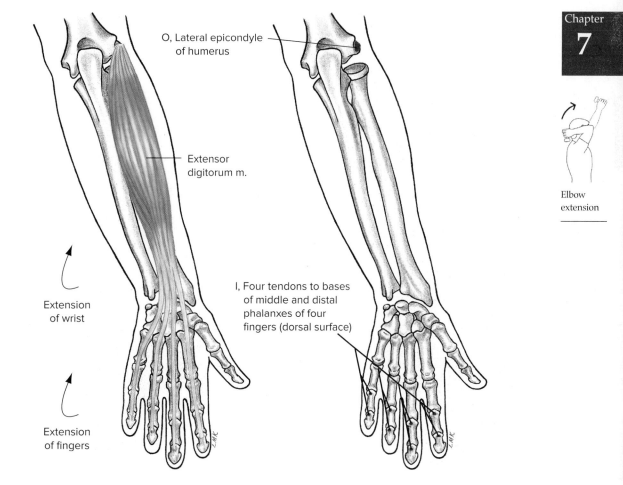

O, Lateral epicondyle of humerus

Extensor digitorum m.

Extension of wrist

Extension of fingers

I, Four tendons to bases of middle and distal phalanxes of four fingers (dorsal surface)

FIG. 7.21 • Extensor digitorum muscle, posterior view. O, Origin; I, Insertion.

# Extensor indicis muscle  FIG. 7.22

(eks-ten´sor in´di-sis)

just medial to the extensor digitorum tendon of the index finger with extension of the index fingers and flexion of the third, fourth, and fifth fingers

## Origin

Between middle and distal one-third of the posterior ulna

## Innervation

Radial nerve, posterior interosseous branch (C6–C8)

## Insertion

Base of the middle and distal phalanxes of the second phalange (dorsal surface)

## Application, strengthening, and flexibility

The extensor indicis muscle is the pointing muscle. That is, it is responsible for extending the index finger, particularly when the other fingers are flexed. It also provides weak assistance to wrist extension and may be developed through exercises similar to those described for the extensor digitorum.

The extensor indicis is stretched by passively taking the index finger into maximal flexion at its metacarpophalangeal, proximal interphalangeal, and distal interphalangeal joints while fully flexing the wrist.

## Action

Extension of the index finger at the metacarpophalangeal joint

Weak wrist extension

Weak supination of the forearm from a pronated position

## Palpation

With forearm pronated on the posterior aspect of the distal forearm and dorsal surface of the hand

O, Between middle and distal one-third of posterior ulna

Extensor indicis m.

Extension

I, Base of middle and distal phalanxes of second phalange (dorsal surface)

FIG. 7.22 • Extensor indicis muscle, posterior view. O, Origin; I, Insertion.

# Extensor digiti minimi muscle  FIG. 7.23

(eks-ten´sor dij´i-ti min´im-i)

## Origin

Lateral epicondyle of the humerus

## Insertion

Base of the middle and distal phalanxes of the fifth
phalange (dorsal surface)

## Action

Extension of the little finger at the metacarpophalan-
geal joint
Weak wrist extension
Weak elbow extension

## Palpation

Passing over the dorsal aspect of the distal radioulnar joint, particularly with relaxed flexion of other
fingers and alternating fifth finger extension and
relaxation; dorsal surface of forearm immediately
medial to the extensor digitorum and lateral to the
extensor carpi ulnaris

## Innervation

Radial nerve, posterior interosseous branch (C6–C8)

## Application, strengthening, and flexibility

The primary function of the extensor digiti minimi
muscle is to assist the extensor digitorum in extending the little finger. Because of its dorsal relationship to the wrist, it also provides weak assistance
in wrist extension. It is strengthened with the same
exercises described for the extensor digitorum.

The extensor digiti minimi is stretched by passively taking the little finger into maximal flexion
at its metacarpophalangeal, proximal interphalangeal, and distal interphalangeal joints while fully
flexing the wrist and elbow.

5th MCP
extension

Wrist
extension

Elbow
extension

Extensor digiti minimi m.

Extension

FIG. 7.23 • Extensor
digiti minimi muscle,
posterior view. O,
Origin; I, Insertion.

O, Lateral epicondyle
of humerus

I, Base of middle and
distal phalanxes
of fifth phalange
(dorsal surface)

# Extensor pollicis longus muscle  FIG. 7.24

(eks-ten´sor pol´i-sis lon´gus)

Thumb
CMC
extension

## Origin

Posterior lateral surface of the lower middle ulna

## Insertion

Base of the distal phalanx of the thumb (dorsal
surface)

Thumb
MCP
extension

## Action

Extension of the thumb at the carpometacarpal, meta-
carpophalangeal, and interphalangeal joints
Extension of the wrist
Abduction of the wrist
Weak supination of the forearm from a pronated
position

Thumb IP
extension

## Palpation

Dorsal aspect of the hand to its insertion on the base
of the distal phalanx; also on the posterior surface
of the lower forearm between radius and ulna just
proximal to the extensor indicis and medial to the
extensor pollicis brevis and abductor pollicis longus
with the forearm pronated and fingers in relaxed
flexion while actively extending the thumb

## Innervation

Radial nerve, posterior interosseous branch (C6–C8)

## Application, strengthening, and flexibility

The primary function of the extensor pollicis
longus muscle is extension of the thumb, although
it does provide weak assistance in wrist extension.

It may be strengthened by extending the flexed
thumb against manual resistance. It is stretched
by passively taking the entire thumb into maximal
flexion at its carpometacarpal, metacarpophalangeal, and interphalangeal joints while fully flexing
the wrist with the forearm in pronation.

The tendons of the extensor pollicis longus and
extensor pollicis brevis, along with the tendon of
the abductor pollicis longus, form the "anatomical snuffbox," the small depression that develops
between these two tendons when they contract.
The name *anatomical snuffbox* originates from
tobacco users placing their snuff in this depression.
Deep in the snuffbox the scaphoid bone can be palpated and is often a site of point tenderness when
it is fractured. Due to its posterolateral relationship
to the wrist it also assists in wrist abduction.

**Chapter 7**

Wrist
extension

Wrist
abduction

Radioulnar
supination

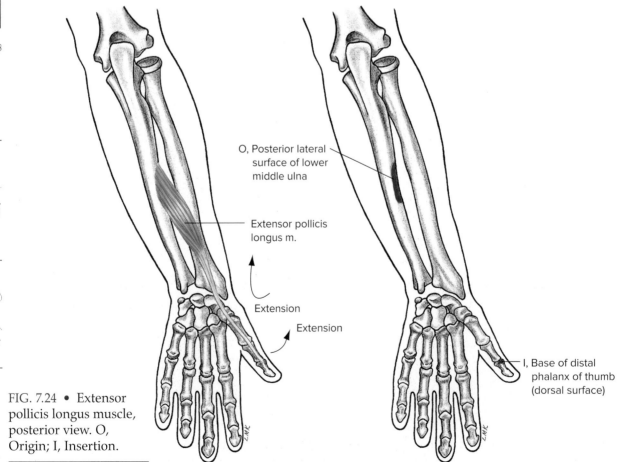

FIG. 7.24 • Extensor
pollicis longus muscle,
posterior view. O,
Origin; I, Insertion.

O, Posterior lateral surface of lower middle ulna

Extensor pollicis longus m.

Extension

Extension

I, Base of distal phalanx of thumb (dorsal surface)

# Extensor pollicis brevis muscle   FIG. 7.25

(eks-ten´sor pol´i-sis bre´vis)

## Origin

Posterior surface of the lower middle radius

## Insertion

Base of the proximal phalanx of the thumb (dorsal surface)

## Action

Extension of the thumb at the carpometacarpal and metacarpophalangeal joints
Wrist abduction
Weak wrist extension

## Palpation

Just lateral to the extensor pollicis longus tendon on the dorsal side of the hand to its insertion on the proximal phalanx with extension of the thumb carpometacarpal and metacarpophalangeal joints and flexion of the interphalangeal joint

## Innervation

Radial nerve, posterior interosseous branch (C6, C7)

## Application, strengthening, and flexibility

The extensor pollicis brevis assists the extensor pollicis longus in extending the thumb. Because of its dorsal relationship to the wrist, it, too, provides weak assistance in wrist extension.

It may be strengthened through the same exercises described for the extensor pollicis longus muscle. It is stretched by passively taking the first carpometacarpal joint and the metacarpophalangeal joint of the thumb into maximal flexion while fully flexing and adducting the wrist. Due to its lateral orientation at the wrist it assists in wrist abduction.

Thumb CMC extension

Thumb MCP extension

Wrist abduction

Wrist extension

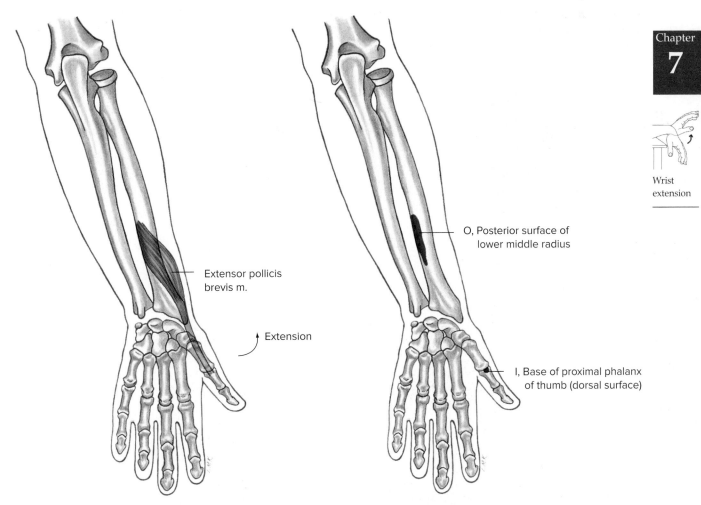

Extensor pollicis brevis m.

Extension

O, Posterior surface of lower middle radius

I, Base of proximal phalanx of thumb (dorsal surface)

FIG. 7.25 • Extensor pollicis brevis muscle, posterior view. O, Origin; I, Insertion.

# Abductor pollicis longus
## muscle FIG. 7.26

(ab-duk´tor pol´i-sis lon´gus)

Thumb
CMC
abduction

### Origin

Posterior aspect of the radius and midshaft of the ulna

### Insertion

Base of the first metacarpal (dorsal lateral surface)

Wrist
abduction

### Action

Abduction of the thumb at the carpometacarpal joint
Abduction of the wrist
Extension of the thumb at the carpometacarpal joint
Weak supination of the forearm from a pronated position
Weak extension of the wrist joint

Thumb
CMC
extension

Radioulnar
supination

Wrist
extension

### Palpation

With forearm in neutral pronation/supination on the lateral aspect of the wrist joint just proximal to the first metacarpal during active thumb and wrist abduction

### Innervation

Radial nerve, posterior interosseous branch (C6, C7)

### Application, strengthening, and flexibility

The primary function of the abductor pollicis longus muscle is abduction of the thumb, although it does provide some assistance in abduction of the wrist. It may be developed by abducting the thumb from the adducted position against a manually applied resistance. Stretching of the abductor pollicis longus is accomplished by fully flexing and adducting the entire thumb across the palm with the wrist fully adducted and in slight flexion.

Abductor pollicis
longus m.

Abduction

O, Posterior aspect of the
radius and midshaft
of ulna

I, Base of first metacarpal
(dorsal lateral surface)

FIG. 7.26 • Abductor pollicis longus muscle, posterior view. O, Origin; I, Insertion.

# Intrinsic muscles of the hand

The intrinsic hand muscles may be grouped according to location as well as according to the parts of the hand they control (Fig. 7.27). The abductor pollicis brevis, opponens pollicis, flexor pollicis brevis, and adductor pollicis make up the thenar eminence—the muscular pad on the palmar surface of the first metacarpal. The hypothenar eminence is the muscular pad that forms the ulnar border on the palmar surface of the hand and is made up of the abductor digiti minimi, flexor digiti minimi brevis, palmaris brevis, and opponens digiti minimi. The intermediate muscles of the hand consist of three palmar interossei, four dorsal interossei, and four lumbrical muscles.

Four intrinsic muscles act on the carpometacarpal joint of the thumb. The opponens pollicis is the muscle that causes opposition in the thumb metacarpal. The abductor pollicis brevis abducts the thumb metacarpal and is assisted in this action by the flexor pollicis brevis, which also flexes the thumb metacarpal. The metacarpal of the thumb is adducted by the adductor pollicis. Both the flexor pollicis brevis and the adductor pollicis flex the proximal phalanx of the thumb.

The three palmar interossei are adductors of the second, fourth, and fifth phalanges. The four dorsal interossei both flex and abduct the index, middle, and ring proximal phalanxes, in addition to assisting with extension of the middle and distal phalanxes of these fingers. The third dorsal

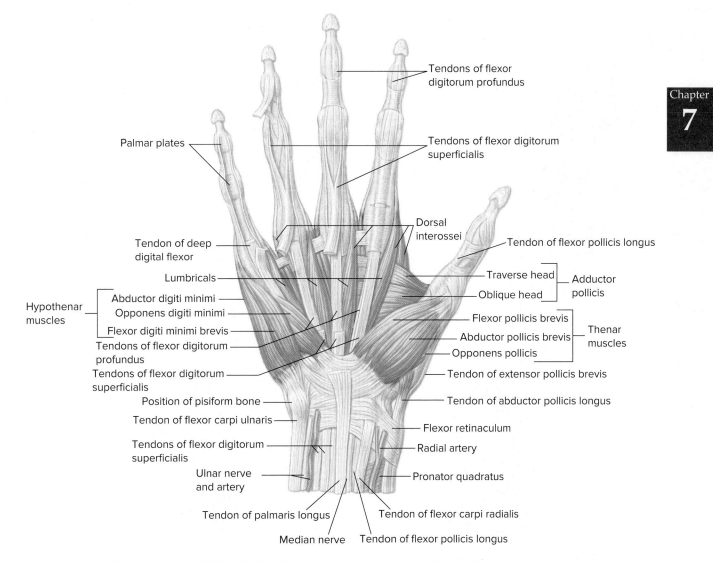

FIG. 7.27 • Intrinsic muscles of the right hand, anterior view.

interossei also adducts the middle finger. The four lumbricals flex the index, middle, ring, and little proximal phalanxes and extend the middle and distal phalanxes of these fingers.

Three muscles act on the little finger. The opponens digiti minimi causes opposition of the little finger metacarpal. The abductor digiti minimi abducts the little finger metacarpal, and the flexor digiti minimi brevis flexes this metacarpal.

Refer to Table 7.3 for further details regarding the intrinsic muscles of the hand.

## TABLE 7.3 · Intrinsic muscles of the hand

| | Muscle | Origin | Insertion | Action | Palpation | Innervation |
|---|---|---|---|---|---|---|
| **Thenar muscles** | Opponens pollicis | Anterior surface of transverse carpal ligament, trapezium | Lateral border of 1st metacarpal | CMC opposition of thumb | Palmar aspect of 1st metacarpal with opposition of fingertips to thumb | Median nerve (C6, C7) |
| | Abductor pollicis brevis | Anterior surface of transverse carpal ligament, trapezium, scaphoid | Base of 1st proximal phalanx | CMC abduction of thumb | Radial aspect of palmar surface of 1st metacarpal with 1st CMC abduction | Median nerve (C6, C7) |
| | Flexor pollicis brevis | Superficial head: trapezium and transverse carpal ligament; Deep head: ulnar aspect of 1st metacarpal | Base of proximal phalanx of 1st metacarpal | CMC flexion and abduction; MCP flexion of thumb | Medial aspect of thenar eminence just proximal to 1st MCP joint with 1st MCP flexion against resistance | Superficial head: median nerve (C6, C7); Deep head: ulnar nerve (C8, T1) |
| | Adductor pollicis | Transverse head: anterior shaft of 3rd metacarpal; Oblique head: base of 2nd and 3rd metacarpals, capitate, trapezoid | Ulnar aspect of base of proximal phalanx of 1st metacarpal | CMC adduction; MCP flexion of thumb | Palmar surface between 1st and 2nd metacarpal with 1st CMC adduction | Ulnar nerve (C8, T1) |
| **Intermediate muscles** | Palmar interossei | Shaft of 2nd, 4th, and 5th metacarpals and extensor expansions | Bases of 2nd, 4th, and 5th proximal phalanxes and extensor expansions | MCP adduction of 2nd, 4th, and 5th phalanges | Cannot be palpated | Ulnar nerve (C8, T1) |
| | Dorsal interossei | Two heads on shafts on adjacent metacarpals | Bases of 2nd, 3rd, and 4th proximal phalanxes and extensor expansions | MCP flexion and abduction; PIP/DIP extension of 2nd, 3rd, and 4th phalanges; MCP adduction of 3rd phalange | Dorsal surface between 1st and 2nd metacarpals, between shafts of 2nd through 5th metacarpals with active abduction/adduction of 2nd, 3rd, and 4th MCP joints | Ulnar nerve, palmar branch (C8, T1) |
| | Lumbricals | Flexor digitorum profundus tendon in center of palm | Extensor expansions on radial side of 2nd, 3rd, 4th, and 5th proximal phalanxes | MCP flexion and PIP/DIP extension of 2nd, 3rd, 4th, and 5th phalanges | Cannot be palpated | 1st and 2nd: median nerve (C6, C7); 3rd and 4th: ulnar nerve (C8, T1) |

TABLE 7.3 (continued) • **Intrinsic muscles of the hand**

2nd–5th MCP flexion

| | Muscle | Origin | Insertion | Action | Palpation | Innervation |
|---|---|---|---|---|---|---|
| Hypothenar muscles | Opponens digiti minimi | Hook of hamate and adjacent transverse carpal ligament | Medial border of 5th metacarpal | MCP opposition of 5th phalange | On radial aspect of hypothenar eminence with opposition of 5th phalange to thumb | Ulnar nerve (C8, T1) |
| | Abductor digiti minimi | Pisiform and flexor carpi ulnaris tendon | Ulnar aspect of base of 5th proximal phalanx | MCP abduction of 5th phalange | Ulnar aspect of hypothenar eminence with 5th MCP abduction | Ulnar nerve (C8, T1) |
| | Flexor digiti minimi brevis | Hook of hamate and adjacent transverse carpal ligament | Ulnar aspect of base of 5th proximal phalanx | MCP flexion of 5th phalange | Palmar surface of 5th metacarpal, lateral to opponens digiti minimi with 5th MCP flexion against resistance | Ulnar nerve (C8, T1) |
| | Palmaris brevis | Transverse carpal ligament and medial margin of palmar aponeurosis | Skin of ulnar border of palm | Tenses the skin on the ulnar side | Ulnar border of the palm of the hand | Ulnar nerve (C8, T1) |

## REVIEW EXERCISES

1. List the planes in which each of the following wrist, hand, and finger joint movements occurs. List the axis of rotation for each movement in each plane.
   a. Abduction
   b. Adduction
   c. Flexion
   d. Extension
2. Discuss why the thumb is the most important part of the hand.
3. How should boys and girls be taught to do push-ups? Justify your answer.
   a. Hands flat on the floor
   b. Fingertips

4. **Muscle analysis chart** • Wrist, hand, and fingers

| Fill in the chart by listing the muscles primarily involved in each movement. | |
|---|---|
| Wrist and hand | |
| Flexion | Extension |
| Adduction | Abduction |
| Fingers, metacarpophalangeal joints | |
| Flexion | Extension |
| Abduction | Adduction |

Muscle analysis chart (continued)

| Fingers, proximal interphalangeal joints | |
|---|---|
| Flexion | Extension |

| Fingers, distal interphalangeal joints | |
|---|---|
| Flexion | Extension |

| Thumb, carpometacarpal, metacarpophalangeal, and interphalangeal joints | |
|---|---|
| Flexion | Extension |
| Abduction | Adduction |

5. List the muscles involved in the little finger as you type on a computer keyboard and reach for the left tab key with the wrists properly stabilized in an ergonomic position.

6. Describe the importance of the intrinsic muscles in the hand as you reach to turn a doorknob.

7. Determine the kind of flexibility exercises that would be indicated for a patient with carpal tunnel syndrome, and explain in detail how they should be performed.

# LABORATORY EXERCISES

1. Locate the following parts of the humerus, radius, ulna, carpals, and metacarpals on a human skeleton and on a subject.
   a. Skeleton
      1. Medial epicondyle
      2. Lateral epicondyle
      3. Lateral supracondylar ridge
      4. Trochlea
      5. Capitulum
      6. Coronoid process
      7. Tuberosity of the radius
      8. Styloid process—radius
      9. Styloid process—ulna
      10. Lister's tubercle
      11. First and third metacarpals
      12. Wrist bones
      13. First phalanx of third metacarpal

   b. Subject
      1. Medial epicondyle
      2. Lateral epicondyle
      3. Lateral supracondylar ridge
      4. Lister's tubercle
      5. Pisiform
      6. Scaphoid (navicular)

2. How and where can the following muscles be palpated on a human subject?
   a. Flexor pollicis longus
   b. Flexor carpi radialis
   c. Flexor carpi ulnaris
   d. Extensor digitorum communis
   e. Extensor pollicis longus
   f. Extensor carpi ulnaris

3. Demonstrate the action and list the muscles primarily responsible for the following movements at the wrist joint:
   a. Flexion
   b. Extension
   c. Abduction
   d. Adduction

4. With a laboratory partner, determine how and why maintaining full flexion of all the fingers is impossible when passively moving the wrist into maximal flexion. Is it also difficult to maintain maximal extension of all the finger joints while passively taking the wrist into full extension?

## 5. Wrist and hand joint exercise movement analysis chart

After analyzing each exercise in the chart, break each into two primary movement phases, such as a lifting phase and a lowering phase. For each phase, determine the wrist and hand joint movements that occur, and then list the wrist and hand joint muscles primarily responsible for causing/controlling those movements or maintaining the position. Beside each muscle in each movement, indicate the type of contraction as follows: I—isometric; C—concentric; E—eccentric.

| Exercise | Initial movement (lifting) phase | | Secondary movement (lowering) phase | |
|---|---|---|---|---|
| | Movement(s) | Agonist(s)—(contraction type) | Movement(s) | Agonist(s)—(contraction type) |
| Push-up | | | | |
| Chin-up | | | | |
| Bench press | | | | |
| Dip | | | | |
| Lat pull | | | | |
| Ball squeeze | | | | |
| Frisbee throw | | | | |

## 6. Wrist and hand joint sport skill analysis chart

Analyze each skill in the chart and list the movements of the right and left wrist and hand joints in each phase of the skill. You may prefer to list the initial positions that the wrist and hand joints are in for the stance phase. After each movement, list the wrist and hand joint muscle(s) primarily responsible for causing/controlling the movement or maintaining the position. Beside each muscle in each movement, indicate the type of contraction as follows: I—isometric; C—concentric; E—eccentric. It may be desirable to study the concepts for analysis in Chapter 8 for the various phases.

| Exercise | | Stance phase | Preparatory phase | Movement phase | Follow-through phase |
|---|---|---|---|---|---|
| Baseball pitch | (R) | | | | |
| | (L) | | | | |
| Volleyball serve | (R) | | | | |
| | (L) | | | | |
| Tennis serve | (R) | | | | |
| | (L) | | | | |

## Wrist and hand joint sport skill analysis chart (continued)

| Exercise | | Stance phase | Preparatory phase | Movement phase | Follow-through phase |
|---|---|---|---|---|---|
| Softball pitch | (R) | | | | |
| | (L) | | | | |
| Tennis backhand | (R) | | | | |
| | (L) | | | | |
| Batting | (R) | | | | |
| | (L) | | | | |
| Bowling | (R) | | | | |
| | (L) | | | | |
| Basketball free throw | (R) | | | | |
| | (L) | | | | |

Chapter

**7**

# References

Cleland JA, Koppenhaver S, Su J: *Netter's orthopaedic clinical examination: an evidence-based approach,* ed 3, Philadelphia, 2016, Elsevier.

Gabbard CP, et al: Effects of grip and forearm position on flex arm hang performance, *Research Quarterly for Exercise and Sport,* July 1983.

Gench BE, Hinson MM, Harvey PT: *Anatomical kinesiology,* Dubuque, IA, 1995, Eddie Bowers.

Guarantors of Brain: *Aids to the examination of the peripheral nervous system,* ed 4, London, 2000, Saunders.

Hamilton N, Weimer W, Luttgens K: *Kinesiology: scientific basis of human motion,* ed 12, New York, 2012, McGraw-Hill.

Hislop HJ, Montgomery J: *Daniels and Worthingham's muscle testing: techniques of manual examination,* ed 8, Philadelphia, 2007, Saunders.

Houglum PA, Bertoti DB: *Brunstom's clinical kinesiology,* ed 6, Philadelphia, 2012, FA Davis.

Lindsay DT: *Functional human anatomy,* St. Louis, 1996, Mosby.

Magee DJ: *Orthopedic physical assessment,* ed 6, St. Louis, 2014, Elsevier Saunders.

Muscolino JE: *The muscular system manual: the skeletal muscles of the human body,* ed 3, St. Louis, 2010, Elsevier Mosby.

Norkin CC, Levangie PK: *Joint structure and function—a comprehensive analysis,* ed 5, Philadelphia, 2011, Davis.

Norkin CC, White DJ: *Measurement of joint motion: a guide to goniometry,* ed 4, Philadelphia, 2009, Davis.

Oatis CA: *Kinesiology: the mechanics and pathomechanics of human movement,* ed 2, Philadelphia, 2008, Lippincott Williams & Wilkins.

Putz R, Pabst R: *Sobotta - Atlas of human anatomy one volume edition,* ed 14, Germany, 2008, Elsevier.

Rasch PJ: *Kinesiology and applied anatomy,* ed 7, Philadelphia, 1989, Lea & Febiger.

Raya G, Akelman E: *The hand: anatomy, examination, and diagnosis,* ed 4, Philadelphia, 2011, Lippincott Williams & Wilkins.

Schünke M, Schulte E, Schumacher U: *Thieme atlas of anatomy: general anatomy and musculoskeletal system,* ed 2, New York, 2014, Thieme Medical.

Seeley RR, Stephens TD, Tate P: *Anatomy & physiology,* ed 8, New York, 2008, McGraw-Hill.

Seeley RR, VanPutte CL, Regan JL, Russo AF: *Seeley's anatomy & physiology,* ed 9, New York, 2011, McGraw-Hall.

Sieg KW, Adams SP: *Illustrated essentials of musculoskeletal anatomy,* ed 4, Gainesville, FL, 2002, Megabooks.

Sisto DJ, et al: An electromyographic analysis of the elbow in pitching, *American Journal of Sports Medicine* 15:260, May–June 1987.

Springer SI: Racquetball and elbow injuries, *National Racquetball* 16:7, March 1987.

Stone RJ, Stone JA: *Atlas of the skeletal muscles,* ed 6, New York, 2009, McGraw-Hill.

Van De Graaff KM: *Human anatomy,* ed 6, Dubuque, IA, 2002, McGraw-Hill.

# Worksheet Exercises

For in- or out-of-class assignments, or for testing, utilize this tear-out worksheet.

## Worksheet 1

Using crayons or colored markers, draw and label on the worksheet the following muscles. Indicate the origin and insertion of each muscle with an "O" and an "I," respectively.

a.  Flexor pollicis longus

b.  Flexor carpi radialis

c.  Flexor carpi ulnaris

d.  Extensor digitorum

e.  Extensor pollicis longus

f.  Extensor carpi ulnaris

g.  Extensor pollicis brevis

h.  Palmaris longus

i.  Extensor carpi radialis longus

j.  Extensor carpi radialis brevis

k.  Extensor digiti minimi

l.  Extensor indicis

m.  Flexor digitorum superficialis

n.  Flexor digitorum profundus

o.  Abductor pollicis longus

# Worksheet Exercises

For in- or out-of-class assignments, or for testing, utilize this tear-out worksheet.

## Worksheet 2

Label and indicate with arrows the following movements of the wrist and hands. For each motion, list the agonist muscle(s), the plane in which the motion occurs, and its axis of rotation.

a. Flexion

b. Extension

c. Abduction (ulnar flexion)

d. Adduction (radial flexion)

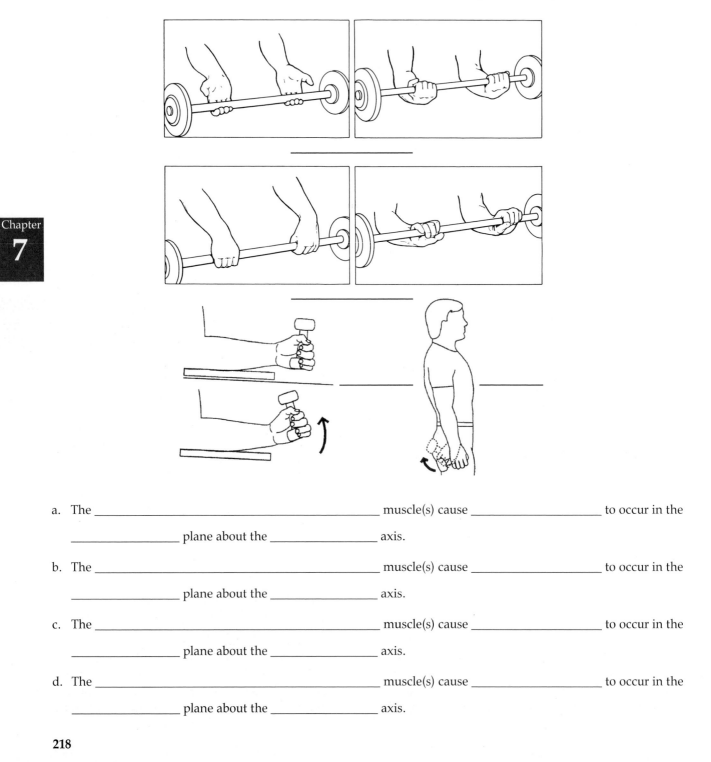

a. The _____ muscle(s) cause _____ to occur in the

_____ plane about the _____ axis.

b. The _____ muscle(s) cause _____ to occur in the

_____ plane about the _____ axis.

c. The _____ muscle(s) cause _____ to occur in the

_____ plane about the _____ axis.

d. The _____ muscle(s) cause _____ to occur in the

_____ plane about the _____ axis.

# THE HIP JOINT AND PELVIC GIRDLE

## Objectives

- To identify on a human skeleton or subject selected bony features of the hip joint and pelvic girdle

- To label on a skeletal chart selected bony features of the hip joint and pelvic girdle

- To draw on a skeletal chart the individual muscles of the hip joint

- To demonstrate, using a human subject, all the movements of the hip joint and pelvic girdle and list their respective planes of movement and axes of motion

- To palpate on a human subject the muscles of the hip joint and pelvic girdle

- To determine, list, and palpate the muscles of the hip joint and their antagonists and appreciate the role of the ligaments in providing stability

- To learn and understand the innervation of the hip joint muscles

- To determine, through analysis, the hip movements and muscles involved in selected skills and exercises

The hip, or acetabular femoral joint, is a relatively stable joint due to its bony architecture, strong ligaments, and large, supportive muscles. It functions in weight bearing and locomotion, which is enhanced significantly by the hip's wide range of motion, which provides the ability to run, crossover cut, side-step cut, jump, and make many other directional changes.

## Bones   FIGS. 8.1 TO 8.3

The hip joint is a ball-and-socket joint that consists of the head of the femur connecting with the acetabulum of the pelvic girdle. The femur projects out laterally from its head toward the greater trochanter and then angles back toward the midline as it runs inferiorly to form the proximal bone of the knee. It is the longest bone in the body. The pelvic girdle consists of a right and left pelvic bone joined together posteriorly by the sacrum. The sacrum can be considered an extension of the spinal column with five fused vertebrae. Extending inferior from the sacrum is the coccyx. The pelvic bones are made up of three bones: the ilium, the ischium, and the pubis. At birth and during growth and development, there are three distinct bones. At maturity, they are fused to form one pelvic bone known as the os coxae.

The pelvic bone can be divided roughly into three areas, starting from the acetabulum:

Upper two-fifths = ilium
Posterior and lower two-fifths = ischium
Anterior and lower one-fifth = pubis

In studying the muscles of the hip and thigh, it is helpful to focus on the important bony landmarks, keeping in mind their purpose as key attachment points for the muscles. The anterior pelvis serves to provide points of origin for muscles generally involved in flexing the hip. Specifically, the tensor fasciae latae arises from the anterior iliac crest, the sartorius originates on the anterior superior iliac spine, and the rectus femoris originates on the anterior inferior iliac spine. Laterally, the gluteus medius and minimus, which abduct the hip, originate just below the iliac crest. Posteriorly, the gluteus maximus originates on the posterior iliac crest as well as the posterior sacrum and coccyx. Posteroinferiorly, the ischial tuberosity serves as the point of origin for the hamstrings, which extend the hip. Medially, the pubis and its inferior ramus serve as the point of origin for the hip adductors, which include the adductor magnus, adductor longus, adductor brevis, pectineus, and gracilis.

The proximal thigh generally serves as a point of insertion for some of the short muscles of the hip and as the origin for three of the knee extensors. Most notably, the greater trochanter is the point of insertion for all of the gluteal muscles and five of the six deep external rotators. Although not palpable, the lesser trochanter serves as the bony landmark upon which the iliopsoas inserts. Anteriorly, the three vasti muscles of the quadriceps originate proximally. Posteriorly, the linea aspera serves as the insertion for the hip adductors.

FIG. 8.1 • Right pelvis and femur, anterior view.

Chapter
8

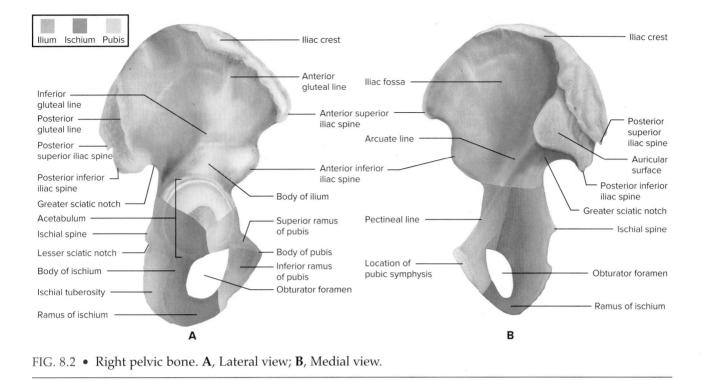

Ilium  Ischium  Pubis

**A (Lateral view labels, left side):**
Inferior gluteal line
Posterior gluteal line
Posterior superior iliac spine
Posterior inferior iliac spine
Greater sciatic notch
Acetabulum
Ischial spine
Lesser sciatic notch
Body of ischium
Ischial tuberosity
Ramus of ischium

**A (right side labels):**
Iliac crest
Anterior gluteal line
Anterior superior iliac spine
Anterior inferior iliac spine
Body of ilium
Superior ramus of pubis
Body of pubis
Inferior ramus of pubis
Obturator foramen

**A**

**B (Medial view, left side labels):**
Iliac fossa
Arcuate line
Pectineal line
Location of pubic symphysis

**B (right side labels):**
Iliac crest
Posterior superior iliac spine
Auricular surface
Posterior inferior iliac spine
Greater sciatic notch
Ischial spine
Obturator foramen
Ramus of ischium

**B**

FIG. 8.2 • Right pelvic bone. **A**, Lateral view; **B**, Medial view.

Distally, the patella serves as a major bony landmark to which all four quadriceps muscles attach and ultimately insert on the tibial tuberosity via the patella tendon. The remainder of the hip muscles insert on the proximal tibia or fibula. The sartorius, gracilis, and semitendinosus insert on the upper anteromedial surface of the tibia just below the medial condyle, after crossing the knee posteromedially. The semimembranosus inserts posteromedially on the medial tibial condyle. Laterally, the biceps femoris inserts primarily on the head of the fibula, with some fibers attaching on the lateral tibial condyle. Anterolaterally, Gerdy's tubercle provides the insertion point for the iliotibial tract of the tensor fasciae latae.

## Joints   FIGS. 8.1 TO 8.7

Anteriorly, the pelvic bones are joined to form the symphysis pubis, an amphiarthrodial joint. Posteriorly, the sacrum is located between the two pelvic bones and forms the sacroiliac joints. Strong ligaments unite these bones to form rigid, slightly movable joints. The bones are large and heavy and for the most part are covered by thick, heavy muscles. Very minimal oscillating-type movements can occur in these joints, as in walking or in hip flexion when lying on one's back. However, movements usually involve the entire pelvic girdle and hip joints. In walking, there is hip flexion and extension with rotation of the pelvic girdle, forward in hip flexion and backward in hip extension. Jogging and running result in faster movements and a greater range of movement.

Sport skills such as kicking a football or soccer ball are other good examples of hip and pelvic movements. Pelvic rotation helps increase the length of the stride in running; in kicking, it can result in a greater range of motion, which translates into a greater distance or more speed to the kick.

Except for the glenohumeral joint, the hip is one of the most mobile joints of the body, largely because of its multiaxial arrangement. Unlike the glenohumeral, the hip joint's bony architecture provides a great deal of stability, resulting in relatively few hip joint subluxations and dislocations.

The hip joint is classified as an enarthrodial-type joint and is formed by the femoral head inserting into the socket provided by the acetabulum of the pelvis. An extremely strong and dense ligamentous capsule, illustrated in Figs. 8.4 and 8.5, reinforces the joint, especially anteriorly. Anteriorly, the iliofemoral, or Y, ligament prevents hip hyperextension. The teres ligament attaches from deep in the acetabulum to a depression in the head of the femur and slightly limits adduction. The pubofemoral ligament is located anteromedially and inferiorly and limits excessive extension and abduction. Posteriorly, the triangular ischiofemoral ligament extends

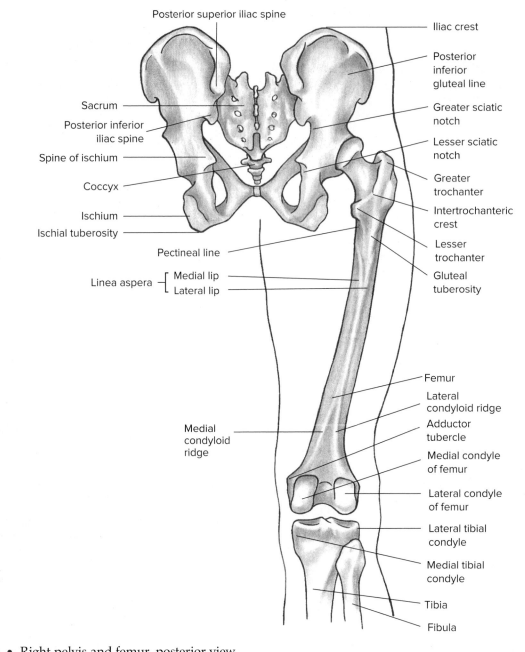

Posterior superior iliac spine

Iliac crest

Posterior inferior gluteal line

Greater sciatic notch

Lesser sciatic notch

Greater trochanter

Intertrochanteric crest

Lesser trochanter

Gluteal tuberosity

Sacrum

Posterior inferior iliac spine

Spine of ischium

Coccyx

Ischium

Ischial tuberosity

Pectineal line

Linea aspera { Medial lip / Lateral lip

Femur

Lateral condyloid ridge

Adductor tubercle

Medial condyle of femur

Lateral condyle of femur

Lateral tibial condyle

Medial tibial condyle

Tibia

Fibula

Medial condyloid ridge

FIG. 8.3 • Right pelvis and femur, posterior view.

from the ischium below to the trochanteric fossa of the femur and limits internal rotation. Table 8.1 provides a listing of the major ligaments of the hip and symphysis pubis.

Similar to the glenoid fossa of the shoulder joint, the acetabulum is lined around most of its periphery with a labrum to enhance stability and provide some shock absorption. Covering the remaining surface of the acetabulum, as well as the femoral head, is articular cartilage that may gradually degenerate with age and/or injury, leading to

osteoarthritis characterized by pain, stiffness, and limited range of motion.

Because of individual differences, there is some disagreement about the exact possible range of each movement in the hip joint, but the ranges are generally 0 to 130 degrees of flexion, 0 to 30 degrees of extension, 0 to 35 degrees of abduction, 0 to 30 degrees of adduction, 0 to 45 degrees of internal rotation, and 0 to 50 degrees of external rotation (Fig. 8.8). Although rarely referred to as distinct motions, the hip, when flexed

**TABLE 8.1 • Major ligaments of the hip and symphysis pubis joints**

| Joint / structure | Ligament | Proximal attachments | Distal attachments | Functional application and joint stability |
|---|---|---|---|---|
| Hip | Inguinal ligament (Poupart's ligament) | Pubic tubercle of the pubis (inferior medial end) | Anterior superior iliac spine (ASIS) of Ilium | Forms the base of the inguinal canal; site of indirect inguinal hernia may occur; marks the superior border of the femoral triangle and the inferior border of the inguinal triangle; supports the muscle running inferior to its fibers (iliopsoas, pectineus) as well as the femoral artery, femoral vein, and femoral nerve |
| | Iliofemoral ligament ("Y" ligament, Ligament of Bigelow) | Anterior inferior iliac spine | Anterior surface of the intertrochanteric line of the femur | Two bands running diagonally from anterior pelvis to femur; strongest ligament in the body; shaped like an inverted "Y"; reinforces the anterior surface of the articular capsule of the hip joint; prevents hip extension and helps maintain posture |
| | Pubofemoral ligament (pubocapsular ligament) | Obturator crest, superior ramus of pubis, and iliopubic eminence | Intertrochanteric line on anterior femur | Reinforces the inferior and anterior capsule of the hip joint; limits excessive extension, abduction, and external rotation |
| | Ischiofemoral ligament (ischiocapsular ligament) | Posteriorly on inferior acetabular margin of ischium, superior margin of ischial tuberosity | Zona obicularis and intertrochanteric line posteriorly | Triangular band of strong fibers that reinforce the posterior articular capsule of the hip joint; limits internal rotation |
| | Ligament of head of the femur (ligamentum teres, round ligament) | Transverse acetabular ligament along ischial and pubic margins of acetabular notch | Fovea capitis of femur | Triangular in its origin, contains within it the acetabular branch of the obturator artery to supply the head of and proximal neck of the femur; taut in excessive hip adduction, flexion, and external rotation |
| | Transverse acetabular ligament | Inferior acetabular rim of pubis | Inferior acetabular rim of ischium | Functions as a tension band between the posteroinferior and anteroinferior aspects of the acetabulum; prevents inferior displacement of head of femur |
| Pubic Symphysis | Inferior pubic ligament (arcuate ligament) | Inferior ramus of one pubic bone beneath the pubic symphysis | Inferior ramus of the opposite pubic bone | Stabilizes the pubic symphysis; maintains superior/inferior glide and separation/compression of the pubic symphysis |
| | Superior pubic ligament | Superior to pubic symphysis | Superior to pubic symphysis | Stabilizes the pubic symphysis; maintains superior/inferior glide and separation/compression of the pubic symphysis |

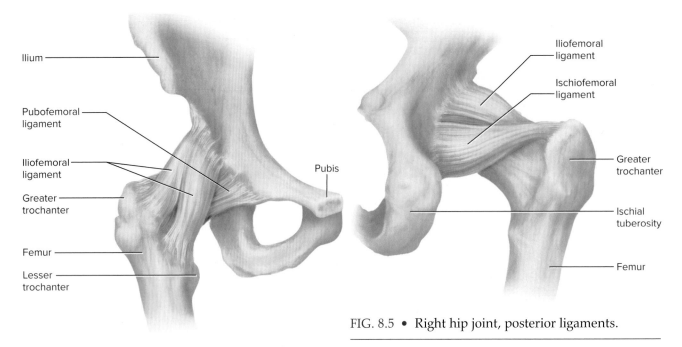

FIG. 8.4 • Right hip joint, anterior ligaments.

FIG. 8.5 • Right hip joint, posterior ligaments.

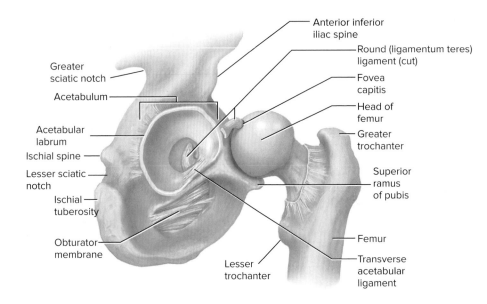

FIG. 8.6 • Right hip joint, lateral view, femur retracted.

to 90 degrees, can adduct and abduct in the transverse plane, similar to the glenohumeral joint. These motions include approximately 40 degrees of horizontal adduction and 60 degrees of horizontal abduction.

The pelvic girdle moves back and forth within three planes for a total of six different movements. To avoid confusion, it is important to analyze the pelvic girdle activity to determine the exact location of the movement. All pelvic girdle rotation actually results from motion at one or more of the following locations: the right hip, the left hip, the lumbar spine. Although it is not essential for movement to occur in all three of these areas, it must occur in at least one for the pelvis to rotate in any direction. Table 8.2 lists the motions at the hips and lumbar spine that can often accompany rotation of the pelvic girdle.

FIG. 8.7 • Right hip joint, coronal section.

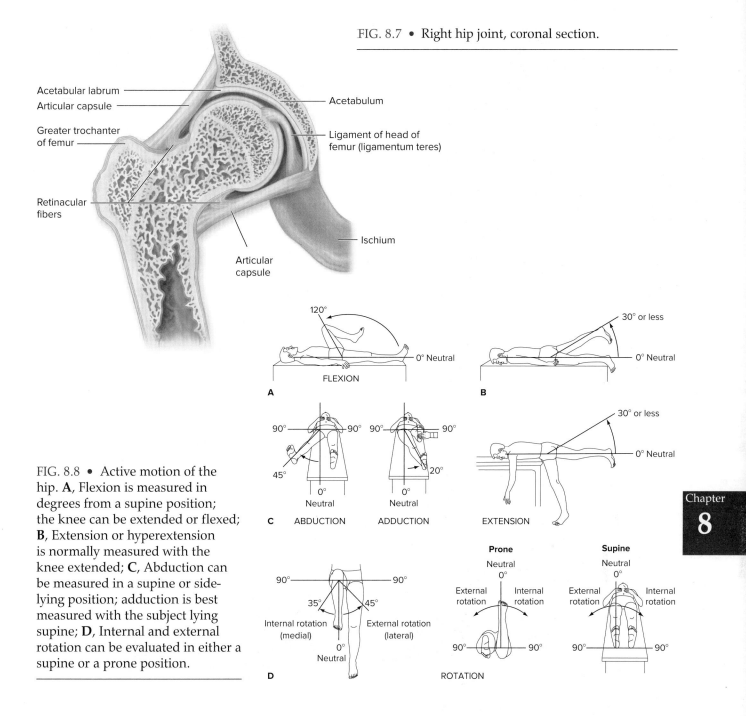

Acetabular labrum

Articular capsule

Greater trochanter
of femur

Retinacular
fibers

Acetabulum

Ligament of head of
femur (ligamentum teres)

Ischium

Articular
capsule

FIG. 8.8 • Active motion of the
hip. **A**, Flexion is measured in
degrees from a supine position;
the knee can be extended or flexed;
**B**, Extension or hyperextension
is normally measured with the
knee extended; **C**, Abduction can
be measured in a supine or side-
lying position; adduction is best
measured with the subject lying
supine; **D**, Internal and external
rotation can be evaluated in either a
supine or a prone position.

120°

0° Neutral

FLEXION

**A**

30° or less

0° Neutral

**B**

90° — 90°       90° — 90°

45°              20°

0°               0°
Neutral          Neutral

**C**   ABDUCTION    ADDUCTION

30° or less

0° Neutral

EXTENSION

90° — 90°

35°   45°

Internal rotation        External rotation
(medial)                 (lateral)

0°
Neutral

**D**

**Prone**
Neutral
0°

External       Internal
rotation       rotation

90° — 90°

**Supine**
Neutral
0°

External       Internal
rotation       rotation

90° — 90°

ROTATION

Chapter
**8**

## TABLE 8.2 • **Hip and lumbar spine motions accompanying pelvic rotation**

| Pelvic rotation | Lumbar spine motion | Right hip motion | Left hip motion |
|---|---|---|---|
| Anterior rotation | Extension | Flexion | Flexion |
| Posterior rotation | Flexion | Extension | Extension |
| Right lateral rotation | Left lateral flexion | Abduction | Adduction |
| Left lateral rotation | Right lateral flexion | Adduction | Abduction |
| Right transverse rotation | Left lateral rotation | Internal rotation | External rotation |
| Left transverse rotation | Right lateral rotation | External rotation | Internal rotation |

## Movements FIGS. 8.9, 8.10

Hip flexion

Hip extension

Hip abduction

Hip external rotation

Hip adduction

Hip internal rotation

Anterior and posterior pelvic rotation occur in the sagittal or anteroposterior plane, whereas right and left lateral pelvic rotation occur in the lateral or frontal plane. Right transverse (clockwise) pelvic rotation and left transverse (counterclockwise) pelvic rotation occur in the horizontal or transverse plane of motion.

**Hip flexion:** movement of the anterior femur from any point toward the anterior pelvis in the sagittal plane

**Hip extension:** movement of the posterior femur from any point toward the posterior pelvis in the sagittal plane

**Hip abduction:** movement of the femur in the frontal plane laterally to the side away from the midline

**Hip adduction:** movement of the femur in the frontal plane medially toward the midline

| Flexion A | Extension B | Abduction C | Adduction D |

| External rotation E | Internal rotation F | Diagonal abduction G | Diagonal adduction H |

Courtesy of Britt Jones (all)

FIG. 8.9 • Movements of the hip.

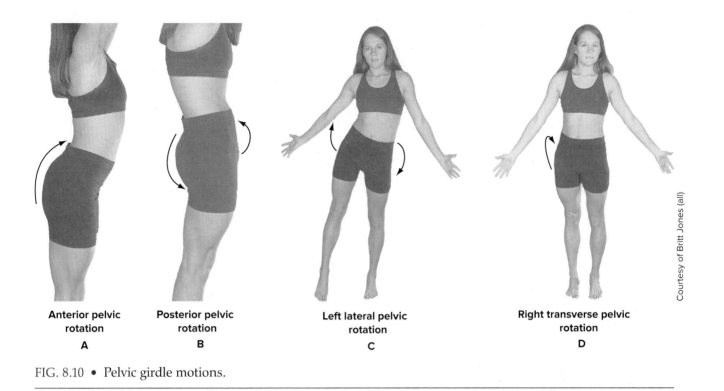

Anterior pelvic rotation
**A**

Posterior pelvic rotation
**B**

Left lateral pelvic rotation
**C**

Right transverse pelvic rotation
**D**

Courtesy of Britt Jones (all)

FIG. 8.10 • Pelvic girdle motions.

**Hip external rotation:** lateral rotary movement of the femur in the transverse plane around its longitudinal axis away from the midline; lateral rotation

**Hip internal rotation:** medial rotary movement of the femur in the transverse plane around its longitudinal axis toward the midline; medial rotation

**Hip diagonal abduction:** movement of the femur in a diagonal plane away from the midline of the body

**Hip diagonal adduction:** movement of the femur in a diagonal plane toward the midline of the body

**Hip horizontal adduction:** movement of the femur in a horizontal or transverse plane toward the pelvis

**Hip horizontal abduction:** movement of the femur in a horizontal or transverse plane away from the pelvis

**Anterior pelvic rotation:** anterior movement of the upper pelvis; the iliac crest tilts forward in a sagittal plane; anterior tilt; downward rotation; accomplished by hip flexion and/or lumbar extension

**Posterior pelvic rotation:** posterior movement of the upper pelvis; the iliac crest tilts backward in a sagittal plane; posterior tilt; upward rotation; accomplished by hip extension and/or lumbar flexion

**Left lateral pelvic rotation:** in the frontal plane, the left pelvis moves inferiorly in relation to the right pelvis; either the left pelvis rotates downward or the right pelvis rotates upward; left lateral tilt; accomplished by left hip abduction, right hip adduction, and/or right lumbar lateral flexion

**Right lateral pelvic rotation:** in the frontal plane, the right pelvis moves inferiorly in relation to the left pelvis; either the right pelvis rotates downward or the left pelvis rotates upward; right lateral tilt; accomplished by right hip abduction, left hip adduction, and/or left lumbar lateral flexion

**Left transverse pelvic rotation:** in a horizontal plane of motion, rotation of the pelvis to the body's left; the right iliac crest moves anteriorly in relation to the left iliac crest, which moves posteriorly; accomplished by right hip external rotation, left hip internal rotation, and/or right lumbar rotation

**Right transverse pelvic rotation:** in a horizontal plane of motion, rotation of the pelvis to the body's right; the left iliac crest moves anteriorly in relation to the right iliac crest, which moves posteriorly; accomplished by left hip external rotation, right hip internal rotation, and/or left lumbar rotation

Some confusion in understanding and learning the pelvic girdle motions can be avoided by always thinking about these movements from the perspective of the person's pelvis that is actually moving. Further understanding may be gained by thinking of the person as moving his or her pelvis as if steering a vehicle, as shown in Fig. 8.11.

Chapter
**8**

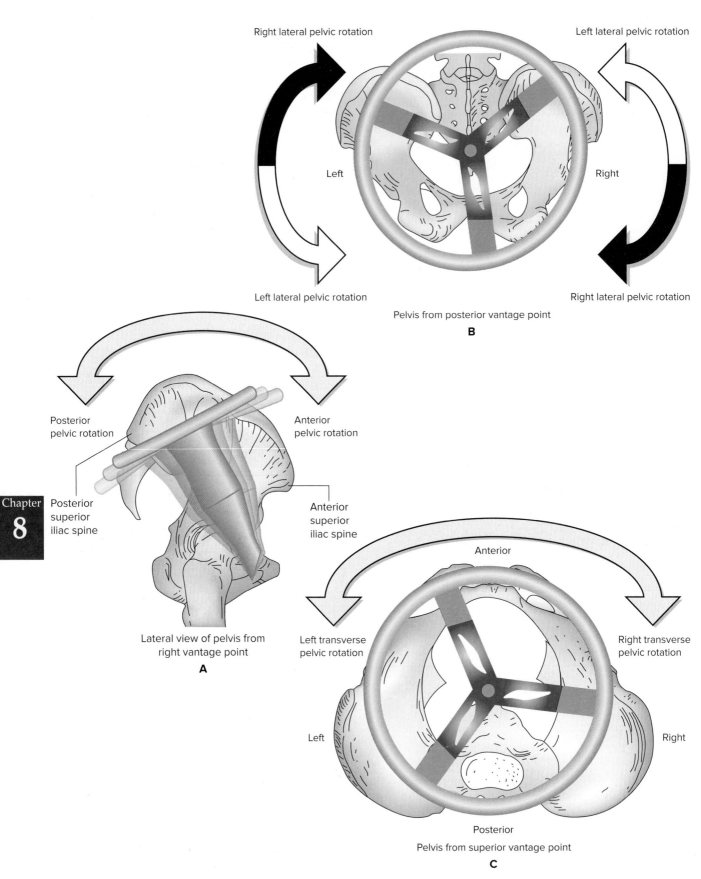

FIG. 8.11 • **A**, Anterior and posterior pelvic rotation; **B**, Right and left lateral pelvic rotation; **C**, Right and left transverse pelvic rotation.

## Muscles  FIGS. 8.12 TO 8.14

At the hip joint, there are seven two-joint muscles that have one action at the hip and another at the knee. The muscles actually involved in hip and pelvic girdle motions depend largely on the direction of the movement and the position of the body in relation to the earth and its gravitational forces. In addition, it should be noted that the body part that moves the most will be the part least stabilized. For example, when one is standing on both feet and contracting the hip flexors, the trunk and pelvis will rotate anteriorly, but when one is lying supine and contracting the hip flexors, the thighs will move forward into flexion on the stable pelvis.

For another example, the hip flexor muscles are used in moving the thighs toward the trunk, but the extensor muscles are used eccentrically when the pelvis and the trunk move downward slowly on the femur and concentrically when the trunk is raised on the femur—this, of course, occurs in rising to the standing position.

In the downward phase of the knee-bend exercise, the movement at the hips and knees is flexion. The muscles primarily involved are the hip and knee extensors in eccentric contraction.

Iliopsoas (deep)
Tensor fasciae latae
Sartorius
Adductor longus
Rectus femoris
Vastus lateralis
Vastus intermedius (deep)
Vastus medialis

Gluteus minimus
Piriformis
Pectineus
Adductor longus / brevis
Vastus lateralis
Vastus intermedius
Gracilis
Adductor magnus
Semimembranosus

Gluteus medius
Gluteus maximus
Gracilis
Semitendinosus
Iliotibial band
Biceps femoris
Gastrocnemius

**A**    **B**

FIG. 8.12 • Superficial and deep muscles of the lower extremity. **A**, Anterior view; **B**, Posterior view.

Additionally, it is important to understand that the action a particular muscle has on the hip may vary depending on the position of the femur in relation to the pelvis at the time. As the hip moves through its rather large ranges of motion, the lines of pull of specific muscles may change significantly. This is best seen with the adductors. If the hip is in flexion, the adductors, upon concentric contraction, tend to cause extension, and if the hip is in extension, they tend to cause flexion.

## Hip joint and pelvic girdle muscles—location

Muscle location largely determines the muscle action. Seventeen or more muscles are found in the area (the six external rotators are counted as one muscle). Most hip joint and pelvic girdle muscles are large and strong.

Anterior
    Primarily hip flexion
        Iliopsoas (iliacus and psoas)
        Pectineus
        Rectus femoris[*†]
        Sartorius[†]

Lateral
    Primarily hip abduction
        Gluteus medius
        Gluteus minimus
        External rotators
        Tensor fasciae latae[†]

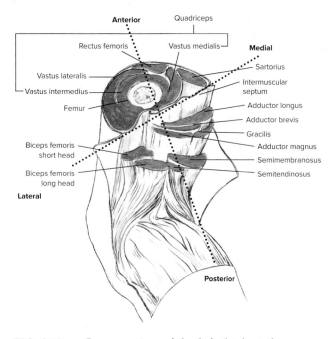

FIG. 8.13 • Cross section of the left thigh at the midsection.

[*]Two-joint muscles; knee actions are discussed in Chapter 9.
[†]Two-joint muscles.

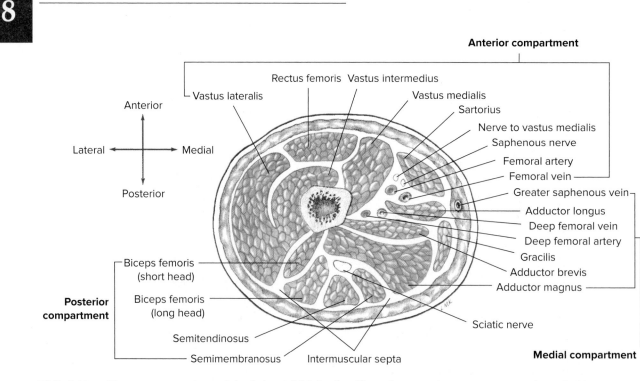

FIG. 8.14 • Transverse section of the left midthigh, detailing the anterior, posterior, and medial compartments.

Posterior
    Primarily hip extension
        Gluteus maximus
        Biceps femoris[*][†]
        Semitendinosus[*][†]
        Semimembranosus[*][†]
        External rotators (six deep)

Medial
    Primarily hip adduction
        Adductor brevis
        Adductor longus
        Adductor magnus
        Gracilis[†]

[*]Two-joint muscles; knee actions are discussed in Chapter 9.
[†]Two-joint muscles.

## Muscle identification

In developing a thorough and practical knowledge of the muscular system, it is essential that individual muscles be understood. Figs. 8.13, 8.15, 8.16, 8.17, and 8.18 illustrate groups of muscles that work together to produce joint movement. While viewing the muscles in these figures, correlate them with Table 8.3.

The muscles of the pelvis that act on the hip joint may be divided into two regions—the iliac and gluteal regions. The iliac region contains the iliopsoas muscle, which flexes the hip. The iliopsoas actually is two different muscles: the iliacus and the psoas major, although some include the psoas minor in discussion of the iliopsoas. The 10 muscles of the

FIG. 8.15 • Muscles of the right anterior pelvic and thigh regions.

FIG. 8.16 • Muscles of the right medial thigh.

Chapter
**8**

**231**

gluteal region function primarily to extend and rotate the hip. Located in the gluteal region are the gluteus maximus, gluteus medius, gluteus minimi, and tensor fasciae latae and the six deep external rotators—piriformis, obturator externus, obturator internus, gemellus superior, gemellus inferior, and quadratus femoris.

The thigh is divided into three compartments by the intermuscular septa (Fig. 8.14). The anterior compartment contains the rectus femoris, vastus medialis, vastus intermedius, vastus lateralis, and sartorius. The hamstring muscle group, consisting of the biceps femoris, semitendinosus, and semimembranosus, is located in the posterior compartment. The medial compartment contains the thigh muscles primarily responsible for adduction of the hip, which are the adductor brevis, adductor longus, adductor magnus, pectineus, and gracilis.

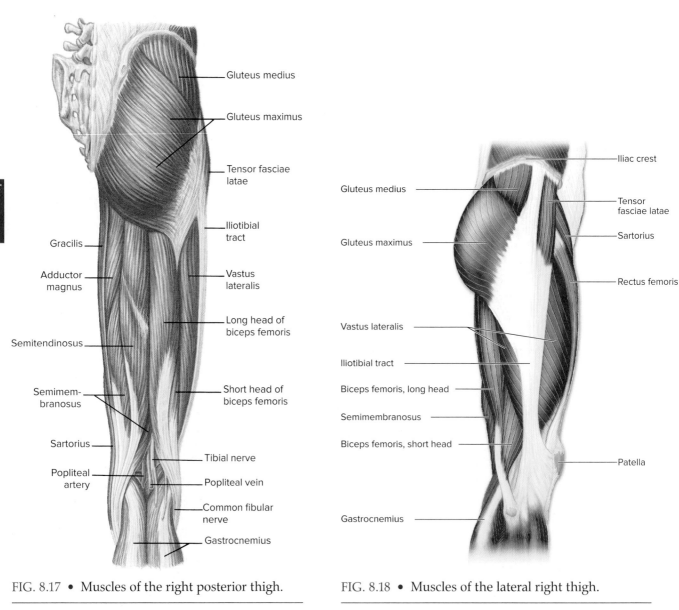

FIG. 8.17 • Muscles of the right posterior thigh.

FIG. 8.18 • Muscles of the lateral right thigh.

TABLE 8.3 · **Agonist muscles of the hip joint**

| | Muscle | Origin | Insertion | Action | Plane of motion | Palpation | Innervation |
|---|---|---|---|---|---|---|---|
| Anterior | Iliacus | Inner surface of the ilium | Lesser trochanter of the femur and shaft just below | Flexion of the hip | Sagittal | Difficult to palpate; deep against posterior abdominal wall; with subject seated and leaning slightly forward to relax abdominal muscles, palpate psoas major deeply between iliac crest and 12th rib about halfway between ASIS and umbilicus with active hip flexion; palpate iliopsoas distal tendon on anterior aspect of hip approximately 1½ inches below center of inguinal ligament with active hip flexion/ extension of supine subject, immediately lateral to pectineus and medial to sartorius | Lumbar nerve and femoral nerve (L2–L4) |
| | | | | Anterior pelvic rotation | | | |
| | | | | External rotation of the hip | Transverse | | |
| | | | | Transverse pelvic rotation contralaterally when ipsilateral femur is stabilized | | | |
| | Psoas major and minor | Lower borders of the transverse processes (L1–L5), sides of the bodies of the last thoracic vertebra (T12), the lumbar vertebrae (L1–L5), intervertebral fibrocartilages, and base of the sacrum | Lesser trochanter of the femur and shaft just below psoas minor; pectineal line (of pubis) and iliopectineal eminence (Psoas minor is involved only in lumbar spine movements.) | Flexion of the hip | Sagittal | | |
| | | | | Anterior pelvic rotation | | | |
| | | | | Flexion of lumbar spine | | | |
| | | | | External rotation of the hip | Transverse | | |
| | | | | Transverse pelvic rotation contralaterally when ipsilateral femur is stabilized | | | |
| | | | | Lateral flexion of lumbar spine | Frontal | | |
| | | | | Lateral pelvic rotation to contralateral side | | | |
| | Rectus femoris | Anterior inferior iliac spine of the ilium and groove (posterior) above the acetabulum | Superior aspect of the patella and patella tendon to the tibial tuberosity | Flexion of the hip | Sagittal | Straight down anterior thigh from anterior inferior iliac spine to patella with resisted hip flexion/knee extension | Femoral nerve (L2–L4) |
| | | | | Extension of the knee | | | |
| | | | | Anterior pelvic rotation | | | |
| | Sartorius | Anterior superior iliac spine and notch just below the spine | Anterior medial surface of the tibia just below the condyle | Flexion of the hip | Sagittal | Located proximally just medial to tensor fascia latae and lateral to iliopsoas; palpate superficially from anterior superior iliac spine to medial tibial condyle during combined resistance of hip flexion/ external rotation/ abduction and knee flexion in a supine position | Femoral nerve (L2, L3) |
| | | | | Flexion of the knee | | | |
| | | | | Anterior pelvic rotation | | | |
| | | | | External rotation of the thigh as it flexes the hip and knee | Transverse | | |
| | | | | Weak internal rotation of the knee | | | |
| | | | | Abduction of the hip | Frontal | | |

TABLE 8.3 (continued) • **Agonist muscles of the hip joint**

| | Muscle | Origin | Insertion | Action | Plane of motion | Palpation | Innervation |
|---|---|---|---|---|---|---|---|
| Medial | Pectineus | Space 1 inch wide on the front of the pubis just above the crest | Rough line leading from the lesser trochanter down to the linea aspera | Flexion of the hip | Sagittal | Difficult to distinguish from other adductors; anterior aspect of hip approximately 1½ inches below center of inguinal ligament; just lateral and slightly proximal to adductor longus and medial to iliopsoas during flexion and adduction of supine subject | Femoral nerve (L2–L4) |
| | | | | Adduction of the hip | Frontal | | |
| | | | | External rotation of the hip | Transverse | | |
| | Adductor brevis | Front of the inferior pubic ramus just below the origin of the adductor longus | Lower 2/3 of the pectineal line of the femur and the upper half of the medial lip of the linea aspera | Adduction of the hip | Frontal | Deep to adductor longus and superficial to adductor magnus; very difficult to palpate and differentiate from adductor longus, which is immediately inferior; proximal portion is just lateral to adductor longus | Obturator nerve (L3, L4) |
| | | | | External rotation as it adducts the hip | Transverse | | |
| | | | | Assists in flexion of the hip | Sagittal | | |
| | Adductor longus | Anterior pubis just below its crest | Middle 1/3 of the linea aspera | Adduction of the hip | Frontal | Most prominent muscle proximally on anteromedial thigh just inferior to the pubic bone with resisted adduction | Obturator nerve (L3, L4) |
| | | | | External rotation as it adducts the hip | Transverse | | |
| | | | | Assists in flexion of the hip | Sagittal | | |
| | Adductor magnus | Edge of the entire pubic ramus and the ischium and ischial tuberosity | Whole length of the linea aspera, medial condyloid ridge, and adductor tubercle | Adduction of the hip | Frontal | Medial aspect of thigh between gracilis and medial hamstrings from ischial tuberosity to adductor tubercle with resisted adduction from abducted position | Anterior: obturator nerve (L2–L4)<br><br>Posterior: sciatic nerve (L4, L5, S1–S3) |
| | | | | External rotation as the hip adducts | Transverse | | |
| | | | | Extension of the hip | Sagittal | | |
| | Gracilis | Antero-medial edge of the descending ramus of the pubis | Anterior medial surface of the tibia below the condyle | Adduction of the hip | Frontal | A thin superficial tendon on anteromedial thigh with knee flexion and resisted adduction; just posterior to adductor longus and medial to the semitendinosus | Obturator nerve (L2–L4) |
| | | | | Weak flexion of the knee | Sagittal | | |
| | | | | Assists with flexion of the hip | | | |
| | | | | Internal rotation of the hip | Transverse | | |
| | | | | Weak internal rotation of the knee | | | |

Chapter 8

TABLE 8.3 (continued) • **Agonist muscles of the hip joint**

| | Muscle | Origin | Insertion | Action | Plane of motion | Palpation | Innervation |
|---|---|---|---|---|---|---|---|
| **Posterior** | Semitendinosus | Ischial tuberosity | Upper anterior medial surface of the tibia just below the condyle | Flexion of the knee | Sagittal | Posteromedial aspect of distal thigh with combined knee flexion and internal rotation against resistance; just distal to ischial tuberosity in a prone position with hip internally rotated during active knee flexion | Sciatic nerve—tibial division (L5, S1, S2) |
| | | | | Extension of the hip | | | |
| | | | | Posterior pelvic rotation | | | |
| | | | | Internal rotation of the hip | Transverse | | |
| | | | | Internal rotation of the flexed knee | | | |
| | Semimembranosus | Ischial tuberosity | Posteromedial surface of the medial tibial condyle | Flexion of the knee | Sagittal | Largely covered by other muscles, tendon can be felt at posteromedial aspect of knee just deep to semitendinosus tendon with combined knee flexion and internal rotation against resistance | Sciatic nerve—tibial division (L5, S1, S2) |
| | | | | Extension of the hip | | | |
| | | | | Posterior pelvic rotation | | | |
| | | | | Internal rotation of the hip | Transverse | | |
| | | | | Internal rotation of the knee | | | |
| | Biceps femoris | Long head: ischial tuberosity. Short head: lower half of the linea aspera, and lateral condyloid ridge | Head of the fibula and lateral condyle of the tibia | Flexion of the knee | Sagittal | Posterolateral aspect of distal thigh with combined knee flexion and external rotation against resistance; just distal to the ischial tuberosity in a prone position with hip internally rotated during active knee flexion | Long head: sciatic nerve—tibial division (S1–S3) Short head: sciatic nerve—peroneal division (L5, S1, S2) |
| | | | | Extension of the hip | | | |
| | | | | Posterior pelvic rotation | | | |
| | | | | External rotation of the hip | Transverse | | |
| | | | | External rotation of the knee | | | |
| | Gluteus maximus | Posterior 1/4 of the crest of the ilium, posterior surface of the sacrum and coccyx near the ilium, and fascia of the lumbar area | Oblique ridge on the lateral surface of the greater trochanter and the iliotibial band of the fasciae latae | Extension of the hip | Sagittal | Running downward and laterally between posterior iliac crest superiorly, anal cleft medially, and gluteal fold inferiorly, emphasized with hip extension, external rotation, and abduction | Inferior gluteal nerve (L5, S1, S2) |
| | | | | Posterior pelvic rotation | | | |
| | | | | External rotation of the hip | Transverse | | |
| | | | | Upper fibers: assist in hip abduction | Frontal | | |
| | | | | Lower fibers: assist in hip adduction | | | |

Chapter
**8**

235

TABLE 8.3 (continued) • **Agonist muscles of the hip joint**

| | Muscle | Origin | Insertion | Action | Plane of motion | Palpation | Innervation |
|---|---|---|---|---|---|---|---|
| Lateral | Gluteus medius | Lateral surface of the ilium just below the crest | Posterior and middle surfaces of the greater trochanter of the femur | Abduction of the hip | Frontal | Slightly in front of and a few inches above the greater trochanter with active elevation of opposite pelvis from a standing position or active abduction when side-lying on contralateral pelvis | Superior gluteal nerve (L4, L5, S1) |
| | | | | Lateral pelvic rotation to ipsilateral side | | | |
| | | | | Anterior fibers: internal rotation of the hip | Transverse | | |
| | | | | Posterior fibers: external rotation of the hip | | | |
| | | | | Anterior fibers: flexion of the hip | Sagittal | | |
| | | | | Anterior fibers: anterior pelvic rotation | | | |
| | | | | Posterior fibers: extension of the hip | | | |
| | | | | Posterior fibers: posterior pelvic rotation | | | |
| | Gluteus minimus | Lateral surface of the ilium just below the origin of the gluteus medius | Anterior surface of the greater trochanter of the femur | Abduction of the hip | Frontal | Deep to the gluteus medius; covered by tensor fasciae latae between anterior iliac crest and greater trochanter during internal rotation and abduction | Superior gluteal nerve (L4, L5, S1) |
| | | | | Lateral pelvic rotation to ipsilateral side | | | |
| | | | | Hip internal rotation as femur is abducted | Transverse | | |
| | | | | Flexion of the hip | Sagittal | | |
| | | | | Anterior pelvic rotation | | | |
| | Tensor fasciae latae | Anterior iliac crest and surface of the ilium just below the crest | One-fourth of the way down the thigh into the iliotibial tract, which in turn inserts onto Gerdy's tubercle of the anterolateral tibial condyle | Abduction of the hip | Frontal | Anterolaterally, between anterior iliac crest and greater trochanter during flexion, internal rotation, and abduction | Superior gluteal nerve (L4, L5, S1) |
| | | | | Lateral pelvic rotation to ipsilateral side | | | |
| | | | | Flexion of the hip | Sagittal | | |
| | | | | Anterior pelvic rotation | | | |
| | | | | Internal rotation of the hip as it flexes | Transverse | | |

Chapter

**8**

TABLE 8.3 (continued) • **Agonist muscles of the hip joint**

| | Muscle | Origin | Insertion | Action | Plane of motion | Palpation | Innervation |
|---|---|---|---|---|---|---|---|
| Deep posterior | Piriformis | Anterior sacrum, posterior portions of the ischium, and obturator foramen | Superior and posterior aspect of the greater trochanter | Hip external rotation | Transverse | With subject prone and gluteus maximus relaxed, palpate deeply between posterior superior greater trochanter and sacrum while passively internally/externally rotating femur | First and second sacral nerves (S1, S2) |
| | Gemellus superior | Ischial spine | Posterior aspect of the greater trochanter immediately below piriformis | Hip external rotation | Transverse | With subject prone and gluteus maximus relaxed, palpate deeply between posterior superior greater trochanter and ischial spine while passively internally/externally rotating femur | Sacral nerve (L5, S1, S2) |
| | Gemellus inferior | Ischial tuberosity | Posterior aspect of the greater trochanter with obturator internus | Hip external rotation | Transverse | With subject prone and gluteus maximus relaxed, palpate deeply between posterior greater trochanter and ischial tuberosity while passively internally/externally rotating femur | Branches from sacral plexus (L4, L5, S1, S2) |
| | Obturator internus | Margin of obturator foramen | Posterior aspect of the greater trochanter with gamellus superior | Hip external rotation | Transverse | With subject prone and gluteus maximus relaxed, palpate deeply between posterior superior greater trochanter and obturator foramen while passively internally/externally rotating femur | Branches from sacral plexus (L4, L5, S1, S2) |
| | Obturator externus | Inferior margin of obturator foramen | Posterior aspect of the greater trochanter immediately below obturator internus | Hip external rotation | Transverse | With subject prone and gluteus maximus relaxed, palpate deeply between inferior posterior greater trochanter and obturator foramen while passively internally/externally rotating femur | Obturator nerve (L3, L4) |
| | Quadratus femoris | Ischial tuberosity | Intertrochanteric ridge of femur | Hip external rotation | Transverse | With subject prone and gluteus maximus relaxed, palpate deeply between inferior posterior greater trochanter and ischial tuberosity while passively internally/externally rotating femur | Branches from sacral plexus (L4, L5, S1) |

Chapter

**8**

# Nerves   FIG. 8.19

The muscles of the hip and pelvic girdle are all innervated from the lumbar and sacral plexus, known collectively as the lumbosacral plexus. The lumbar plexus is formed by the anterior rami of spinal nerves L1 through L4 and some fibers from T12. The lower abdomen and the anterior and medial portions of the lower extremity are innervated by nerves arising from the lumbar plexus. The sacral plexus is formed by the anterior rami of L4, L5, and S1 through S4. The lower back, pelvis, perineum, posterior surface of the thigh and leg, and dorsal and plantar surfaces of the foot are innervated by nerves arising from the sacral plexus.

FIG. 8.19 • Nerves of the lumbosacral plexus. **A**, Close-up; **B**, Anterior view; **C**, Posterior view.

The major nerves of significance arising from the lumbar plexus to innervate the muscles of the hip are the femoral and obturator nerves. The femoral nerve (Fig. 8.20) arises from the posterior division of the lumbar plexus and innervates the anterior muscles of the thigh, including the iliopsoas, rectus femoris, vastus medialis, vastus intermedius, vastus lateralis, pectineus, and sartorius. It also provides sensation to the anterior and medial thigh and the medial leg and foot. The obturator nerve (Fig. 8.21) arises from the anterior division of the lumbar plexus and provides innervation to the hip adductors, such as the adductor brevis, adductor longus, adductor magnus, and gracilis, as well

as the obturator externus. The obturator nerve provides sensation to the medial thigh.

The nerves arising from the sacral plexus that innervate the muscles of the hip are the superior gluteal, inferior gluteal, sciatic, and branches from the sacral plexus. The superior gluteal nerve arises from L4, L5, and S1 to innervate the gluteus medius, gluteus minimus, and tensor fasciae latae. The inferior gluteal nerve arises from L5, S1, and S2 to supply the gluteus maximus. Branches from the sacral plexus innervate the piriformis (S1, S2), gemellus superior (L5, S1, S2), gemellus inferior and obturator internus (L4, L5, S1, S2), and quadratus femoris (L4, L5, S1).

FIG. 8.20 • Muscular and cutaneous distribution of the femoral nerve.

The sciatic nerve is composed of the tibial and common peroneal (fibular) nerves, which are wrapped together in a connective tissue sheath until reaching approximately midway down the posterior thigh. The sciatic nerve tibial division (Fig. 8.22) innervates the semitendinosus, semimembranosus, biceps femoris (long head), and adductor magnus. The sciatic nerve supplies sensation to the anterolateral and posterolateral lower leg as well as to most of the dorsal and plantar aspects of the foot. The tibial division provides sensation to the posterolateral lower leg and plantar aspect of the foot, while the peroneal division provides sensation to the anterolateral lower leg and dorsum of the foot. Both of these nerves continue down the lower extremity to provide motor and sensory function to the muscles of the lower leg; this will be addressed in Chapters 9 and 10. See Fig. 8.23 for a full listing of the hip muscular innervations.

FIG. 8.21 • Muscular and cutaneous distribution of the obturator nerve.

FIG. 8.22 • Muscular and cutaneous distribution of the tibial nerve.

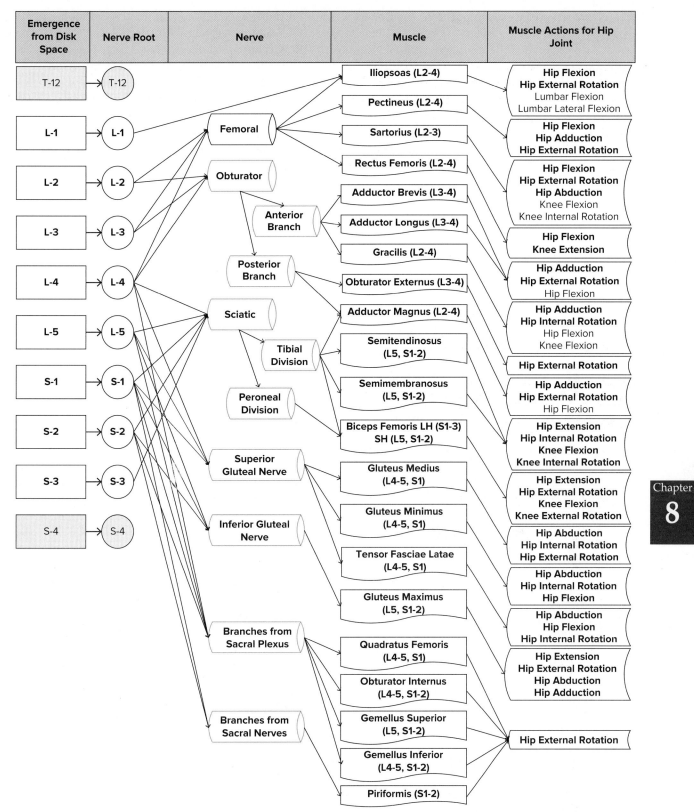

FIG. 8.23 • Innervation for hip joint muscles and their actions. The neuromuscular structures associated with the primary movers for the hip joint and their actions are bolded.

# Iliopsoas muscle    FIG. 8.24

(il´e-o-so´as)

Hip flexion

Hip external rotation

Lumbar flexion

Lumbar lateral flexion

## Origin

Iliacus: inner surface of the ilium

Psoas major and minor: lower borders of the transverse processes (L1–L5), sides of the bodies of the last thoracic vertebra (T12), lumbar vertebrae (L1–L5), intervertebral fibrocartilages, and base of the sacrum

## Insertion

Iliacus and psoas major: lesser trochanter of the femur and the shaft just below

Psoas minor: pectineal line (of pubis) and iliopectineal eminence

Psoas minor muscle

Psoas major muscle

Iliacus muscle

Flexion

External rotation

O, Lower borders of the transverse processes of lumbar vertebrae 1–5 (psoas major and minor)

O, Inner surface of the ilium (iliacus)

I, Lesser trochanter of femur and shaft just below (iliacus and psoas major)

O, Sides of the bodies of the last thoracic (T12) and all the lumbar vertebrae (L1–L5), intervertebral fibrocartilages, and base of sacrum (psoas major and minor)

I, Pectineal line (of pubis) and iliopectineal eminence (psoas minor)

FIG. 8.24 • Iliopsoas muscle, anterior view. O, Origin; I, Insertion.

## Action

Flexion of the hip

Anterior pelvic rotation

External rotation of the hip

Transverse pelvis rotation contralaterally when ipsilateral femur is stabilized

Flexion of lumbar spine (psoas major and minor)

Lateral flexion of lumbar spine (psoas major and minor)

Lateral pelvic rotation to contralateral side (psoas major and minor)

## Palpation

Difficult to palpate; deep against posterior abdominal wall; with subject seated and leaning slightly forward to relax abdominal muscles, palpate psoas major deeply between iliac crest and 12th rib about halfway between ASIS and umbilicus with active hip flexion; palpate iliopsoas distal tendon on anterior aspect of hip approximately 1½ inches below center of inguinal ligament with active hip flexion/extension of supine subject, immediately lateral to pectineus and medial to sartorius

## Innervation

Lumbar nerve and femoral nerve (L2–L4)

## Application, strengthening, and flexibility

The iliopsoas is commonly referred to as if it were one muscle, but it is actually composed of the iliacus and the psoas major. Some anatomy texts make this distinction and list each muscle individually. The psoas minor attaches on the pubis above the hip joint and therefore does not act on the hip joint. Most authorities do not include it in discussion of the iliopsoas.

The iliopsoas muscle is powerful in actions such as raising the lower extremity from the floor while in a supine position. The psoas major's origin in the lower back tends to move the lower back anteriorly or, in the supine position, pulls up the lower back as it raises the thighs. For this reason, lower back problems are often aggravated by this activity, and bilateral 6-inch leg raises are usually not recommended. The abdominals are the muscles that can be used to prevent this lower back strain by pulling up on the front of the pelvis, thus flattening the back. Leg raising is primarily hip flexion and not abdominal action. The back may be injured by strenuous and prolonged leg-raising exercises due to the iliopsoas pulling the lumbar spine into hyperextension and increasing the lordotic curve, particularly in the absence of adequate stabilization by the abdominals. The iliopsoas contracts strongly, both concentrically and eccentrically, in sit-ups, particularly if the hip is not flexed. The more flexed and/or abducted the hips are, the less the iliopsoas will be activated with abdominal strengthening exercises.

The iliopsoas may be exercised by supporting the arms on a dip bar or parallel bars and then flexing the hips to lift the legs. This may be done initially with the knees flexed in a tucked position to lessen the resistance. As the muscle becomes more developed, the knees can be straightened, which increases the resistance arm length to add more resistance. This concept of increasing or decreasing the resistance by modifying the resistance arm is explained further in Chapter 3. See Appendix 2 for more commonly used exercises for the iliopsoas and other muscles in this chapter.

To stretch the iliopsoas, which often becomes tight with excessive straight-leg sit-ups and contributes to anterior pelvic tilting, the hip must be extended so that the femur is behind the plane of the body. In order to somewhat isolate the iliopsoas, full knee flexion should be avoided. Slight additional stretch may be applied by internally rotating the hip while it is extended.

# Rectus femoris muscle   FIG. 8.25

(rek´tus fem´or-is)

Hip flexion

Knee extension

## Origin

Anterior inferior iliac spine of the ilium and groove (posterior) above the acetabulum

## Insertion

Superior aspect of the patella and patellar tendon to the tibial tuberosity

## Action

Flexion of the hip
Extension of the knee
Anterior pelvic rotation

## Palpation

Straight down anterior thigh from anterior inferior iliac spine to patella with resisted knee extension and hip flexion

## Innervation

Femoral nerve (L2–L4)

## Application, strengthening, and flexibility

Pulling from the anterior inferior iliac spine of the ilium, the rectus femoris muscle has the same tendency to anteriorly rotate the pelvis (down in front and up in back). Only the abdominal muscles can prevent this from occurring. In speaking of the hip flexor group in general, it may be said that many people permit the pelvis to be permanently tilted forward as they get older. The relaxed abdominal wall does not hold the pelvis up; therefore, an increased lumbar curve results.

Generally, a muscle's ability to exert force decreases as it shortens. This explains why the rectus femoris muscle is a powerful extensor of the knee when the hip is extended but is weaker when the hip is flexed. This muscle is exercised, along with the vastus group, in running, jumping, hopping, and skipping. In these movements, the hips are extended powerfully by the gluteus maximus and the hamstring muscles, which counteract the tendency of the rectus femoris muscle to flex the hip while it extends the knee. It can be remembered as one of the quadriceps muscle group. The rectus femoris is developed by performing hip flexion exercises or knee extension exercises against manual resistance.

The rectus femoris is best stretched in a sidelying position by having a partner take the knee into full flexion and simultaneously take the hip into extension.

FIG. 8.25 • Rectus femoris muscle, anterior view. O, Origin; I, Insertion.

# Sartorius muscle    FIG. 8.26

(sar-to´ri-us)

## Origin

Anterior superior iliac spine and notch just below the
   spine

## Insertion

Anterior medial surface of the tibia just below the
   condyle

## Action

Flexion of the hip
Flexion of the knee
External rotation of the thigh as it flexes the hip and knee
Abduction of the hip
Anterior pelvic rotation
Weak internal rotation of the knee

## Palpation

Located proximally just medial to the tensor fasciae
   latae and lateral to the iliopsoas; palpate superficially
   from the anterior superior iliac spine to the medial
   tibial condyle during combined resistance of hip
   flexion/external rotation/abduction and knee flexion
   in a supine position

## Innervation

Femoral nerve (L2, L3)

## Application, strengthening, and flexibility

Pulling from the anterior superior iliac spine and
the notch just below it, the tendency again is to tilt
the pelvis anteriorly (down in front) as this muscle
contracts. The abdominal muscles must prevent
this tendency by posteriorly rotating the pelvis
(pulling up in front), thus flattening the lower
back.

The sartorius, a two-joint muscle, is effective as
a hip flexor or as a knee flexor. Sometimes referred
to as the tailor's muscle, it is active in all the hip and
knee movements used in assuming the sitting position of a tailor. It is weak when both flexions take
place at the same time. Observe that, in attempting to cross the knees when in a sitting position,
one customarily leans well back, thus raising the
origin to lengthen this muscle, making it more
effective in flexing and crossing the knees. With
the knees held extended, the sartorius becomes a
more effective hip flexor. It is the longest muscle
in the body and is strengthened when hip flexion
activities are performed as described for developing the iliopsoas. Stretching may be accomplished
by a partner passively taking the hip into extreme
extension, adduction, and internal rotation with
the knee extended.

Hip flexion

Knee
flexion

Hip
external
rotation

Hip
abduction

Knee
internal
rotation

Hip flexion

External
rotation

O, Anterior superior iliac
spine and notch just
below spine of ilium

Sartorius m.

Knee flexion

I, Anterior medial
surface of the
tibia just below
the condyle

FIG. 8.26 • Sartorius muscle, anterior view. O, Origin; I, Insertion.

# Pectineus muscle    FIG. 8.27

(pek-tin´e-us)

### Origin

Space 1 inch wide on the front of the pubis just above the crest (pectineal line)

### Insertion

Rough line leading from the lesser trochanter down to the linea aspera (pectineal line of femur)

### Action

Flexion of the hip
Adduction of the hip
External rotation of the hip
Anterior pelvic rotation

### Palpation

Difficult to distinguish from other adductors; anterior aspect of hip approximately 1½ inches below center of inguinal ligament; just lateral and slightly proximal to adductor longus and medial to iliopsoas during flexion and adduction of supine subject

### Innervation

Femoral nerve (L2–L4)

### Application, strengthening, and flexibility

As the pectineus contracts, it also tends to rotate the pelvis anteriorly. The abdominal muscles pulling up on the pelvis in front prevent this tilting action.

The pectineus muscle is exercised together with the iliopsoas muscle in leg raising and lowering. Hip flexion exercises and hip adduction exercises against resistance may be used for strengthening this muscle.

The pectineus is stretched by fully abducting the extended and internally rotated hip.

Hip flexion

Hip adduction

Hip external rotation

Pectineus m.

External rotation

Adduction

Flexion

O, Space 1 inch wide on front of pubis above crest (pectineal line)

I, Rough line leading from lesser trochanter down to linea aspera (pectineal line of femur)

FIG. 8.27 • Pectineus muscle, anterior view. O, Origin; I, Insertion.

# Adductor brevis muscle    FIG. 8.28

(ad-duk´tor bre´vis)

## Origin

Front of the inferior pubic ramus just below the origin
of the adductor longus

## Insertion

Lower two-thirds of the pectineal line of the femur
and upper half of the medial lip of the linea aspera

## Action

Adduction of the hip
External rotation as it adducts the hip
Assists in flexion of the hip
Assists in anterior pelvic rotation

## Palpation

Deep to the adductor longus and superficial to the
adductor magnus; very difficult to palpate and differentiate from adductor longus, which is immediately inferior; proximal portion is just lateral to
adductor longus

## Innervation

Obturator nerve (L3, L4)

## Application, strengthening, and flexibility

The adductor brevis muscle, along with the other
adductor muscles, provides powerful movement of
the thighs toward each other. Squeezing the thighs
together toward each other against resistance is
effective in strengthening the adductor brevis.
Abducting the extended and internally rotated hip
provides stretching of the adductor brevis.

Hip
adduction

Hip
external
rotation

Hip flexion

FIG. 8.28 • Adductor brevis muscle, anterior view. O, Origin; I, Insertion.

## Adductor longus muscle  FIG. 8.29
### (ad-duk´tor lon´gus)

Hip
adduction

### Origin
Anterior pubis just below its crest

### Insertion
Middle third of the linea aspera

Hip
external
rotation

### Action
Adduction of the hip
External rotation as it adducts the hip
Assists in flexion of the hip
Assists in anterior pelvic rotation

### Palpation
Most prominent muscle proximally on anteromedial
   thigh just inferior to the pubic bone with resisted
   adduction

Hip flexion

### Innervation
Obturator nerve (L3, L4)

### Application, strengthening, and flexibility
The muscle may be strengthened by using the scissors exercise, which requires the subject to sit on the floor with the legs spread wide while a partner puts his or her legs or arms inside each lower leg to provide resistance. As the subject attempts to adduct his or her legs together, the partner provides manual resistance throughout the range of motion. This exercise may be used for either one or both legs. The adductor longus is stretched in the same manner as the adductor brevis.

Chapter
**8**

Adductor longus m.

Adduction

O, Anterior pubis
just below its crest

I, Middle third
of linea aspera

FIG. 8.29 • Adductor longus muscle, anterior view. O, Origin; I, Insertion.

# Adductor magnus muscle    FIG. 8.30

(ad-duk´tor mag´nus)

### Origin

Edge of the entire ramus of the pubis and the ischium and ischial tuberosity

### Insertion

Whole length of the linea aspera, medial condyloid ridge, and adductor tubercle

### Action

Adduction of the hip

External rotation as the hip adducts

Extension of the hip

### Palpation

Medial aspect of thigh between gracilis and medial hamstrings from ischial tuberosity to adductor tubercle with resisted adduction from abducted position

### Innervation

Anterior: obturator nerve (L2–L4)

Posterior: sciatic nerve (L4, L5, S1–S3)

## Application, strengthening, and flexibility

The upper portion that originates on the ischiopubic ramus is sometimes listed separately as the adductor minimus, leaving the portion originating on the ischial tuberosity to be listed as the adductor magnus. The adductor minimus portion is referred to as the adductor portion, with the larger portion being referred to as the hamstring portion.

The adductor magnus muscle is used in the breaststroke kick in swimming and in horseback riding. Since the adductor muscles (adductor magnus, adductor longus, adductor brevis, and gracilis) are not heavily used in ordinary movement, some prescribed activity for them should be provided. Some modern exercise equipment is engineered to provide resistance for hip adduction movement. Hip adduction exercises such as those described for the adductor brevis and the adductor longus may be used for strengthening the adductor magnus as well. The adductor magnus is stretched in the same manner as the adductor brevis and adductor longus.

Hip adduction

Hip external rotation

Hip extension

Chapter

**8**

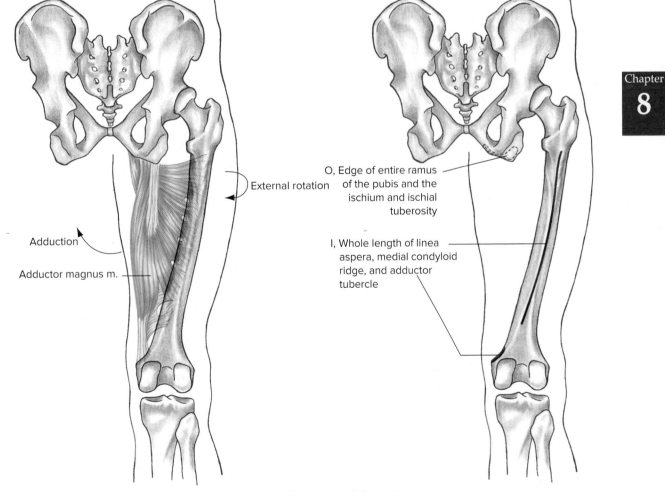

External rotation

Adduction

Adductor magnus m.

O, Edge of entire ramus of the pubis and the ischium and ischial tuberosity

I, Whole length of linea aspera, medial condyloid ridge, and adductor tubercle

FIG. 8.30 • Adductor magnus muscle, posterior view. O, Origin; I, Insertion.

# Gracilis muscle   FIG. 8.31

(gras´il-is)

Hip
adduction

Knee
flexion

## Origin

Anteromedial edge of the descending ramus of the pubis

## Insertion

Anterior medial surface of the tibia just below the condyle

## Action

Adduction of the hip
Weak flexion of the knee
Internal rotation of the hip
Assists with flexion of the hip
Weak internal rotation of the knee

Hip
internal
rotation

## Palpation

A thin tendon on anteromedial thigh with knee flexion and resisted adduction just posterior to adductor longus and medial to the semitendinosus

Hip flexion

## Innervation

Obturator nerve (L2–L4)

## Application, strengthening, and flexibility

Also known as the adductor gracilis, this muscle performs the same function as the other adductors but adds some weak assistance to knee flexion.

The adductor muscles as a group (adductor magnus, adductor longus, adductor brevis, and gracilis) are called into action in horseback riding and in doing the breaststroke kick in swimming. Proper development of the adductor group prevents soreness after participation in these sports. The gracilis is strengthened with the same exercises as described for the other hip adductors. The gracilis may be stretched in a manner similar to the adductors, except that the knee must be extended.

Knee
internal
rotation

FIG. 8.31 • Gracilis muscle, anterior view. O, Origin; I, Insertion.

# Semitendinosus muscle   FIG. 8.32

(sem´i-ten-di-no´sus)

## Origin
Ischial tuberosity

## Insertion
Anterior medial surface of the tibia just below the condyle

## Action
Flexion of the knee
Extension of the hip
Internal rotation of the hip
Internal rotation of the flexed knee
Posterior pelvic rotation

## Palpation
Posteromedial aspect of distal thigh with combined knee flexion and internal rotation against resistance just distal to the ischial tuberosity in a prone position with hip internally rotated during active knee flexion

## Innervation
Sciatic nerve—tibial division (L5, S1, S2)

## Application, strengthening, and flexibility
This two-joint muscle is most effective when contracting to either extend the hip or flex the knee.

When there is extension of the hip and flexion of the knee at the same time, both movements are weak. When the trunk is flexed forward with the knees straight, the hamstring muscles have a powerful pull on the rear pelvis and tilt it down in back by full contraction. If the knees are flexed when this movement takes place, one can observe that the work is done chiefly by the gluteus maximus muscle.

On the other hand, when the muscles are used in powerful flexion of the knees, as in hanging by the knees from a bar, the flexors of the hip come into play to raise the origin of these muscles and make them more effective as knee flexors. By full extension of the hips in this movement, the knee flexion movement is weakened. These muscles are used in ordinary walking as extensors of the hip and allow the gluteus maximus to relax in the movement.

The semitendinosus is best developed through knee flexion exercises against resistance. Commonly known as hamstring curls or leg curls, they may be performed in a prone position on a knee table or standing with ankle weights attached. This muscle is emphasized when performing hamstring curls while attempting to maintain the knee joint in internal rotation. This internally rotated position brings its insertion in alignment with its origin.

The semitendinosus is stretched by maximally extending the knee while flexing the internally rotated and slightly abducted hip.

Knee flexion

Hip extension

Hip internal rotation

Knee internal rotation

**Chapter 8**

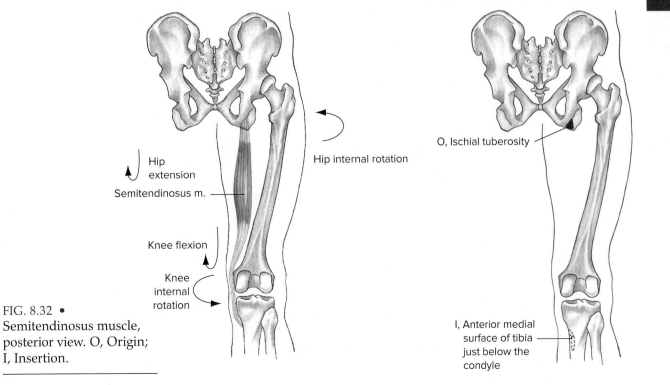

Hip extension

Semitendinosus m.

Knee flexion

Knee internal rotation

Hip internal rotation

O, Ischial tuberosity

I, Anterior medial surface of tibia just below the condyle

FIG. 8.32 •
Semitendinosus muscle, posterior view. O, Origin; I, Insertion.

Knee
flexion

# Semimembranosus muscle   FIG. 8.33

(sem´i-mem´bra-no´sus)

Hip
extension

## Origin

Ischial tuberosity

## Insertion

Posteromedial surface of the medial tibial condyle

Hip
internal
rotation

## Action

Flexion of the knee
Extension of the hip
Internal rotation of the hip
Internal rotation of the flexed knee
Posterior pelvic rotation

## Palpation

Largely covered by other muscles, the tendon can be
    felt at the posteromedial aspect of the knee just deep

Knee
internal
rotation

to the semitendinosus tendon with combined knee
flexion and internal rotation against resistance

## Innervation

Sciatic nerve—tibial division (L5, S1, S2)

## Application, strengthening, and flexibility

Both the semitendinosus and semimembranosus
are responsible for internal rotation of the knee,
along with the popliteus muscle, which is discussed
in Chapter 9. Because of the manner in which they
cross the joint, the muscles are very important in
providing dynamic medial stability to the knee
joint.

The semimembranosus is best developed by
performing leg curls. Internal rotation of the knee
throughout the range accentuates the activity of
this muscle. The semimembranosus is stretched
in the same manner as the semitendinosus.

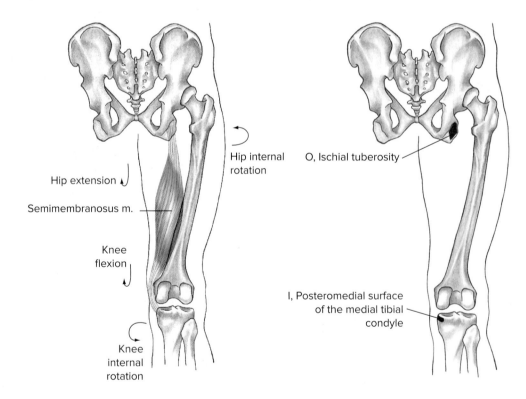

FIG. 8.33 • Semimembranosus muscle, posterior view. O, Origin; I, Insertion.

# Biceps femoris muscle   FIG. 8.34

(bi´seps fem´or-is)

## Origin

Long head: ischial tuberosity
Short head: lower half of the linea aspera, and lateral
condyloid ridge

## Insertion

Lateral condyle of the tibia and head of the fibula

## Action

Flexion of the knee
Extension of the hip
External rotation of the hip
External rotation of the flexed knee
Posterior pelvic rotation

## Palpation

Posterolateral aspect of distal thigh with combined
knee flexion and external rotation against resistance
and just distal to the ischial tuberosity in a prone
position with hip externally rotated during active
knee flexion

## Innervation

Long head: sciatic nerve—tibial division
 (S1–S3);
Short head: sciatic nerve—peroneal division
 (L5, S1, S2)

## Application, strengthening, and flexibility

The semitendinosus, semimembranosus, and
biceps femoris muscles are known as the hamstrings. These muscles, together with the gluteus
maximus muscle, are used in extension of the hip
when the knees are straight or nearly so. Thus, in
running, jumping, skipping, and hopping, these
muscles are used together. The hamstrings are used
without the aid of the gluteus maximus, however,
when hanging from a bar by the knees. Similarly,
the gluteus maximus is used without the aid of the
hamstrings when the knees are flexed while the
hips are actively extended. This occurs when rising
from a kneebend position to a standing position.

The biceps femoris is best developed through
hamstring curls as described for the semitendinosus,
but it is emphasized more if the knee is maintained
in external rotation throughout the range of motion,
which brings the origin and insertion more in line
with each other. The biceps femoris is stretched by
maximally extending the knee while flexing the
externally rotated and slightly adducted hip.

Knee
flexion

Hip
extension

Hip
external
rotation

Knee
external
rotation

Chapter
8

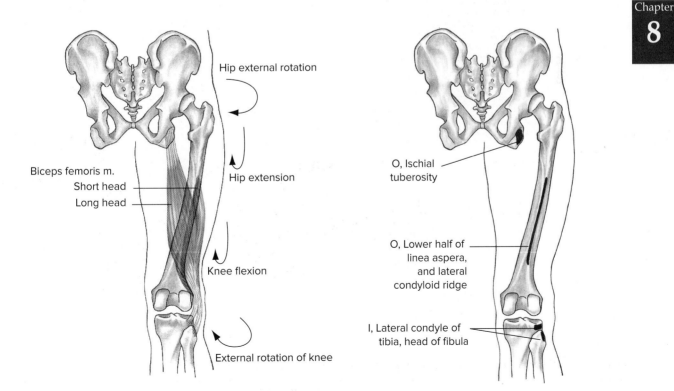

FIG. 8.34 • Biceps femoris muscle, posterior view. O, Origin; I, Insertion.

# Gluteus maximus muscle   FIG. 8.35

(gluˊte-us maksˊi-mus)

Hip extension

## Origin

Posterior one-fourth of the crest of the ilium, posterior surface of the sacrum and coccyx near the ilium, and fascia of the lumbar area

Hip external rotation

## Insertion

Oblique ridge (gluteal tuberosity) on the lateral surface of the greater trochanter and the iliotibial band of the fasciae latae

## Action

Extension of the hip
External rotation of the hip
Upper fibers: assist in hip abduction
Lower fibers: assist in hip adduction
Posterior pelvic rotation

Hip abduction

## Palpation

Running downward and laterally between posterior iliac crest superiorly, anal cleft medially, and gluteal fold inferiorly, emphasized with hip extension, external rotation, and abduction

Hip adduction

## Innervation

Inferior gluteal nerve (L5, S1, S2)

## Application, strengthening, and flexibility

The gluteus maximus muscle comes into action when movement between the pelvis and the femur approaches and goes beyond 15 degrees of extension. As a result, it is not used extensively in ordinary walking. It is important in extension of the thigh with external rotation.

Strong action of the gluteus maximus muscle is seen in running, hopping, skipping, and jumping. Powerful extension of the thigh is secured in the return to standing from a squatting position, especially with weighted barbells placed on the shoulders.

Hip extension exercises from a forward-leaning or prone position may be used to develop this muscle. This muscle is most emphasized when the hip starts from a flexed position and moves to full extension and abduction, with the knee flexed 30 degrees or more to reduce the hamstrings' involvement in the action.

The gluteus maximus is stretched in the supine position with full hip flexion to the ipsilateral axilla and then to the contralateral axilla with the knee in flexion. Simultaneous internal hip rotation accentuates this stretch.

Chapter
8

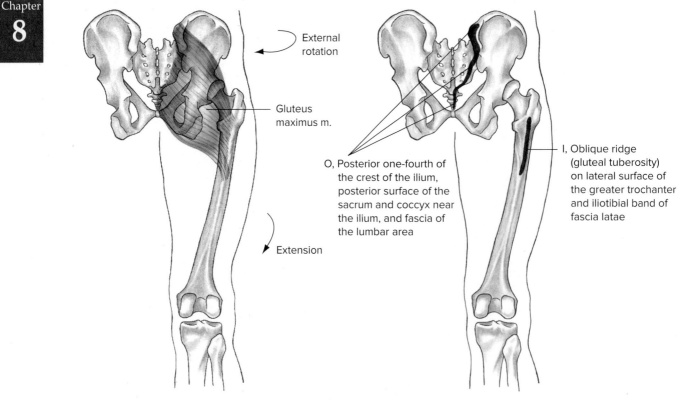

External rotation

Gluteus maximus m.

O, Posterior one-fourth of the crest of the ilium, posterior surface of the sacrum and coccyx near the ilium, and fascia of the lumbar area

Extension

I, Oblique ridge (gluteal tuberosity) on lateral surface of the greater trochanter and iliotibial band of fascia latae

FIG. 8.35 • Gluteus maximus muscle, posterior view. O, Origin; I, Insertion.

# Gluteus medius muscle    FIG. 8.36

(glu´te-us me´di-us)

## Origin

Lateral surface of the ilium just below the crest

## Insertion

Posterior and middle surfaces of the greater trochanter of the femur

## Action

Abduction of the hip

Lateral pelvic rotation to ipsilateral side

Anterior fibers: internal rotation, flexion of the hip, and anterior pelvic rotation

Posterior fibers: external rotation, extension of the hip, and posterior pelvic rotation

## Palpation

Slightly in front of and a few inches above the greater trochanter with active elevation of opposite pelvis from a standing position or active abduction when side-lying on contralateral pelvis

## Innervation

Superior gluteal nerve (L4, L5, S1)

## Application, strengthening, and flexibility

Typical action of the gluteus medius and gluteus minimus muscles is seen in walking. As the weight of the body is suspended on one leg, these muscles prevent the opposite pelvis from sagging. Weakness in the gluteus medius and gluteus minimus can result in a Trendelenburg gait, which is characterized by the trunk lurching to the side of the weakness when the contralateral pelvis drops. With this weakness, the individual's opposite pelvis will sag on weight bearing because the hip abductors on the weight-bearing side are not strong enough to maintain the opposite side at or near level.

Hip external rotation exercises performed against resistance can provide some strengthening of the gluteus medius, but it is best strengthened by performing side-lying leg raises or hip abduction exercises as described for the tensor fasciae latae. The gluteus medius is best stretched by moving the hip into extreme adduction in front of the opposite extremity and then behind it.

Hip abduction

Hip internal rotation

Hip flexion

Hip external rotation

Chapter 8

Hip extension

FIG. 8.36 • Gluteus medius muscle, posterior view. O, Origin; I, Insertion.

# Gluteus minimus muscle   FIG. 8.37

(glu´te-us min´i-mus)

### Origin

Lateral surface of the ilium just below the origin of the
gluteus medius

### Insertion

Anterior surface of the greater trochanter of the
femur

### Action

Abduction of the hip
Lateral pelvic rotation to ipsilateral side
Internal rotation as the femur abducts
Flexion of the hip
Anterior pelvic rotation

### Palpation

Deep to the gluteus medius; covered by tensor fasciae
latae between anterior iliac crest and greater trochanter during internal rotation and abduction

### Innervation

Superior gluteal nerve (L4, L5, S1)

## Application, strengthening, and flexibility

Both the gluteus minimus and the gluteus medius
are used in powerfully maintaining proper hip
abduction while running. As a result, both of
these muscles are exercised effectively in running,
hopping, and skipping, in which weight is transferred forcefully from one foot to the other. As the
body ages, the gluteus medius and gluteus minimus muscles tend to lose their effectiveness. The
spring of youth, as far as the hips are concerned,
resides in these muscles. To have great drive in
the legs, these muscles must be fully developed.

The gluteus minimus is best strengthened by
performing hip abduction exercises similar to those
described for the tensor fasciae latae and gluteus
medius muscles. It may also be developed by performing hip internal rotation exercises against
manual resistance. Stretching of this muscle is
accomplished by extreme hip adduction with slight
external rotation.

FIG. 8.37 • Gluteus minimus muscle, posterior view. O, Origin; I, Insertion.

# Tensor fasciae latae muscle   FIG. 8.38

(ten´sor fas´i-e la´te)

## Origin

Anterior iliac crest and surface of the ilium just below the crest

## Insertion

One-fourth of the way down the thigh into the iliotibial tract, which in turn inserts onto Gerdy's tubercle of the anterolateral tibial condyle

## Action

Abduction of the hip
Flexion of the hip
Lateral pelvic rotation to ipsilateral side
Tendency to rotate the hip internally as it flexes
Anterior pelvic rotation

## Palpation

Anterolaterally, between anterior iliac crest and greater trochanter during internal rotation, flexion, and abduction

## Innervation

Superior gluteal nerve (L4, L5, S1)

## Application, strengthening, and flexibility

The tensor fasciae latae muscle aids in preventing external rotation of the hip as it is flexed by other flexor muscles.

The tensor fasciae latae muscle is used when flexion and internal rotation take place. This is a weak movement but is important in helping direct the leg forward so that the foot is placed straight forward in walking and running. Thus, from the supine position, raising the leg with definite internal rotation of the femur will call it into action.

The tensor fasciae latae may be developed by performing hip abduction exercises against gravity and resistance while in a side-lying position. This is done simply by abducting the hip that is up and then slowly lowering it back to rest against the other leg. Stretch may be applied by remaining on the side and having a partner passively move the downside hip into full extension, adduction, and external rotation.

Hip abduction

Hip flexion

Hip internal rotation

Hip flexion

Tensor fasciae latae m.

Hip abduction

O, Anterior iliac crest and surface of ilium just below crest

I, One-fourth of the way down the thigh into the iliotibial tract, which in turn inserts onto Gerdy's tubercle of the anterolateral tibial condyle

FIG. 8.38 • Tensor fasciae latae muscle, anterior view. O, Origin; I, Insertion.

Chapter
8

# The six deep lateral rotator muscles—piriformis, gemellus superior, gemellus inferior, obturator externus, obturator internus, quadratus femoris   FIG. 8.39

(pi-ri-for´mis); (je-mel´us su-pe´ri-or); (je-mel´us in-fe´ri-or); (ob-tu-ra´tor eks-ter´nus); (ob-tu-ra´tor in-ter´nus); (kwad-ra´tus fem´or-is)

## Origin

Anterior sacrum, posterior portions of the ischium, and obturator foramen; Ischial spine; Ischial tuberosity; Margin of obturator foramen; Inferior margin of obturator foramen; Ischial tuberosity

## Insertion

Superior and posterior aspect of the greater trochanter; Posterior aspect of the greater trochanter immediately below piriformis; Posterior aspect of the greater trochanter with obturator internus; Posterior aspect of the greater trochanter with gemellus superior; Posterior aspect of the greater trochanter immediately below obturator internus; Intertrochanteric ridge of femur

## Action

External rotation of the hip

## Palpation

Although not directly palpable, deep palpation is possible between the posterior superior greater trochanter and obturator foramen with the subject prone during relaxation of the gluteus maximus while passively using the lower leg flexed at the knee to passively internally and externally rotate the femur or alternately contract/relax the external rotators slightly

## Innervation

Piriformis: first or second sacral nerve (S1, S2)
Gemellus superior: sacral nerve (L5, S1, S2)
Gemellus inferior: branches from sacral plexus (L4, L5, S1, S2)
Obturator externus: obturator nerve (L3, L4)
Obturator internus: branches from sacral plexus (L4, L5, S1, S2)
Quadratus femoris: branches from sacral plexus (L4, L5, S1)

## Application, strengthening, and flexibility

The six lateral rotators are used powerfully in movements of external rotation of the femur, as in sports in which the individual takes off on one leg from preliminary internal rotation. Throwing a baseball and swinging a baseball bat, in which there is rotation of the hip, are typical examples.

Standing on one leg and forcefully turning the body away from that leg is accomplished by contraction of these muscles, and this may be repeated for strengthening purposes. A partner may provide resistance as development progresses. The six deep lateral rotators may be stretched in the supine position with a partner passively internally rotating and slightly flexing the hip.

Of special note, the sciatic nerve usually passes just inferior to the piriformis muscle but may pass through it. As a result, tightness in the piriformis muscle may contribute to compression on the sciatic nerve. The piriformis may be stretched by having the subject lie on the uninvolved side with a partner passively taking the hip into full internal rotation combined with hip adduction and slight to moderate hip flexion.

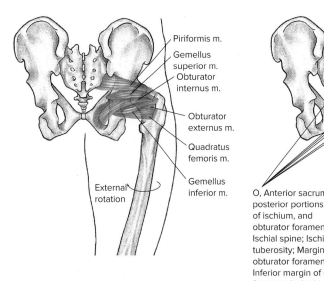

FIG. 8.39 • The six deep lateral rotator muscles, posterior view: piriformis, gemellus superior, gemellus inferior, obturator externus, obturator internus, and quadratus femoris. O, Origin; I, Insertion.

# REVIEW EXERCISES

1. List the planes in which each of the following hip joint movements occurs. List the respective axis of rotation for each movement in each plane.
   a. Flexion
   b. Extension
   c. Adduction
   d. Abduction
   e. External rotation
   f. Internal rotation
2. How is walking different from running in terms of the use of the hip joint muscle actions and the range of motion?
3. Research common hip disorders such as osteoarthritis, femoroacetabular impingement, snapping hip syndrome, groin strains, hamstring strains, greater trochanteric bursitis, and slipped capital femoral epiphysis. Report your findings in class.

4. **Muscle analysis chart** · Hip joint

| Complete the chart by listing the muscles primarily involved in each movement. | |
|---|---|
| Flexion | Extension |
| Abduction | Adduction |
| External rotation | Internal rotation |

5. **Antagonistic muscle action chart** · Hip joint and pelvic girdle

| Complete the chart by listing the muscle(s) or parts of muscles that are antagonist in their actions to the muscles in the left column. | |
|---|---|
| Agonist | Antagonist |
| Gluteus maximus | |
| Gluteus medius | |
| Gluteus minimus | |
| Biceps femoris | |
| Semimembranosus/Semitendinosus | |
| Adductor magnus/Adductor brevis | |
| Adductor longus | |
| Gracilis | |
| Lateral rotators | |
| Rectus femoris | |
| Sartorius | |
| Pectineus | |
| Iliopsoas | |
| Tensor fasciae latae | |

# LABORATORY EXERCISES

1. Locate the following parts of the pelvic girdle and hip joint on a human skeleton and on a subject:
   a. Skeleton
      1. Ilium
      2. Ischium
      3. Pubis
      4. Symphysis pubis
      5. Acetabulum
      6. Rami (ascending and descending)
      7. Obturator foramen
      8. Ischial tuberosity
      9. Anterior superior iliac spine
      10. Posterior superior iliac spine
      11. Greater trochanter
      12. Lesser trochanter
   b. Subject
      1. Crest of ilium
      2. Anterior superior iliac spine
      3. Posterior superior iliac spine
      4. Ischial tuberosity
      5. Greater trochanter

2. How and where can the following muscles be palpated on a human subject?
   a. Gracilis
   b. Sartorius
   c. Gluteus maximus
   d. Gluteus medius
   e. Gluteus minimus
   f. Biceps femoris
   g. Rectus femoris
   h. Semimembranosus
   i. Semitendinosus
   j. Adductor magnus
   k. Adductor longus
   l. Adductor brevis

3. Be prepared to indicate on a human skeleton, using a long rubber band, where each muscle has its origin and insertion.

4. Distinguish between hip flexion and trunk flexion by performing each individually and then both together.

5. Demonstrate the movement and list the muscles primarily responsible for the following hip movements:
   a. Flexion
   b. Extension
   c. Adduction
   d. Abduction
   e. External rotation
   f. Internal rotation

6. How may the walking gait be affected by a weakness in the gluteus medius muscle? Have a laboratory partner demonstrate the gait pattern associated with gluteus medius weakness. What is the name of this dysfunctional gait?

7. How might bilateral iliopsoas tightness affect the posture and movement of the lumbar spine in the standing position? Demonstrate and discuss this effect with a laboratory partner.

8. How might bilateral hamstring tightness affect the posture and movement of the lumbar spine in the standing position? Demonstrate and discuss this effect with a laboratory partner.

Chapter
8

9. Hip joint exercise movement analysis chart

After analyzing each exercise in the chart, break each into two primary movement phases, such as a lifting phase and a lowering phase. For each phase, determine which hip joint movements occur, and then list the hip joint muscles primarily responsible for causing/controlling those movements. Beside each muscle in each movement, indicate the type of contraction as follows: I—isometric; C—concentric; E—eccentric.

| Exercise | Initial movement (lifting) phase | | Secondary movement (lowering) phase | |
|---|---|---|---|---|
| | Movement(s) | Agonist(s)–(contraction type) | Movement(s) | Agonist(s)–(contraction type) |
| Push-up | | | | |
| Squat | | | | |
| Dead lift | | | | |
| Hip sled | | | | |
| Forward lunge | | | | |
| Rowing exercise | | | | |
| Stair machine | | | | |

## 10. Hip joint sport skill analysis chart

Analyze each skill in the chart, and list the movements of the right and left hip joints in each phase of the skill. You may prefer to list the initial position the hip joint is in for the stance phase. After each movement, list the hip joint muscle(s) primarily responsible for causing/controlling the movement. Beside each muscle in each movement, indicate the type of contraction as follows: I—isometric, C—concentric; E—eccentric. It may be desirable to review the concepts for analysis in Chapter 12 for the various phases.

| Exercise | | Stance phase | Preparatory phase | Movement phase | Follow-through phase |
|---|---|---|---|---|---|
| Baseball pitch | (R) | | | | |
| | (L) | | | | |
| Football punting | (R) | | | | |
| | (L) | | | | |
| Walking | (R) | | | | |
| | (L) | | | | |
| Softball pitch | (R) | | | | |
| | (L) | | | | |
| Soccer pass | (R) | | | | |
| | (L) | | | | |
| Batting | (R) | | | | |
| | (L) | | | | |
| Bowling | (R) | | | | |
| | (L) | | | | |
| Basketball jump shot | (R) | | | | |
| | (L) | | | | |

Chapter
**8**

# References

Cerezal L, Kassarjian A, Canga A, et al. Anatomy, biomechanics, imaging, and management of ligamentum teres injuries. *Radiographics: Review Publication Radiological Society of North America* 30:1637–1651, 2010.

Cleland JA, Koppenhaver S, Su J: *Netter's orthopaedic clinical examination: an evidence-based approach*, ed 3, Philadelphia, 2016, Elsevier.

Field D: *Anatomy: palpation and surface markings*, ed 3, Oxford, 2001, Butterworth-Heinemann.

Hamilton N, Weimer W, Luttgens K: *Kinesiology: scientific basis of human motion*, ed 12, New York, 2012, McGraw-Hill.

Hislop HJ, Montgomery J: *Daniels and Worthingham's muscle testing: techniques of manual examination*, ed 8, Philadelphia, 2007, Saunders.

Houglum PA, Bertoti DB: *Brunstom's clinical kinesiology*, ed 6, Philadelphia, 2012, FA Davis.

Kendall FP, McCreary EK, Provance PG, Rodgers MM, Romani WA: *Muscles: testing and function, with posture and pain*, ed 5, Baltimore, 2005, Lippincott Williams & Wilkins.

Lindsay DT: *Functional human anatomy*, St. Louis, 1996, Mosby.

Lysholm J, Wikland J: Injuries in runners, *American Journal of Sports Medicine* 15:168, September–October 1986.

Magee DJ: *Orthopedic physical assessment*, ed 6, St. Louis, 2014, Elsevier Saunders.

Muscolino JE: *The muscular system manual: the skeletal muscles of the human body*, ed 3, St. Louis, 2010, Elsevier Mosby.

Nam D, Osbahr DC, Choi D, Ranawat AS, Kelly BT, Coleman SH: Defining the origins of the iliofemoral, ischiofemoral, and pubofemoral ligaments of the hip capsuloligamentous complex utilizing computer navigation. *HSS Journal,* 7(3):239–243, 2011.

Noahes TD, et al: Pelvic stress fractures in long distance runners, *American Journal of Sports Medicine* 13:120, March–April 1985.

Oatis CA: *Kinesiology: the mechanics and pathomechanics of human movement*, ed 2, Philadelphia, 2008, Lippincott Williams & Wilkins.

Prentice WE: *Principles of athletic training: a competency-based approach*, ed 15, New York, 2014, McGraw-Hill.

Putz R, Pabst R: *Sobotta - atlas of human anatomy one volume edition*, ed 14, Germany, 2008, Elsevier.

Saladin KS: *Anatomy & physiology: the unity of form and function*, ed 5, New York, 2010, McGraw-Hill.

Schünke M, Schulte E, Schumacher U: *Thieme atlas of anatomy: general anatomy and musculoskeletal system*, ed 2, New York, 2014, Thieme Medical.

Seeley RR, Stephens TD, Tate P: *Anatomy & physiology*, ed 8, New York, 2008, McGraw-Hill.

Seeley RR, VanPutte CL, Regan JL, Russo AF: *Seeley's anatomy & physiology*, ed 9, New York, 2011, McGraw-Hill.

Shier D, Butler J, Lewis R: *Hole's human anatomy and physiology*, ed 12, New York, 2010, McGraw-Hill.

Sieg KW, Adams SP: *Illustrated essentials of musculoskeletal anatomy*, ed 4, Gainesville, FL, 2002, Megabooks.

Stone RJ, Stone JA: *Atlas of the skeletal muscles*, ed 6, New York, 2009, McGraw-Hill.

Thibodeau GA, Patton KT: *Anatomy & physiology*, ed 9, St. Louis, 1993, Mosby.

Van De Graaff KM: *Human anatomy*, ed 6, Dubuque, IA, 2002, McGraw-Hill.

Chapter

8

# Worksheet Exercises

For in- or out-of-class assignments, or for testing, utilize this tear-out worksheet.

## Worksheet 1

Using crayons or colored markers, draw and label on the worksheet the following muscles. Indicate the origin and insertion of each muscle with an "O" and an "I," respectively, and draw in the origin and insertion on the contralateral side of the skeleton.

a. Iliopsoas

b. Rectus femoris

c. Sartorius

d. Pectineus

e. Adductor brevis

f. Adductor longus

g. Adductor magnus

h. Gracilis

# Worksheet Exercises

For in- or out-of-class assignments, or for testing, utilize this tear-out worksheet.

## Worksheet 2

Using crayons or colored markers, draw and label on the worksheet the following muscles. Indicate the origin and insertion of each muscle with an "O" and an "I," respectively, and draw in the origin and insertion on the contralateral side of the skeleton.

- a. Semitendinosus
- b. Semimembranosus
- c. Biceps femoris
- d. Gluteus maximus
- e. Gluteus medius
- f. Gluteus minimis
- g. Tensor fasciae latae

# THE KNEE JOINT

## Objectives

- To identify on a human skeleton selected bony features of the knee

- To appreciate and explain the role of the cartilaginous and ligamentous structures of the knee joint in providing stability

- To draw and label on a skeletal chart muscles and ligaments of the knee joint

- To palpate the superficial knee joint structures and muscles on a human subject

- To demonstrate and palpate with a fellow student all the movements of the knee joint and list their respective planes of motion and axes of rotation

- To name and explain the actions and importance of the quadriceps and hamstring muscles

- To list and organize the muscles that produce the movements of the knee joint and list their antagonists

- To learn and understand the innervation of the knee joint muscles

- To determine, through analysis, the knee movements and muscles involved in selected skills and exercises

The knee joint is the largest diarthrodial joint in the body and is very complex. It is primarily a hinge joint. The combined functions of weight bearing and locomotion place considerable stress, strain, compression, and torsion on the knee joint. Powerful knee joint extensor and flexor muscles, combined with a strong ligamentous structure, provide a strong functioning joint in most instances.

## Bones   FIG. 9.1

The enlarged femoral condyles articulate on the enlarged condyles of the tibia, somewhat in a horizontal line. Because the femur projects downward at an oblique angle toward the midline, its medial condyle is slightly larger than the lateral condyle.

The top of the medial and lateral tibial condyles, known as the medial and lateral tibial plateaus, serve as receptacles for the femoral condyles. The tibia is the medial bone in the leg and bears much more of the body's weight than the fibula. The fibula serves as the attachment for some very important knee joint structures, although it does not articulate with the femur or patella and is not part of knee joint articulation.

The patella is a sesamoid (floating) bone contained within the quadriceps muscle group and the patellar tendon. Its location allows it to serve the quadriceps in a fashion similar to the work of a pulley by creating an improved angle of pull. This results in a greater mechanical advantage when performing knee extension.

Key bony landmarks of the knee include the superior and inferior poles of the patella, the tibial tuberosity, Gerdy's tubercle, the medial and lateral femoral condyles, the upper anterior medial surface of the tibia, and the head of the fibula. The three vasti muscles of the quadriceps originate on the proximal femur and insert along with the rectus

Chapter
**9**

Femur shaft

Adductor tubercle
Lateral epicondyle
Patellar surface
Base of patella
Apex of patella
Intercondylar eminence
Gerdy's tubercle
Proximal tibiofibular joint
Lateral surface
Fibula

Medial epicondyle
Medial condyle
Medial condyle
Tibial tuberosity
Anterior crest
Tibia

Adductor tubercle
Lateral epicondyle
Lateral condyle
Intercondylar fossa
Articular facets
Lateral condyle
Apex
Head of fibula
Fibula

A                    B

FIG. 9.1 • Bones of the right knee—femur, patella, tibia, and fibula. **A,** Anterior view; **B,** Posterior view.

femoris on the superior pole of the patella. Their specific insertion into the patella varies in that the vastus medialis and vastus lateralis insert into the patella from a superomedial and superolateral angle, respectively. The superficial rectus femoris and the vastus intermedius, which lies directly beneath it, both attach to the patella from the superior direction. From here their insertion is ultimately on the tibial tuberosity by way of the large patellar tendon, which runs from the inferior patellar pole to the tibial tuberosity. Gerdy's tubercle, located on the anterolateral aspect of the lateral tibial condyle, is the insertion point for the iliotibial tract of the tensor fasciae latae.

The upper anteromedial surface of the tibia just below the medial condyle serves as the insertion for the sartorius, gracilis, and semitendinosus. The semimembranosus inserts posteromedially on the medial tibial condyle. The head of the fibula is the primary location of the biceps femoris insertion, although some of its fibers insert on the lateral tibial condyle. The popliteus origin is located on the lateral aspect of the lateral femoral condyle.

Additionally, the tibial collateral ligament originates on the medial aspect of the upper medial femoral condyle and inserts on the medial surface of the tibia. Laterally, the shorter fibula collateral originates on the lateral femoral condyle very close to the popliteus origin and inserts on the head of the fibula.

# Joints FIGS. 9.2, 9.3

The knee joint proper, or tibiofemoral joint, is classified as a ginglymus joint because it functions like a hinge. It moves between flexion and extension without side-to-side movement into abduction or adduction. However, it is sometimes referred to as a trochoginglymus joint because of the internal and external rotation movements that can occur during flexion. Some authorities argue that it should be classified as a condyloid or "double condyloid" joint due to its bicondylar structure. The patellofemoral joint is classified as an arthrodial joint due to the gliding nature of the patella on the femoral condyles.

The ligaments provide static stability to the knee joint, and contractions of the quadriceps and hamstrings produce dynamic stability. The surfaces between the femur and tibia are protected by articular cartilage, as is true of all diarthrodial joints. In addition to the articular cartilage covering the ends of the bones, specialized cartilages (see Fig. 9.2), known as the menisci, form cushions between the bones. These menisci are attached to the tibia and deepen the tibial plateaus, thereby enhancing stability.

The medial semilunar cartilage, or, more technically, the medial meniscus, is located on the medial tibial plateau to form a receptacle for the medial femoral condyle. The lateral semilunar cartilage (lateral meniscus) sits on the lateral tibial plateau to receive the lateral femoral condyle. Both of these menisci are thicker on the outside border and taper down to be very thin on the inside border. They can slip about slightly and are held in place by various small ligaments. The medial meniscus is the larger of the two and has a much more open C appearance than the rather closed C configuration of the lateral meniscus. Either or both of the menisci may be torn in several different areas from a variety of mechanisms, resulting in varying degrees of severity and problems. These injuries often occur due to the significant compression and shear forces that develop as the knee rotates while flexing or extending during quick directional changes in running.

Two very important ligaments of the knee are the anterior and posterior cruciate ligaments, so named because they cross within the knee between the tibia and the femur. These ligaments are vital in maintaining the anterior and posterior stability of the knee joint, respectively, as well as its rotatory stability (see Fig. 9.2).

The anterior cruciate ligament (ACL) tear is one of the most common serious injuries to the knee and has been shown to be significantly more common in females than males during similar sports such as basketball and soccer. The mechanism of this injury often involves noncontact rotary forces associated with planting, cutting, and landing in a valgus position. Studies have also shown that the ACL may be disrupted in a hyperextension mechanism or solely by a violent contraction of the quadriceps that pulls the tibia forward on the femur. Recent studies suggest that ACL injury prevention programs incorporating detailed conditioning exercises and techniques designed to improve neuromuscular coordination and control among the hamstrings and quadriceps, maintain proper knee alignment, and utilize proper landing techniques may be effective in reducing the likelihood of injury.

Fortunately, the posterior cruciate ligament (PCL) is not often injured. Injuries of the posterior cruciate usually come about through direct contact with an opponent or with the playing surface. Many of the PCL injuries that do occur are partial tears with minimal involvement of other knee structures. In many cases, even with complete tears, athletes may remain fairly competitive at a high level after a brief nonsurgical treatment and rehabilitation program.

On the medial side of the knee is the tibial (medial) collateral ligament (MCL; see Fig. 9.2), which maintains medial stability by resisting valgus forces or preventing the knee joint from being abducted. Injuries to the tibial collateral occur quite commonly, particularly in contact or collision sports in which a teammate or an opponent falls against the lateral aspect of the knee or leg, causing medial opening of the knee joint and/or external rotation resulting in stress to the medial ligamentous structures. Its deeper fibers are attached to the medial meniscus, which may be affected with injuries to the ligament.

On the lateral side of the knee, the fibular (lateral) collateral ligament (LCL) joins the fibula and the femur. Injuries to this ligament are infrequent.

In addition to the four major ligaments which have been discussed, there are numerous other ligaments, most of which are detailed in Fig. 9.2 and listed in Table 9.1. These ligaments are generally of lesser importance, not typically injured significantly in isolation, and will not be discussed further.*

The knee joint (see Fig. 9.3) is well supplied with synovial fluid from the synovial cavity, which lies under the patella and between the surfaces of the tibia and the femur. Commonly, this synovial cavity

Chapter

**9**

---

*More detailed discussion of the knee is found in anatomy texts and athletic training manuals.

**A**      **Anterior view with patella removed**

**B**      **Posterior view**

**C**      **Superior view with femur removed**

FIG. 9.2 • Ligaments and menisci of the right knee. **A,** Anterior view with patella removed; **B,** Posterior view; **C,** Superior view with femur removed.

## TABLE 9.1 • Major ligaments of the knee joint

| Joint/Structure | Ligament | Proximal attachments | Distal attachments | Functional application and joint stability |
|---|---|---|---|---|
| Tibiofemoral joint | Tibial (medial) collateral ligament (MCL) | Medial epicondyle of femur just below adductor tubercle | Medial surface of the medial condyle of the tibia deep to pes anserinus | Composed of a superficial and deep layer, provides medial stability against valgus stress and external rotation |
| | Fibular (lateral) collateral ligament (LCL) | Lateral epicondyle of femur | Head of fibula | Extracapsular ligament; not part of the fibrous capsule of the knee; passes superficial to the popliteus tendon origin; provides lateral stability against varus stress |
| | Anterior cruciate ligament (ACL) | Medial surface of the lateral femoral condyle | Anterior aspect of the intercondylar eminence of the tibia | Limits the anterior displacement of the tibia relative to the femur in all positions of joint movement and limits excessive rotation |
| | Posterior cruciate ligament (PCL) | Anterolateral surface of the medial femoral condyle in intercodylar notch | Posterior aspect of the tibia approximately 1 cm below joint line | Limits posterior displacement of the tibia relative to the femur in all positions of joint movement |
| | Posterior Oblique Ligament (POL) | Adductor tubercle of femur | Posteromedial tibia and posterior aspect of the joint capsule | A thickening of the medial capsular ligament which provides posteromedial stability against valgus stress; prevents posteromedial tibial translation |
| | Anterior meniscofemoral ligament (Ligament of Humphrey) | Posterior horn of the lateral meniscus | Lateral surface of the medial femoral condyle | Passes anterior and medial to posterior cruciate ligament; stabilizes and protects the posterolateral corner of the knee and posterior horn of lateral meniscus |
| | Posterior meniscofemoral ligament (Ligament of Wrisberg) | Posterior part of the lateral meniscus | Lateral surface of the medial condyle of femur | Passes superior and medial behind the posterior cruciate ligament; stabilizes and protects the posterolateral corner of the knee and posterior horn of lateral meniscus |
| | Arcuate popliteal ligament | Lateral femoral epicondyle and oblique popliteal ligament | Posterior aspect of fibular head | Y-shaped; stabilizes the posterolateral corner of the knee joint, prevents varus hyperextension |
| | Transverse ligament (anterior meniscomeniscal) | Anterior horn of the medial meniscus | Anterior convex margin of the lateral meniscus | Prevents anterior displacement of medial meniscus |
| Patellofemoral joint | Medial patellofemoral ligament | Arises from the capsular fibers of the vastus medialis obliquus muscle at the adductor tubercle | Superomedial aspect of the patella | Maintain the patella within the trochlea groove, prevents lateral subluxation |
| Proximal tibiofibular joint | Anterior ligament of fibular head | Head of fibula | Lateral condyle of tibia | Two to three broad, flat bands, which pass obliquely downward from the anterior aspect of lateral tibial condyle to the fibular head; stabilizes the proximal tibiofibular joint |
| | Posterior ligament of fibular head | Head of fibula | Lateral condyle of tibia | One to two thick broad bands which run obliquely inferolateral from posterior lateral tibial condyle to posterior head of the fibula; reinforced by the popliteus tendon; stabilize the proximal tibiofibular joint |

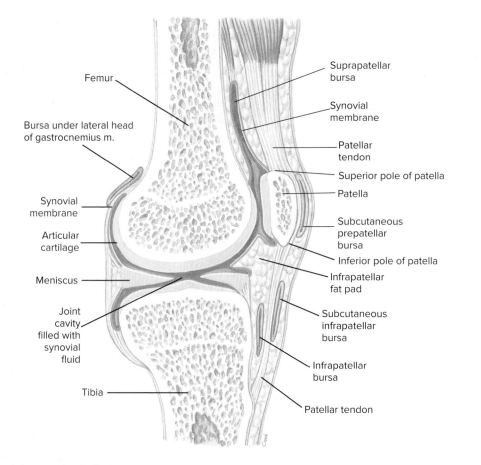

FIG. 9.3 • Knee joint, sagittal view.

is called the capsule of the knee. Just posterior to the patellar tendon is the infrapatellar fat pad, which is often an insertion point for synovial folds of tissue known as **plica**. A plica is an anatomical variant among some individuals that may be irritated or inflamed with injuries or overuse of the knee. More than 10 bursae are located around the knee, some of which are connected to the synovial cavity. Bursae are located where they can absorb shock or reduce friction.

The knee can usually extend to 180 degrees, or a straight line, although it is not uncommon for some knees to hyperextend up to 10 degrees or more. When the knee is in full extension, it can move from there to about 150 degrees of flexion. With the knee flexed 30 degrees or more, approximately 30 degrees of internal rotation and 45 degrees of external rotation can occur (Fig. 9.4).

Due to the shape of the medial femoral condyle, the knee must "screw home" to fully extend. As the knee approaches full extension, the tibia must externally rotate approximately 10 degrees to achieve proper alignment of the tibial and femoral condyles. In full extension, due to the

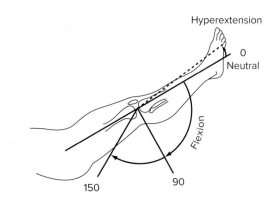

FIG. 9.4 • Active motion of the knee. Flexion is measured in degrees from the zero starting position, which is an extended straight leg with the subject either prone or supine. Hyperextension is measured in degrees opposite the zero starting point.

close congruency of the articular surfaces, there is no appreciable rotation of the knee. During initial flexion from a fully extended position, the knee "unlocks" by the tibia's rotating internally, to a degree, from its externally rotated position to achieve flexion.

## Movements FIG. 9.5

Flexion and extension of the knee occur in the sagittal plane, whereas internal and external rotation occur in the horizontal plane. The knee will not allow rotation unless it is flexed 20 to 30 degrees or more.

**Flexion:** bending or decreasing the angle between the femur and the leg; characterized by the heel moving toward the buttocks

**Extension:** straightening or increasing the angle between the femur and the leg

**External rotation:** rotary movement of the leg laterally away from the midline

**Internal rotation:** rotary movement of the leg medially toward the midline

## Muscles FIG 8.12

Some of the muscles involved in knee joint movements were discussed in Chapter 8 because of their biarticular arrangement with the hip and the knee joints. These will not be covered fully in this chapter. The knee joint muscles that have already been addressed are

**Knee extensor:** rectus femoris

**Knee flexors:** sartorius, biceps femoris, semitendinosus, semimembranosus, and gracilis

The gastrocnemius muscle, discussed in Chapter 10, also assists minimally with knee flexion.

The muscle group that extends the knee is located in the anterior compartment of the thigh and is known as the quadriceps. It consists of four muscles: the rectus femoris, the vastus lateralis, the vastus intermedius, and the vastus medialis. All four muscles work together to pull the patella superiorly, which in turn pulls the leg into extension at the knee by its attachment to the tibial tuberosity via the patellar tendon.

The central line of pull for the entire quadriceps runs from the anterior superior iliac spine (ASIS) to the center of the patella. The line of pull of the patellar tendon runs from the center of the patella to the center of the tibial tuberosity. The angle formed by the intersection of these two lines at the patella is known as the **Q angle** or quadriceps angle (Fig. 9.6). Normally, in the anatomical position, this angle will be 15 degrees or less for males and 20 degrees or less for females. Generally, females have higher angles due to a wider pelvis. Dynamic Q angles vary significantly during planting and cutting activities. Higher Q angles generally predispose people, in varying degrees, to a variety of potential knee problems, including patellar subluxation or dislocation, patellar compression syndrome, chondromalacia, and ligamentous injuries.

The hamstring muscle group is located in the posterior compartment of the thigh and is responsible for knee flexion. The hamstrings consist of three muscles: the semitendinosus, the semimembranosus, and the biceps femoris. The semimembranosus

Knee flexion

Knee extension

Knee external rotation

Knee internal rotation

| Flexion | Extension | Internal rotation | External rotation |
| A | B | C | D |

Courtesy of Britt Jones (all)

FIG. 9.5 • Movements of the right knee.

and semitendinosus muscles (medial hamstrings) are assisted by the popliteus in internally rotating the knee, whereas the biceps femoris (lateral hamstring) is responsible for knee external rotation. The tensor fascia latae insertion is anterolaterally on Gerdy's tubercle which places its line of pull across the knee from a posterolateral aspect, particularly when the knee is flexed. As a result, some view it as contributing to a very minor degree to knee external rotation, but most anatomists do not acknowledge it for this action.

Two-joint muscles are most effective when either the origin or the insertion is stabilized to prevent movement in the direction of the muscle when it contracts. Additionally, muscles are able to exert greater force when lengthened than when shortened. All the hamstring muscles, as well as the rectus femoris, sartorius, and gracilis, are biarticular (two-joint) muscles.

As an example, the sartorius muscle becomes more effective in flexing at the knee when the pelvis is rotated posteriorly and stabilized by the abdominal muscles, thus increasing its total length by moving its origin farther from its insertion. This is exemplified by trying to flex the knee and cross the legs in the sitting position. One usually leans backward to flex the legs at the knees. This is also illustrated by kicking a football. The kicker invariably leans well backward to raise and fix the origin of the rectus femoris muscle to make it more effective as an extensor of the leg at the knee. And when youngsters hang by the knees, they flex the hips to fix or raise the origin of the hamstrings to make the latter more effective flexors of the knees.

The sartorius, gracilis, and semitendinosus all join together distally to form a tendinous expansion known as the **pes anserinus**, which attaches to the anteromedial aspect of the proximal tibia below the level of the tibial tuberosity. This attachment and the line of pull these muscles have posteromedially to the knee enable them to assist with knee flexion, particularly once the knee is flexed and the hip is externally rotated. The medial and lateral heads of the gastrocnemius attach posteriorly on the medial and lateral femoral condyles, respectively. This relationship to the knee provides the gastrocnemius with a line of pull. As a result, it may assist to minor degree with knee flexion. In a similar manner, the plantaris originates on the inferior aspect of the lateral supracondylar ridge of the distal femur just medial to the lateral gastrocnemius and then runs deep to it, but superficial to the soleus, to insert on the posterior calcaneus medial to the Achilles.

### Knee joint muscles—location

Muscle location is closely related to muscle function with the knee. While viewing the muscles, in Fig. 8.12, correlate them with Table 9.2.

Anterior
    Primarily knee extension
        Rectus femoris*
        Vastus medialis
        Vastus intermedius
        Vastus lateralis

Posterior
    Primarily knee flexion
        Biceps femoris*
        Semimembranosus*
        Semitendinosus*
        Sartorius*
        Gracilis*
        Popliteus
        Gastrocnemius*
        Plantaris*

FIG. 9.6 • Q angle, represented by the angle between the line from the anterior superior iliac spine to the central patella and the line from the central patella to the tibial tuberosity.

*Two-joint muscles; hip actions are discussed in Chapter 8, and ankle actions are discussed in Chapter 10.

TABLE 9.2 · **Agonist muscles of the knee joint**

| | Muscle | Origin | Insertion | Action | Plane of motion | Palpation | Innervation |
|---|---|---|---|---|---|---|---|
| Anterior muscles | Rectus femoris | Anterior inferior iliac spine of the ilium and groove (posterior) above the acetabulum | Superior aspect of the patella and patellar tendon to tibial tuberosity | Extension of the knee / Flexion of the hip / Anterior pelvic rotation | Sagittal | Straight down anterior thigh from anterior inferior iliac spine to patella with resisted hip flexion/knee extension | Femoral nerve (L2–L4) |
| | Vastus intermedius | Upper two-third of anterior surface of femur | Upper border of the patella and patellar tendon to tibial tuberosity | Extension of the knee | Sagittal | Anteromedial distal 1/3 of thigh just above the superomedial patella and deep to the rectus femoris, with extension of the knee, particularly full extension against resistance | Femoral nerve (L2–L4) |
| | Vastus lateralis (externus) | Intertrochanteric line, anterior and inferior borders of the greater trochanter, gluteal tuberosity, upper half of the linea aspera, and entire lateral intermuscular septum | Lateral border of the patella and patellar tendon to tibial tuberosity | Extension of the knee | Sagittal | Slightly distal to greater trochanter down the anterolateral aspect of the thigh to the superolateral patella, with extension of the knee, particularly full extension against resistance | Femoral nerve (L2–L4) |
| | Vastus medialis (internus) | Whole length of linea aspera and medial condyloid ridge | Medial half of upper border of patella and patellar tendon to tibial tuberosity | Extension of the knee | Sagittal | Anteromedial side of the thigh just above the superomedial patella, with extension of the knee, particularly full extension against resistance | Femoral nerve (L2–L4) |

Chapter
**9**

TABLE 9.2 (continued) • **Agonist muscles of the knee joint**

| Muscle | Origin | Insertion | Action | Plane of motion | Palpation | Innervation |
|---|---|---|---|---|---|---|
| **Posterior muscles** | | | | | | |
| Biceps femoris | Long head: ischial tuberosity. Short head: lower half of the linea aspera and lateral condyloid ridge | Head of the fibula and lateral condyle of the tibia | Flexion of the knee | Sagittal | Posterolateral aspect of distal thigh with combined knee flexion and external rotation against resistance; just distal to the ischial tuberosity in a prone position with hip internally rotated during active knee flexion | Long head: sciatic nerve—tibial division (S1–S3) Short head: sciatic nerve—peroneal division (L5, S1, S2) |
| Biceps femoris | | | Extension of the hip | Sagittal | | |
| Biceps femoris | | | Posterior pelvic rotation | Sagittal | | |
| Biceps femoris | | | External rotation of the knee | Transverse | | |
| Biceps femoris | | | External rotation of the hip | Transverse | | |
| Popliteus | Posterior surface of lateral condyle of femur | Upper posterior medial surface of tibia | Internal rotation of the knee as it flexes | Transverse | With subject sitting, knee flexed 90 degrees, palpate deep to the gastrocnemius medially on the posterior proximal tibia and proceed superolaterally toward lateral epicondyle of tibia just deep to fibular collateral ligament, while subject internally rotates knee | Tibial nerve (L5, S1) |
| Popliteus | | | Flexion of the knee | Sagittal | | |
| Semi-membranosus | Ischial tuberosity | Postero-medial surface of the medial tibial condyle | Extension of the hip | Sagittal | Largely covered by other muscles, tendon can be felt at posteromedial aspect of knee just deep to semitendinosus tendon with combined knee flexion and internal rotation against resistance | Sciatic nerve—tibial division (L5, S1, S2) |
| Semi-membranosus | | | Flexion of the knee | Sagittal | | |
| Semi-membranosus | | | Posterior pelvic rotation | Sagittal | | |
| Semi-membranosus | | | Internal rotation of the hip | Transverse | | |
| Semi-membranosus | | | Internal rotation of the knee | Transverse | | |
| Semi-tendinosus | Ischial tuberosity | Upper anterior medial surface of the tibia just below the condyle | Extension of the hip | Sagittal | Posteromedial aspect of the distal thigh with combined knee flexion and internal rotation against resistance; just distal to ischial tuberosity in a prone position with hip internally rotated during active knee flexion | Sciatic nerve—tibial division (L5, S1, S2) |
| Semi-tendinosus | | | Flexion of the knee | Sagittal | | |
| Semi-tendinosus | | | Posterior pelvic rotation | Sagittal | | |
| Semi-tendinosus | | | Internal rotation of the hip | Transverse | | |
| Semi-tendinosus | | | Internal rotation of the knee | Transverse | | |

**Note:** The sartorius and gracilis assist, although not primarily, with knee flexion and internal rotation, and are discussed in detail in Chapter 8. Additionally, a few anatomists include the tensor fascia latae as providing minor assistance with knee external rotation. The gastrocnemius and plantaris, discussed in Chapter 10, assist to some degree with knee flexion.

Chapter 9

## Nerves

The femoral nerve (Fig. 8.20) innervates the knee extensors—rectus femoris, vastus medialis, vastus intermedius, and vastus lateralis. The knee flexors, consisting of the semitendinosus, semimembranosus, biceps femoris (long head), and popliteus, are innervated by the tibial division of the sciatic nerve (Fig. 8.22). The biceps femoris short head is supplied by the peroneal nerve (Fig. 10.10). See Fig. 9.7 for a full listing of the knee muscular innervations.

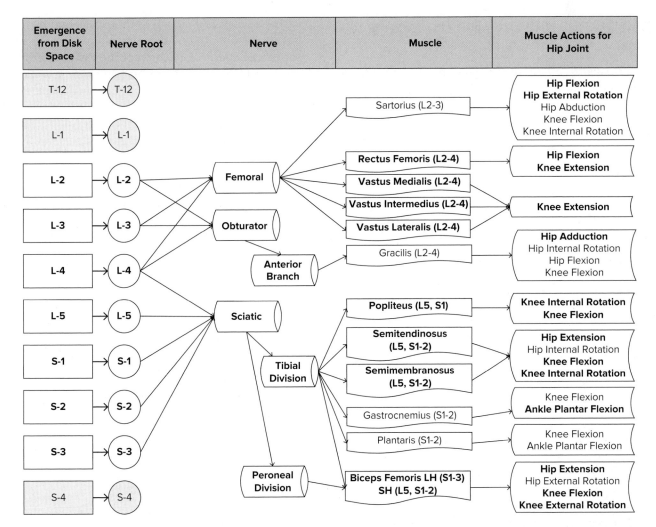

FIG. 9.7 • Innervation for knee joint muscles and their actions. The neuromuscular structures associated with the primary movers for the knee joint and their actions are bolded.

Chapter

**9**

275

## Quadriceps muscles   FIG. 9.8

(kwod ´ri-seps)

The ability to jump is essential in nearly all sports. Individuals who have good jumping ability always have strong quadriceps muscles that extend the leg at the knee. The quadriceps function as a decelerator when it is necessary to decrease speed for changing direction or to prevent falling when landing. This deceleration function is also evident in stopping the body when coming down from a jump. The contraction that occurs in the quadriceps during braking or decelerating actions is eccentric. This eccentric action of the quadriceps controls the slowing of movements initiated in previous phases of the sport skill.

The muscles are the rectus femoris (the only biarticular muscle of the group), vastus lateralis (the largest muscle of the group), vastus intermedius, and vastus medialis. All attach to the patella and by the patellar tendon to the tuberosity of the tibia. All are superficial and palpable, except the vastus intermedius, which is under the rectus femoris. The vertical jump is a simple test that may be used to indicate the strength or power of the quadriceps. It is generally desired that this muscle group be 25% to 33% stronger than the hamstring muscle group (knee flexors).

Development of the strength and endurance of the quadriceps, or "quads," is essential for maintenance of patellofemoral stability, which is often a problem in many physically active individuals. This problem is exacerbated by the quads' being particularly prone to atrophy when injuries occur. The muscles of the quadriceps may be developed by resisted knee extension activities from a seated position; however, full-range knee extensions may be contraindicated with certain patellofemoral conditions. Performing functional weight-bearing activities such as step-ups or squats is particularly useful for strengthening and endurance.

Rectus femoris

Vastus intermedius m.

Vastus lateralis m.

Vastus medialis m.

Tibial tuberosity

FIG. 9.8 •
Quadriceps muscle
group, anterior
view.

## Rectus femoris muscle   FIG. 8.25

(rek ´tus fem ´o-ris)

### Origin

Anterior inferior iliac spine of the ilium and groove (posterior) above the acetabulum

### Insertion

Superior aspect of the patella and patellar tendon to the tibial tuberosity

### Action

Flexion of the hip
Extension of the knee
Anterior pelvic rotation

### Palpation

Straight down anterior thigh from anterior inferior iliac spine to patella, with resisted knee extension and hip flexion

### Innervation

Femoral nerve (L2–L4)

### Application, strengthening, and flexibility

When the hip is flexed, the rectus femoris becomes shorter, which reduces its effectiveness as an extensor of the knee. The work is then done primarily by the three vasti muscles.

Also see the rectus femoris discussion in Chapter 8, p. 244 (Fig. 8.25).

# Vastus lateralis (externus) muscle

FIG. 9.9

(vas´tus lat-er-a´lis)

## Origin

Intertrochanteric line, anterior and inferior borders of the greater trochanter, gluteal tuberosity, upper half of the linea aspera, and entire lateral intermuscular septum

## Insertion

Lateral border of the patella and patellar tendon to the tibial tuberosity

## Action

Extension of the knee

## Palpation

Slightly distal to the greater trochanter down the anterolateral aspect of the thigh to the superolateral patella, with extension of the knee, particularly full extension against resistance

## Innervation

Femoral nerve (L2–L4)

## Application, strengthening, and flexibility

All three of the vasti muscles function with the rectus femoris in knee extension. They are typically used in walking and running and must be used to keep the knee straight, as in standing. The vastus lateralis has a slightly superolateral pull on the patella and, as a result, is occasionally blamed in part for common lateral patellar subluxation and dislocation problems.

The vastus lateralis is strengthened through knee extension activities against resistance. See Appendix 3 for more commonly used exercises for the vastus lateralis and other muscles in this chapter. Stretching occurs by pulling the knee into maximum flexion, such as by standing on one leg and pulling the heel of the other leg up to the buttocks.

Knee extension

Chapter

9

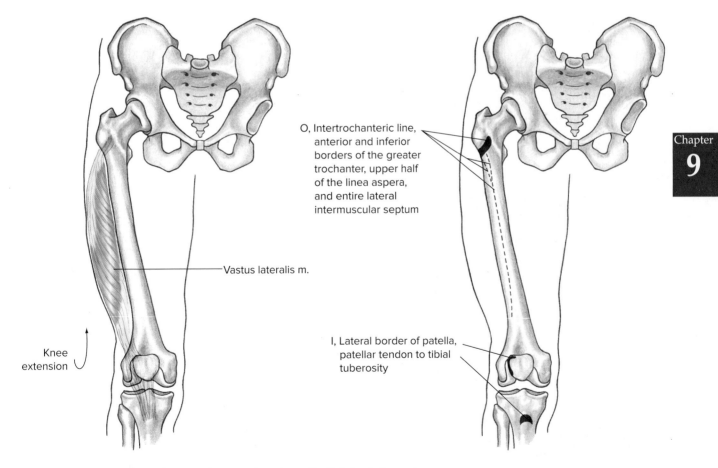

O, Intertrochanteric line, anterior and inferior borders of the greater trochanter, upper half of the linea aspera, and entire lateral intermuscular septum

Vastus lateralis m.

Knee extension

I, Lateral border of patella, patellar tendon to tibial tuberosity

FIG. 9.9 • Vastus lateralis muscle, anterior view. O, Origin; I, Insertion.

# Vastus intermedius muscle  FIG. 9.10

(vas ´tus in ´ter-me ´di-us)

## Origin

Upper two-thirds of the anterior surface of the femur

## Insertion

Upper border of the patella and patellar tendon to the
tibial tuberosity

## Action

Extension of the knee

## Palpation

Anteromedial distal one-third of thigh just above the
superomedial patella and deep to the rectus femoris,
with extension of the knee, particularly full extension against resistance

## Innervation

Femoral nerve (L2–L4)

## Application, strengthening, and flexibility

The three vasti muscles all contract in knee extension. They are used together with the rectus femoris in running, jumping, hopping, skipping, and
walking. The vasti muscles are primarily responsible for extending the knee while the hip is flexed
or being flexed. Thus, in doing a knee bend with
the trunk bent forward at the hip, the vasti are
exercised with less involvement of the rectus femoris. The natural activities mentioned above develop
the quadriceps.

If done properly, squats with a barbell of varying
weights on the shoulders, depending on strength,
are an excellent exercise for developing the quadriceps. Caution should be used, along with strict
attention to proper technique, to avoid injuries to
the knees and lower back. Leg press exercises and
knee extensions with weight machines are other
good exercises. Full knee flexion stretches all of the
quadriceps musculature.

Vastus intermedius m.

Knee extension

O, Upper two-thirds
of anterior surface
of femur

I, Upper border of patella
and patellar tendon to
tibial tuberosity

FIG. 9.10 • Vastus intermedius muscle, anterior view. O, Origin; I, Insertion.

Chapter
9

# Vastus medialis (internus) muscle

FIG. 9.11

(vas´tus me-di-a´lis)

## Origin

Whole length of the linea aspera and the medial condyloid ridge

## Insertion

Medial half of the upper border of the patella and patellar tendon to the tibial tuberosity

## Action

Extension of the knee

## Palpation

Anteromedial side of the thigh just above the superomedial patella, with extension of the knee, particularly full extension against resistance

## Innervation

Femoral nerve (L2–L4)

## Application, strengthening, and flexibility

The vastus medialis is thought to be very important in maintaining patellofemoral stability because of the oblique attachment of its distal fibers to the superior medial patella. This portion of the vastus medialis is referred to as the vastus medialis obliquus (VMO). The vastus medialis is strengthened similarly to the other quadriceps muscles by squats, knee extensions, and leg presses, but the VMO is not really emphasized until the last 10 to 20 degrees of knee extension. Full knee flexion stretches all the quadriceps muscles.

Knee extension

FIG. 9.11 • Vastus medialis muscle, anterior view. O, Origin; I, Insertion.

## Hamstring muscles  FIG. 9.12

The hamstring muscle group, consisting of the biceps femoris, semimembranosus, and semitendinosus, is covered in complete detail in Chapter 8, but further discussion is included here because of its importance in knee function.

Muscle strains involving the hamstrings are very common in football and other sports that require explosive running. This muscle group is often referred to as the running muscle because of its function in acceleration. The hamstring muscles are antagonists to the quadriceps muscles at the knee and are named for their cordlike attachments at the knee. All the hamstring muscles originate on the ischial tuberosity of the pelvic bone, with the exception of the short head of the biceps femoris, which originates on the lower half of the linea aspera and lateral condyloid ridge. The semitendinosus and semimembranosus insert on the anteromedial and posteromedial sides of the tibia, respectively. The biceps femoris inserts on the lateral tibial condyle and head of the fibula—hence the saying "Two to the inside and one to the outside." The short head of the biceps femoris originates on the linea aspera of the femur.

Special exercises to improve the strength and flexibility of this muscle group are important in decreasing knee injuries. Inability to touch the floor with the fingers when the knees are straight is largely a result of a lack of flexibility of the hamstrings. The hamstrings may be strengthened by performing knee or hamstring curls on a knee table against resistance. Some advocate strengthening the hamstrings with Nordic hamstring curls which are done by kneeling on a padded surface with a partner holding the heels while slowing lowering the torso as slowly as possible until losing control and then catching yourself in the push-up position. Tight or inflexible hamstrings are also contributing factors in painful conditions involving the lower back and knee. The flexibility of these muscles may be improved by performing slow, static stretching exercises, such as flexing the hip slowly while maintaining knee extension in a long sitting position.

The hamstrings are primarily knee flexors in addition to serving as hip extensors. Rotation of the knee can occur when it is in a flexed position. Once the knee is flexed approximately 20 degrees, it may rotate by actions of the hamstring muscles. The biceps femoris externally rotates the leg at the knee. The semitendinosus and semimembranosus perform internal rotation. Rotation of the knee permits pivoting movements and change in direction of the body. This rotation of the knee is vital in accommodating forces developing at the hip or ankle during directional changes in order to make the total movement more functional as well as more fluid in appearance. See Figs. 8.32, 8.33, and 8.34 on pages 252, 253, and 254 for the semitendinosus, semimembranosus, and biceps femoris, respectively.

Semitendinosis m.

(long head)
Biceps femoris m.
(short head)

Semimembranosus m.

FIG. 9.12 • The hamstring muscle group, posterior view.

Knee
internal
rotation

Knee
external
rotation

Hip
internal
rotation

Chapter
9

Hip
external
rotation

## Popliteus muscle  FIG. 9.13

(pop´li-te´us)

### Origin

Posterior surface of the lateral condyle of the femur

### Insertion

Upper posteromedial surface of the tibia

### Action

Flexion of the knee
Internal rotation of the knee as it flexes

### Palpation

With subject sitting, knee flexed 90 degrees, palpate deep to the gastrocnemius medially on the posterior proximal tibia and proceed superolaterally toward lateral epicondyle of tibia just deep to fibular collateral ligament, while subject internally rotates knee.

### Innervation

Tibial nerve (L5, S1)

### Application, strengthening, and flexibility

The popliteus muscle is the only true flexor of the leg at the knee. All other flexors are two-joint muscles. The popliteus is vital in providing posterolateral stability to the knee. It assists the medial hamstrings in internal rotation of the leg at the knee and is crucial in internally rotating the knee to unlock it from the "screwed home" full extension position.

Hanging from a bar with the legs flexed at the knee strenuously exercises the popliteus muscle. Also, the less strenuous activities of walking and running exercise this muscle. Specific efforts to strengthen this muscle combine knee internal rotation and flexion exercises against resistance. Stretching of the popliteus is difficult but may be done through passive full knee extension without flexing the hip. Passive maximum external rotation with the knee flexed approximately 20 to 30 degrees also stretches the popliteus.

Knee internal rotation

Knee flexion

O, Posterior surface of lateral condyle of femur

I, Upper posteromedial surface of tibia

Popliteus m.

Flexion

Internal rotation

FIG. 9.13 • Popliteus muscle, posterior view. O, Origin; I, Insertion.

# REVIEW EXERCISES

1. List the planes in which each of the following movements occurs. List the axis of rotation for each movement in each plane.
   a. Extension of the leg at the knee
   b. Flexion of the leg at the knee
   c. Internal rotation of the leg at the knee
   d. External rotation of the leg at the knee
2. Research the acceptability of deep knee bends and duck-walk activities in a physical education program, and report your findings in class.
3. Prepare a report on the knee on one of the following topics: anterior cruciate ligament injuries, meniscal injuries, medial collateral ligament injuries, patellar tendon tendonitis, plica syndrome, anterior knee pain, osteochondritis dissecans, patella subluxation/dislocation, knee bracing, quadriceps rehabilitation.
4. Research preventive and rehabilitative exercises to strengthen the knee joint, and report your findings in class.
5. Which muscle group about the knee would be most important for an athlete with a torn anterior cruciate ligament to develop? Why? For an athlete with a torn posterior cruciate ligament? Why?

6. **Muscle analysis chart** • Knee joint

| Complete the chart by listing the muscles primarily involved in each movement. | |
|---|---|
| Flexion | Extension |
| | |
| Internal rotation | External rotation |
| | |

7. **Antagonistic muscle action chart** • Knee joint

Complete the antagonistic muscle action chart by listing the muscle(s) or parts of muscles that are antagonist in their actions to the muscles in the left column.

| Agonist | Antagonist |
|---|---|
| Biceps femoris | |
| Semitendinosus | |
| Semimembranosus | |
| Popliteus | |
| Rectus femoris | |
| Vastus lateralis | |
| Vastus intermedius | |
| Vastus medialis | |

# LABORATORY EXERCISES

1. Locate the following bony landmarks on a human skeleton and on a subject:
   a. Skeleton
      1. Head and neck of femur
      2. Greater trochanter
      3. Shaft of femur
      4. Lesser trochanter
      5. Linea aspera
      6. Adductor tubercle
      7. Medial femoral condyle
      8. Lateral femoral condyle
      9. Patella
      10. Fibula head
      11. Medial tibial condyle
      12. Lateral tibial condyle
      13. Tibial tuberosity
      14. Gerdy's tubercle
   b. Subject
      1. Greater trochanter
      2. Adductor tubercle
      3. Medial femoral condyle
      4. Lateral femoral condyle
      5. Patella
      6. Fibula head
      7. Medial tibial condyle
      8. Lateral tibial condyle
      9. Tibial tuberosity
      10. Gerdy's tubercle
      11. Inferior patellar pole
2. How and where can the following muscles be palpated on a human subject?

*Note:* Palpate the previously studied hip joint muscles as they are performing actions at the knee.
   a. Gracilis
   b. Sartorius
   c. Biceps femoris
   d. Semitendinosus
   e. Semimembranosus
   f. Rectus femoris
   g. Vastus lateralis
   h. Vastus intermedius
   i. Vastus medialis
   j. Popliteus

3. Be prepared to indicate on a human skeleton, by using a long rubber band, the origin and insertion of the muscles listed in Question 2.
4. Demonstrate the following movements, and list the muscles primarily responsible for each.
   a. Extension of the leg at the knee
   b. Flexion of the leg at the knee
   c. Internal rotation of the leg at the knee
   d. External rotation of the leg at the knee
5. With a laboratory partner, determine how and why maintaining the position of full knee extension limits the ability to maximally flex the hip both actively and passively. Does maintaining excessive hip flexion limit the ability to accomplish full knee extension?
6. With a laboratory partner, determine how and why maintaining the position of full knee flexion limits the ability to maximally extend the hip both actively and passively. Does maintaining excessive hip extension limit the ability to accomplish full knee flexion?
7. Compare and contrast the bony, ligamentous, articular, and cartilaginous aspects of the medial knee joint with those of the lateral knee joint.

8. Knee joint exercise movement analysis chart

After analyzing each exercise in the chart, break each into two primary movement phases, such as a lifting phase and a lowering phase. For each phase, determine what knee joint movements occur, and then list the knee joint muscles primarily responsible for causing/controlling those movements. Beside each muscle in each movement, indicate the type of contraction as follows: I—isometric; C—concentric; E—eccentric.

| Exercise | Initial movement (lifting) phase | | Secondary movement (lowering) phase | |
|---|---|---|---|---|
| | Movement(s) | Agonist(s)—(contraction type) | Movement(s) | Agonist(s)—(contraction type) |
| Push-up | | | | |
| Squat | | | | |
| Dead lift | | | | |
| Hip sled | | | | |
| Forward lunge | | | | |
| Rowing exercise | | | | |
| Stair machine | | | | |

## 9. Knee joint sport skill analysis chart

Analyze each skill in the chart, and list the movements of the right and left knee joints in each phase of the skill. You may prefer to list the initial position the knee joint is in for the stance phase. After each movement, list the knee joint muscle(s) primarily responsible for causing/controlling the movement. Beside each muscle in each movement, indicate the type of contraction as follows: I—isometric; C—concentric; E—eccentric. It may be desirable to review the concepts for analysis in Chapter 12 for the various phases.

| Exercise | | Stance phase | Preparatory phase | Movement phase | Follow-through phase |
|---|---|---|---|---|---|
| Baseball pitch | (R) | | | | |
| | (L) | | | | |
| Football punt | (R) | | | | |
| | (L) | | | | |
| Walking | (R) | | | | |
| | (L) | | | | |
| Softball pitch | (R) | | | | |
| | (L) | | | | |
| Soccer pass | (R) | | | | |
| | (L) | | | | |
| Batting | (R) | | | | |
| | (L) | | | | |
| Bowling | (R) | | | | |
| | (L) | | | | |
| Basketball jump shot | (R) | | | | |
| | (L) | | | | |

**Chapter 9**

## References

Baker BE, et al: Review of meniscal injury and associated sports, *American Journal of Sports Medicine* 13:1, January–February 1985.

Field D: *Anatomy: palpation and surface markings*, ed 3, Oxford, 2001, Butterworth-Heinemann.

Garrick JG, Regna RK: Prophylactic knee bracing, *American Journal of Sports Medicine* 15:471, September–October 1987.

Hamilton N, Weimer W, Luttgens K: *Kinesiology: scientific basis of human motion*, ed 12, New York, 2012, McGraw-Hill.

Hislop HJ, Montgomery J: *Daniels and Worthingham's muscle testing: techniques of manual examination*, ed 8, Philadelphia, 2007, Saunders.

Kelly DW, et al: Patellar and quadriceps tendon ruptures—jumping knee, *American Journal of Sports Medicine* 12:375, September–October 1984.

Lysholm J, Wikland J: Injuries in runners, *American Journal of Sports Medicine* 15:168, September–October 1986.

Magee DJ: *Orthopedic physical assessment*, ed 6, St. Louis, 2014, Elsevier Saunders.

Muscolino JE: *The muscular system manual: the skeletal muscles of the human body*, ed 3, St. Louis, 2010, Elsevier Mosby.

Oatis CA: *Kinesiology: the mechanics and pathomechanics of human movement*, ed 2, Philadelphia, 2008, Lippincott Williams & Wilkins.

Prentice WE: *Principles of athletic training: a competency based approach*, ed 15, New York, 2014, McGraw-Hill.

Seeley RR, Stephens TD, Tate P: *Anatomy & physiology*, ed 8, New York, 2008, McGraw-Hill.

Shier D, Butler J, Lewis R: *Hole's human anatomy and physiology*, ed 12, New York, 2010, McGraw-Hill.

Sieg KW, Adams SP: *Illustrated essentials of musculoskeletal anatomy*, ed 4, Gainesville, FL, 2002, Megabooks.

Stone RJ, Stone JA: *Atlas of the skeletal muscles*, ed 6, New York, 2009, McGraw-Hill.

Van De Graaff KM: *Human anatomy*, ed 6, Dubuque, IA, 2002, McGraw-Hill.

Wroble RR, et al: Pattern of knee injuries in wrestling, a six-year study, *American Journal of Sports Medicine* 14:55, January–February 1986.

# Worksheet Exercises

For in- or out-of-class assignments, or for testing, utilize this tear-out worksheet.

## Worksheet 1

Using crayons or colored markers, draw and label on the worksheet the following muscles. Indicate the origin and insertion of each muscle with an "O" and an "I," respectively, and draw in the origin and insertion on the contralateral side of the skeleton.

a. Rectus femoris

b. Vastus lateralis

c. Vastus intermedius

d. Vastus medialis

e. Biceps femoris

f. Semitendinosus

g. Semimembranosus

h. Popliteus

# Worksheet Exercises

For in- or out-of-class assignments, or for testing, utilize this tear-out worksheet.

## Worksheet 2

Label and indicate with arrows the following movements of the knee joint. For each motion, complete the sentence by supplying the plane in which it occurs and the axis of rotation, as well as the muscles causing the motion.

a. Flexion occurs in the _____ plane about the _____ axis and is accomplished by concentric contractions of the _____
_____ muscles.

b. Extension occurs in the _____ plane about the _____
axis and is accomplished by concentric contractions of the _____
_____ muscles.

c. Internal rotation occurs in the _____ plane about the _____
axis and is accomplished by concentric contractions of the _____
_____ muscles.

d. External rotation occurs in the _____ plane about the _____
axis and is accomplished by concentric contractions of the _____
_____ muscles.

A

B

Courtesy of Britt Jones (all)

C

D

# THE ANKLE AND FOOT JOINTS

## Objectives

- To identify on a human skeleton the most important bone features, ligaments, and arches of the ankle and foot

- To draw and label on a skeletal chart the muscles of the ankle and foot

- To determine, list, and palpate the muscles of the ankle and foot joints and their antagonists

- To demonstrate and palpate the movements of the ankle and foot and list their respective planes of motion and axes of rotation

- To palpate the superficial joint structures including the ligaments on a human subject and appreciate their role in providing joint stability

- To list and organize the muscles that produce movement of the ankle and foot and list their antagonists

- To learn and understand the innervation of the ankle and foot joint muscles

- To determine, through analysis, the ankle and foot movements and muscles involved in selected skills and exercises.

The complexity of the foot is evidenced by the 26 bones, 19 large muscles, many small (intrinsic) muscles, and more than 100 ligaments that make up its structure.

Support and propulsion are the two functions of the foot. Proper functioning and adequate development of the muscles of the foot and practice of proper foot mechanics are essential for everyone. In our modern society, foot trouble is one of the most common ailments. Quite often, people develop poor foot mechanics or gait abnormalities secondary to improper footwear or other relatively minor problems. Poor foot mechanics early in life inevitably leads to foot discomfort and pathological conditions in later years.

Walking may be divided into stance and swing phases (Fig. 10.1). The **stance** phase is further divided into three components—heel-strike, midstance, and toe-off. Midstance may be further separated into loading response, midstance, and terminal stance. Normally, **heel-strike** is characterized by landing on the heel with the foot in supination and the leg in external rotation, followed immediately by pronation and internal rotation of the foot and leg, respectively, during **midstance**. The foot returns to supination and the leg returns to external rotation immediately prior to and during **toe-off**. The **swing** phase occurs when the foot leaves the ground and the leg moves forward to another point of contact. The swing phase may be divided into initial swing, midswing, and terminal swing. Problems often arise when the foot is too rigid and does not pronate adequately or when the foot remains in pronation past midstance. If the foot remains too rigid and does not pronate adequately, then impact forces will not be absorbed through the gait, resulting in shock being transmitted up the kinetic chain. If the foot overpronates or remains in pronation too much past midstance, then propulsive forces are diminished and additional stresses are placed on the kinetic chain. Walking differs from running in that one foot is always in contact with the ground and there is a point at which both feet contact the ground whereas in running there is a point at which neither foot is in contact with the ground, and both feet are never in contact with the ground at the same time.

The fitness revolution that has occurred during the past four decades has resulted in great improvements in shoes available for sports and recreational activities. In the past, a pair of sneakers would suffice for most activities. Now there are basketball, baseball, football, jogging, soccer, tennis, walking, and cross-training shoes. Good shoes are important, but there is no substitute for adequate muscle development, strength, and proper foot mechanics.

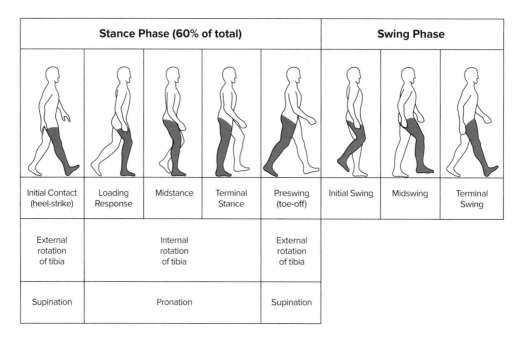

| Stance Phase (60% of total) | | | | | Swing Phase | | |
|---|---|---|---|---|---|---|---|
| Initial Contact (heel-strike) | Loading Response | Midstance | Terminal Stance | Preswing (toe-off) | Initial Swing | Midswing | Terminal Swing |
| External rotation of tibia | Internal rotation of tibia | | | External rotation of tibia | | | |
| Supination | Pronation | | | Supination | | | |

FIG. 10.1 • Walking gait cycle.

## Bones

Each foot has 26 bones, which collectively form the shape of an arch. They connect with the thigh and the remainder of the body through the fibula and tibia (Figs. 10.2 and 10.3). Body weight is transferred from the tibia to the talus and the calcaneus. It should be noted that the talus is one of the few bones involved in locomotion that has no muscle attachments.

The anterior portion of the talus is wider than its posterior portion, and this is a factor in making the ankle more stable in dorsiflexion than in plantar flexion.

In addition to the talus and calcaneus, there are five other bones in the rear foot and mid-foot, known as the tarsals. Between the talus and the three cuneiform bones lies the navicular. The cuboid is located between the calcaneus and the fourth and fifth metatarsals. Distal to the tarsals are the five metatarsals, which in turn correspond to each of the five toes. The toes are known as the phalanges. There are three individual bones in each phalange, except for the great toe, which has only two. Each of these bones is known as a pha-lanx. Finally, there are two sesamoid bones located beneath the first metatarsophalangeal joint and contained within the flexor hallucis brevis tendons.

The distal ends of the tibia and fibula are enlarged and protrude horizontally and inferiorly. These bony protrusions, known as malleoli, serve as a sort of pulley for the tendons of the muscles that run directly posterior to them. Specifically, the peroneus brevis and peroneus longus are immediately behind the lateral malleolus. The muscles immediately posterior to the medial malleolus may be remembered by the phrase "Tom, Dick, and Harry" with the "T" standing for the tibialis posterior, the "D" for the flexor digitorum longus, and the "H" for the flexor hallucis longus. This bony arrangement increases the mechanical advantage

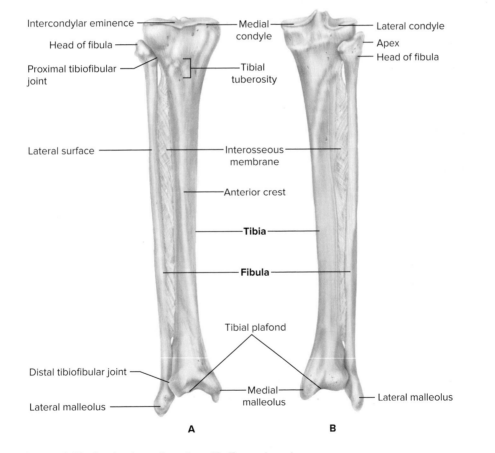

FIG. 10.2 • Right tibia and fibula. **A,** Anterior view; **B,** Posterior view.

**FIG. 10.3 • Right foot. A,** Superior (dorsal) view; **B,** Inferior (plantar) view.

of these muscles in performing their actions of eversion and inversion, respectively. The base of the fifth metatarsal is enlarged and prominent to serve as an attachment point for the peroneus brevis and tertius.

The inner surface of the medial cuneiform and the base of the first metatarsal provide insertion points for the tibialis anterior, while the undersurfaces of the same bones serve as the insertion for the peroneus longus. The tibialis posterior has multiple insertions on the lower inner surfaces of the navicular, cuneiform, and second through fifth metatarsal bases. The tops and undersurfaces of the bases of the second through fifth distal phalanxes are the insertion points for the extensor digitorum longus and the flexor digitorum longus, respectively. Similarly, the top and undersurface of the base of the first distal phalanx provide insertions for the extensor hallucis longus and flexor hallucis longus, respectively.

The posterior surface of the calcaneus is very prominent and serves as the attachment point for the Achilles tendon of the gastrocnemius–soleus complex.

## Joints

The tibia and fibula form the tibiofibular joint, a syndesmotic amphiarthrodial joint (see Fig. 10.2). The bones are joined at both the proximal and distal tibiofibular joints. In addition to the ligaments supporting both of these joints, there is a strong, dense interosseus membrane between the shafts of these two bones. Although only minimal movement is possible between these bones, the distal joint does become sprained occasionally in heavy contact sports such as football. A common component of this injury involves the ankle, or talocrural, joint being in dorsiflexion, which, by making the ankle more stable, allows the ligamentous stress to be transferred to the syndesmosis joint when the dorsiflexed ankle is forced into external rotation. This injury, a sprain of the syndesmosis joint, is commonly referred to as a high ankle sprain and primarily involves the anterior inferior tibiofibular ligament. Secondarily, and with more severe injuries, the posterior tibiofibular ligament, interosseus ligament, and interosseus membrane may be involved.

The ankle joint, technically known as the talocrural joint, is a hinge or ginglymus-type joint (Fig. 10.4). Specifically, it is the joint made up of the talus, the distal tibia, and the distal fibula. The ankle joint allows approximately 50 degrees of plantar flexion and 15 to 20 degrees of dorsiflexion (Fig. 10.5). Greater range of dorsiflexion, particularly in weight bearing, is possible when the knee is flexed, which reduces the tension of the biarticular gastrocnemius muscle. The fibula rotates on its axis 3 to 5 degrees externally with dorsiflexion of the ankle and 3 to 5 degrees internally during plantar flexion. The syndesmosis joint widens by approximately 1 to 2 millimeters during full dorsiflexion.

Inversion and eversion, though commonly thought to be ankle joint movements, technically occur in the subtalar (talocalcaneal) and transverse tarsal joints (talonavicular and calcaneocuboid joints), also known as Chopart's joint. These joints, classified as gliding or arthrodial, combine to allow approximately 20 to 30 degrees of inversion and 5 to 15 degrees of eversion. There is minimal movement within the remainder of the intertarsal and tarsometatarsal arthrodial joints, especially on the medial side.

The phalanges join the metatarsals to form the metatarsophalangeal joints, which are classified as condyloid-type joints. The metatarsophalangeal (MP) joint of the great toe flexes 45 degrees and extends 70 degrees, whereas the interphalangeal (IP) joint can flex from 0 degrees of full extension to 90 degrees of flexion. The MP joints of the four lesser toes allow approximately 40 degrees of flexion and 40 degrees of extension. The MP joints also abduct and adduct minimally. The proximal interphalangeal (PIP) joints in the lesser toes flex from 0 degrees of extension to 35 degrees of flexion. The distal interphalangeal (DIP) joints flex 60 degrees and extend 30 degrees. There is much variation from joint to joint and from person to person in all these joints.

Ankle sprains are one of the most common injuries among physically active people. Sprains involve the stretching or tearing of one or more ligaments. There are far too many ligaments in the foot and ankle to discuss in this text, but a few of the ankle ligaments are shown in Fig. 10.4 and more details are provided in Table 10.1. Far and away the most common ankle sprain results from excessive inversion, usually while in some degree of plantar flexion. This most commonly results in damage to the anterior talofibular ligament, particularly when in greater amounts of plantar flexion. When closer to neutral plantar flexion/dorsiflexion, the true inversion mechanism places more stress on the calcaneofibular ligament. Less common are excessive eversion forces causing injury to the deltoid ligament on the medial aspect of the ankle.

Ligaments in the foot and the ankle assist in stabilizing and maintaining the position of an arch. All 26 bones in the foot are connected with ligaments. This brief discussion focuses on the longitudinal and transverse arches.

There are two longitudinal arches (Fig. 10.6). The medial longitudinal arch, important for shock absorption, is located on the medial side of the foot and extends from the calcaneus to the talus, the navicular, the three cuneiforms, and the distal ends of the three medial metatarsals. The medial longitudinal arch, often implicated in a variety of foot problems, is primarily supported dynamically by the tibialis posterior and tibialis anterior muscles. The lateral longitudinal arch, important in balance, is located on the lateral side of the foot and extends from the calcaneus to the cuboid and the distal ends of the fourth and fifth metatarsals. Individual long arches can be high, medium, or low, but a low arch is not necessarily a weak arch.

The transverse arch (see Fig. 10.6) assists in adapting the foot to the ground and extends across the foot from the first metatarsal to the fifth metatarsal. The distal transverse or metatarsal arch, also a common source of foot problems, is supported by the intrinsic muscles of the foot such as the lumbricals, adductor hallucis, and flexor digiti minimi. These muscles may be strengthened through a towel crunch exercise in which the metatarsophalangeal joints are flexed to grab a towel. The long plantar ligament, longest of all the ligaments of the tarsals, originates on the plantar surface of the calcaneus anterior to the calcaneal tuberosity and inserts on the plantar surface of the cuboid with superficial fibers continuing forward to the bases of the second, third, and fourth metatarsal bones (see Fig. 10.4, *A* and *B*). The plantar fascia, sometimes referred to as the plantar aponeurosis, is a broad structure extending from the medial calcaneal tuberosity to the proximal phalanges of the toes. It assists in stabilizing the medial longitudinal arch and in propelling the body forward during the latter part of the stance phase. A common painful condition involving the plantar fascia is known as plantar fasciitis.

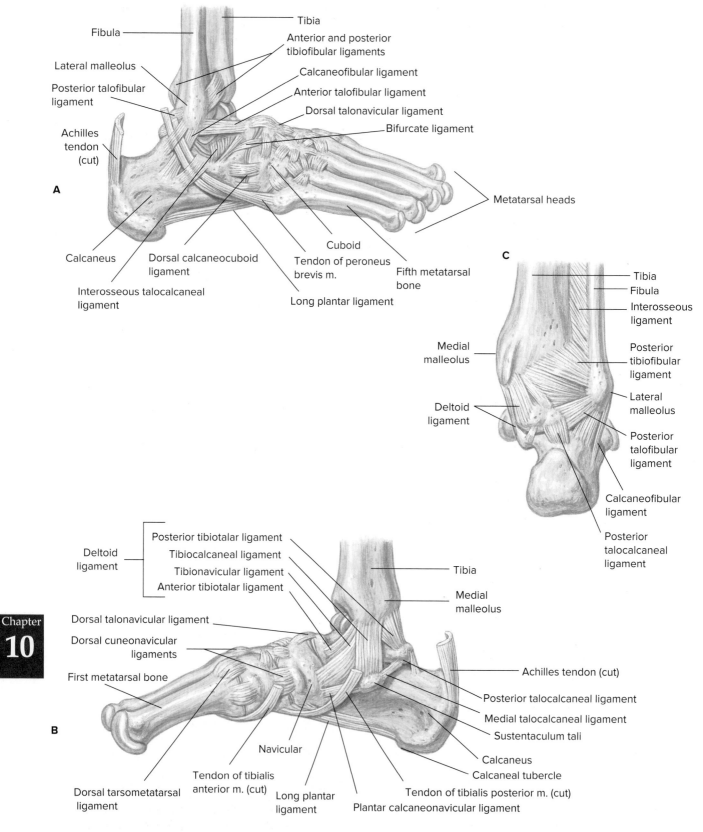

FIG. 10.4 • Right ankle joint. **A,** Lateral view; **B,** Medial view; **C,** Posterior view.

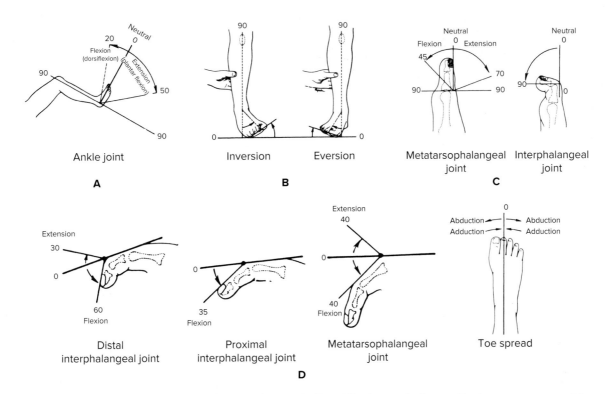

FIG. 10.5 • Active motion of the ankle, foot, and toes. **A,** Dorsiflexion and plantar flexion are measured in degrees from the right-angle neutral position or in percentages of motion as compared to the opposite ankle; **B,** Inversion and eversion normally are estimated in degrees or expressed in percentages as compared to the opposite foot; **C,** Flexion and extension of the great toe; **D,** ROM for the lateral four toes.

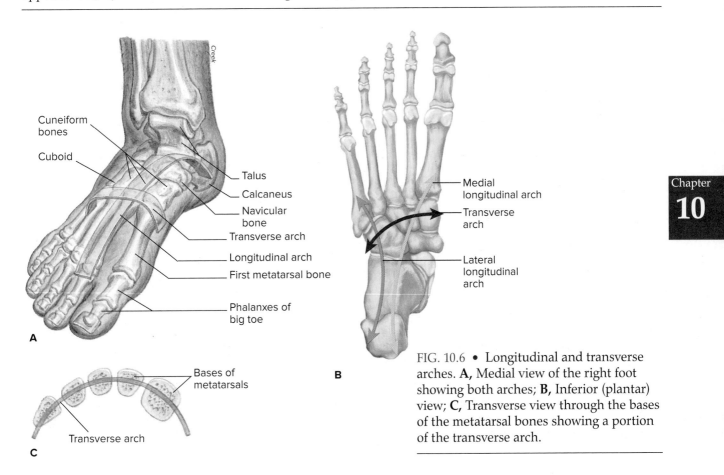

FIG. 10.6 • Longitudinal and transverse arches. **A,** Medial view of the right foot showing both arches; **B,** Inferior (plantar) view; **C,** Transverse view through the bases of the metatarsal bones showing a portion of the transverse arch.

**TABLE 10.1 • Major ligaments of the ankle and foot joints**

| Joint/structure | Ligament | | Proximal attachments | Distal attachments | Functional application and joint stability |
|---|---|---|---|---|---|
| Tibiofibular joint | Anterior tibiofibular ligament | | Anterolateral tubercle of the tibia | Longitudinal tubercle on the anterior aspect of the lateral malleolus | The fibers runs inferiorly and laterally, with the most distal fibers being the longest; prevents excessive fibular movement, especially into external rotation, and checks external talar rotation |
| | Posterior tibiofibular ligament | | Widely on the posterior tubercle of the tibia | Posterior lateral malleolus | Superficial fibers run obliquely, distally and laterally; holds the fibula close in the fibular groove of the tibia |
| | Transverse tibiofibular ligament (deep posterior tibiofibular ligament) | | Posterior tibial margin | Osteochondral junction on the posterior and medial margins of the distal fibula | Thick, strong structure with twisting fibers; prevents posterior talar translation |
| | Interosseous ligament | | Lateral surface of the distal tibia | Anteroinferior triangular segment of the medial aspect of the distal fibular shaft | A thickening of the distal aspect of the interosseous membrane, prevents excessive spreading of the fibula from the tibia with excessive dorsiflexion |
| Ankle (talocrural joint) | Deltoid ligament | Anterior tibiotalar ligament | Anterior portion of medial malleolus of the tibia | Talus anteriorly | Limits excessive eversion, particularly when in plantar flexion; also limits excessive plantarflexion and external rotation |
| | | Tibionavicular ligament | Anterior portion of medial malleolus of the tibia | Tuberosity of the navicular bone | Reinforces the ankle and subtalar joints; limits excessive eversion and plantar flexion |
| | | Tibiocalcaneal ligament | Mid portion of medial malleolus of the tibia | Whole length of the sustentaculum tali of the calcaneus | Reinforces the ankle and subtalar joints; limits excessive eversion and dorsiflexion |
| | | Posterior tibiotalar ligament | Posterior portion of medial malleolus of the tibia | Prominent tubercle on posterior surface of talus, medial to the groove for the tendon of the flexor hallucis longus | Limits excessive eversion, particularly when in dorsiflexion, also limits excessive dorsiflexion and external rotation |
| | Lateral collateral ligament of ankle joint (more commonly referred to by individual ligament names) | Anterior talofibular ligament (ATFL) | Anteroinferior border of the fibula, just distal to the anterior joint line of the ankle | Anterolateral surface of the talar neck, just distal to the articular surface | Weakest of lateral ligaments; limits inversion in plantar flexion; limits anterior translation of talus |
| | | Posterior talofibular ligament (PTFL) | Posterior border of lateral malleolus of the fibula | Posterolateral tubercle of the talus | Strongest lateral ligament; limits posterior talar displacement translation and talar external rotation |
| | | Calcaneofibular ligament (CFL) | Anterior border of lateral malleolus, approximately 1 cm to distal tip | Lateral side of the calcaneus on tubercle distal to subtalar joint and deep to peroneal tendon sheaths | Reinforces the ankle and subtalar joints; limits excessive inversion in neutral or dorsiflexed position |

Chapter
**10**

TABLE 10.1 (continued) • **Major ligaments of the ankle and foot joints**

| Joint/ structure | Ligament | | Proximal attachments | Distal attachments | Functional application and joint stability |
|---|---|---|---|---|---|
| Subtalar joint (talcalcaneal joint) | Lateral talocalcaneal ligament | | Lateral talar surface | Lateral surface of the calcaneus | Just anterior and medial to calcaneofibular ligament; reinforces the ankle joint; limits excessive inversion |
| | Cervical ligament | | Lateral talar process | Lateral superior surface of the calcaneus | Reinforces the subtalar joint; on a deeper plane than the calcaneofibular ligament; limits excessive inversion |
| | Medial talocalcaneal ligament | | Medial talar tubercle | Posterior sustentaculum tali and medial surface calcaneus | Reinforces the subtalar joint, limits excessive eversion |
| | Interosseous talocalcaneal ligament | | Groove between articular facets on inferior talus | Corresponding depression on the superior calcaneus | Binds the calcaneus and talus firmly, taut in eversion |
| | Talonavicular ligament (dorsal talonavicular ligament) | | Neck of talus | Superior surface of the navicular | Stabilizes talocalcaneonavicular joint |
| Calcaneotarsal joint | Bifurcate ligament | Calcaneocuboid ligament | Anterior process of the calcaneus | Medial cuboid | Stabilizes calcaneocuboid joint; limits adduction, plantar flexion and internal rotation of foot |
| | | Dorsal calcaneonavicular ligament | Anterior process of the calcaneus | Lateral navicular | Stabilizes calcaneonavicular joint; limits adduction and internal rotation of foot |
| | Plantar calcaneocuboid ligament (short plantar ligament) | | Anterior tubercle of calcaneus and the depression in front of it | Plantar surface of the cuboid posterior to the groove for the peroneus longus tendon | Deep to the long plantar ligament; supports the longitudinal arch of the foot |
| | Plantar calcaneonavicular ligament (spring ligament) | | Sustentaculum tali on the calcaneus | Plantar surface of navicular | Supports the head of the talus and the medial longitudinal arch |
| | Long plantar ligament (long calcaneocuboid ligament) | | Plantar surface of the calcaneus, anterior to tuberosity | Plantar surface of the cuboid and bases of all lateral metatarsals | Longest of all tarsal ligaments; supports the lateral arch of the foot |

Chapter
**10**

295

TABLE 10.1 (continued) • **Major ligaments of the ankle and foot joints**

| Joint/structure | Ligament | Proximal attachments | Distal attachments | Functional application and joint stability |
|---|---|---|---|---|
| Intertarsal joints | Dorsal cuboideonavicular ligament | Dorsal surface of cuboid | Dorsal surface of navicular | Maintain dorsal stability between cuboid and navicular |
| | Plantar cuboideonavicular ligament | Plantar surface of cuboid | Plantar surface of navicular | Maintain plantar stability between cuboid and navicular |
| | Dorsal cuneocuboid ligament | Dorsal surface of lateral cuneiform | Dorsal surface of cuboid | Maintain dorsal stability between lateral cuneiform and cuboid. |
| | Interosseous cuneocuboid ligament | Central portion of cuboid | Distal aspect of lateral cuneiform | Maintain stability between lateral cuneiform and cuboid |
| | Plantar cuneocuboid ligament | Lateral surface of the cuneiforms | Plantar surface of cuboid | Maintain plantar stability between cuneiforms and cuboid |
| | Plantar intercuneiform ligaments | Plantar surface of the medial, intermediate, and lateral cuneiforms | Plantar surface of the medial, intermediate, and lateral cuneiforms | Maintains plantar stability between cuneiforms |
| | Dorsal cuneonavicular ligament | Dorsal surface of the navicular | Dorsal surfaces of medial, intermediate, and lateral cuneiforms | Maintains dorsal stability between navicular and cuneiforms |
| | Plantar cuneonavicular ligament | Plantar surface of the navicular | Plantar surfaces of medial, intermediate, and lateral cuneiforms | Maintains plantar stability between navicular and cuneiforms |
| | Dorsal intercuneiform ligaments | Dorsal surfaces of medial, intermediate and lateral cuneiforms | Dorsal surfaces of medial, intermediate, and lateral cuneiforms | Maintain dorsal stability between cuneiforms |
| | Interosseous intercuneiform ligaments | Intraarticular surfaces of cuneiform bones | Intraarticular surfaces of cuneiform bones | Maintain stability between cuneiforms |
| | Plantar intercuneiform ligaments | Plantar surfaces of medial, intermediate, and lateral cuneiforms | Plantar surfaces of medial, intermediate, and lateral cuneiforms | Maintain plantar stability between cuneiforms |

TABLE 10.1 (continued) • **Major ligaments of the ankle and foot joints**

| Joint/ structure | Ligament | Proximal attachments | Distal attachments | Functional application and joint stability |
|---|---|---|---|---|
| Intermeta-tarsal joints | Dorsal intermetatarsal ligaments | Dorsal surface of bases of 2nd, 3rd, and 4th metatarsals | Dorsal surface of bases of 3rd, 4th, and 5th metatarsals | Limits excessive movement between the four lateral metatarsals |
| | Interosseous metatarsal ligaments | Interconnects bases of metatarsal bones | Interconnects bases of metatarsals | |
| | Plantar intermetatarsal ligaments | Plantar surface of bases of 2nd, 3rd, and 4th metatarsals | Plantar surface of bases of 3rd, 4th, and 5th metatarsals | |
| | Transverse metatarsal ligaments | Heads of metatarsals | Heads of metatarsals | Connects the heads of the metatarsals |
| Tarsometa-tarsal joints | Cuneometatarsal interos-seous ligaments | Lateral surface of the medial cuneiform; lateral cuneiform; lateral angle of the lateral cuneiform | Adjacent angle of the sec-ond metatarsal; adjacent side of the base of the third metatarsal | Limits motion between the cuneiforms/cuboid and bases of metatarsals |
| | Dorsal tarsometatarsal ligaments | Dorsal aspect of each cuneiform and cuboid | Dorsal aspect of each metatarsal base | |
| | Plantar tarsometatarsal ligaments | Plantar aspect of each cuneiform and cuboid | Plantar aspect of each metatarsal base | |
| Metatarso-phalangeal joints | Tibial collateral ligaments | Dorsal tubercle on tibial side of the metatarsal head | Proximal phalanx bases (medial aspect) | Prevents excessive tibial opening of joint |
| | Fibula collateral ligaments | Dorsal tubercle on fibular side of the metatarsal head | Proximal phalanx bases (lateral aspect) | Prevents excessive fibula opening of joint |
| | Plantar ligaments of metatarsophalangeal articulations | Plantar surface of metatarsophalan-geal articulations between collateral ligaments | Bases of proximal phalanxes | Deep surfaces of the ligaments form parts of the articular facets for the heads of the metatarsals; stabilizes metatarsophalangeal joints and limits extension |

Chapter

**10**

TABLE 10.1 (continued) • **Major ligaments of the ankle and foot joints**

| Joint/ structure | Ligament | Proximal attachments | Distal attachments | Functional application and joint stability |
|---|---|---|---|---|
| Proximal interphalangeal joints | Tibial collateral ligaments | Proximal phalanx heads (medial aspect) | Middle phalanx bases (medial aspect) | Limits excessive tibial opening of joint |
| | Fibular collateral ligaments | Proximal phalanx heads (lateral aspect) | Middle phalanx bases (lateral aspect) | Limits excessive fibular opening of joint |
| | Plantar (accessory) ligaments | Proximal phalanx heads (plantar surface) | Middle phalanx bases (plantar surface) | Limits extension of the joint |
| Distal interphalangeal joints | Tibial collateral ligaments | Middle phalanx heads (medial aspect) | Distal phalanx bases (medial aspect) | Limits excessive tibial opening of joint |
| | Fibular collateral ligaments | Middle phalanx heads (lateral aspect) | Distal phalanx bases (lateral aspect) | Limits excessive fibular opening of joint |
| | Plantar (accessory) ligaments | Middle phalanx heads (plantar surface) | Distal phalanx bases (palmar surface) | Limits extension of the joint |

Chapter

**10**

## Movements   FIG. 10.7

**Dorsiflexion (flexion):** dorsal flexion; movement of the top of the ankle and foot toward the anterior tibia

**Plantar flexion (extension):** movement of the ankle and foot away from the anterior tibia

**Eversion:** turning the ankle and foot outward; abduction, away from the midline; weight is on the medial edge of the foot

**Inversion:** turning the ankle and foot inward; adduction, toward the midline; weight is on the lateral edge of the foot

**Toe flexion:** movement of the toes toward the plantar surface of the foot

**Toe extension:** movement of the toes away from the plantar surface of the foot

**Pronation:** a combination of ankle dorsiflexion, subtalar eversion, and forefoot abduction (toe-out)

**Supination:** a combination of ankle plantar flexion, subtalar inversion, and forefoot adduction (toe-in)

Ankle dorsal flexion

Ankle plantar flexion

Transverse tarsal and subtalar eversion

Transverse tarsal and subtalar inversion

**Dorsiflexion**
**A**

**Plantar flexion**
**B**

**Transverse tarsal and subtalar eversion**
**C**

**Transverse tarsal and subtalar inversion**
**D**

Courtesy of Britt Jones (all)

FIG. 10.7 • Movements of the right ankle and foot.

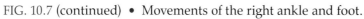

Great toe
MTP and
IP flexion

Great toe
MTP and
IP
extension

**Flexion of the toes**
**E**

**Extension of the toes**
**F**

Courtesy of Britt Jones (all)

**Pronation**
**G**

**Supination**
**H**

FIG. 10.7 (continued) • Movements of the right ankle and foot.

# Ankle and foot muscles  FIGS 10.8, 10.9

The large number of muscles in the ankle and foot may be easier to learn if grouped according to location and function. In general, the muscles located on the anterior aspect of the ankle and foot are the dorsal flexors and/or toe extensors. Those on the posterior aspect are plantar flexors and/or toe flexors. Specifically, the gastrocnemius and the soleus collectively are known as the **triceps surae**, due to their three heads, which together join to the Achilles tendon. Muscles that are evertors are located more to the lateral side, whereas the invertors are located medially.

The lower leg is divided into four compartments, each containing specific muscles (Fig. 10.9). Tightly surrounding and binding each compartment is a dense fascia, which facilitates venous return and prevents excessive swelling of the muscles during exercise. The anterior compartment contains the

FIG. 10.8 • Right Lower leg, ankle, and foot muscles. **A,** Lateral view; **B,** Medial view.

dorsiflexor group, consisting of the tibialis anterior, peroneus tertius, extensor digitorum longus, and extensor hallucis longus. The lateral compartment contains the peroneus longus and peroneus brevis—the two most powerful evertors. The posterior compartment is divided into deep and superficial compartments. The gastrocnemius, soleus, and plantaris are located in the superficial posterior compartment, while the deep posterior compartment is composed of the flexor digitorum longus, flexor hallucis longus, popliteus, and tibialis posterior. All the muscles of the superficial posterior compartment are primarily plantar flexors. The plantaris, absent in some humans, is a vestigial biarticular muscle that contributes minimally to ankle plantar flexion and knee flexion. It originates on the inferior aspect of the lateral supracondylar

line of the distal femur posteriorly, and runs just medial to the lateral head of the gastrocnemius and then deep to it, but superficial to the soleus, to insert on the middle one-third of the posterior calcaneal surface just medial to the Achilles tendon. The deep posterior compartment muscles, except for the popliteus, are plantar flexors but also function as invertors. Although most common with the anterior compartment, any of these components are subject to a condition known as compartment syndrome. This condition may be acute or chronic and may occur secondarily to injury, trauma, or overuse. Symptoms include sharp pain, particularly with increased movement actively or passively; swelling; and weakness in the muscles of the involved compartment. Depending on the severity, emergency surgery may be indicated to

C

D

FIG. 10.8 (continued) • Right Lower leg, ankle, and foot muscles. **A,** Lateral view; **B,** Medial view.

x

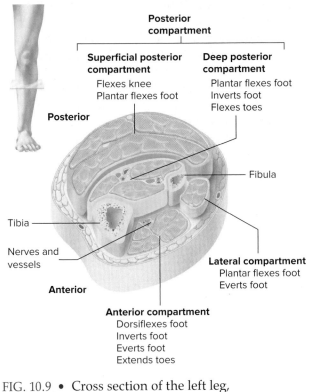

**Posterior compartment**

**Superficial posterior compartment**
Flexes knee
Plantar flexes foot

**Deep posterior compartment**
Plantar flexes foot
Inverts foot
Flexes toes

**Posterior**

Fibula

Tibia

Nerves and vessels

**Lateral compartment**
Plantar flexes foot
Everts foot

**Anterior**

**Anterior compartment**
Dorsiflexes foot
Inverts foot
Everts foot
Extends toes

FIG. 10.9 • Cross section of the left leg, demonstrating the muscular compartments.

release the fascia in order to prevent permanent tissue damage, although many compartment syndromes may be adequately addressed with proper acute management.

Due to the heavy demands placed on leg musculature in the running activities of most sports, both acute and chronic injuries are common. "Shin splints" is a common term used to describe a painful condition of the leg that is often associated with running activities. This condition is not a specific diagnosis but rather is attributed to a number of specific musculotendinous injuries. Most often the tibialis posterior, medial soleus, or tibialis anterior is involved, but the extensor digitorum longus may also be involved. Shin splints often occur as a result of an inappropriate level of flexibility, strength, and endurance for the specific demands of the activity and may be prevented in part by stretching the plantar flexors and strengthening the dorsiflexors.

Additionally, painful cramps caused by acute muscle spasm in the gastrocnemius and soleus occur somewhat commonly and may be relieved through active and passive dorsiflexion. Also, a very disabling injury involves the complete rupture of the strong Achilles tendon, which connects these two plantar flexors to the calcaneus.

## Ankle and Foot Muscles by Function

*Note:* A number of the ankle and foot muscles are capable of helping produce more than one movement.

Plantar flexors
  Gastrocnemius
  Flexor digitorum longus
  Flexor hallucis longus
  Peroneus (fibularis) longus
  Peroneus (fibularis) brevis
  Plantaris
  Soleus
  Tibialis posterior
Evertors
  Peroneus (fibularis) longus
  Peroneus (fibularis) brevis
  Peroneus (fibularis) tertius
  Extensor digitorum longus
Dorsiflexors
  Tibialis anterior
  Peroneus (fibularis) tertius
  Extensor digitorum longus (extensor of the lesser toes)
  Extensor hallucis longus (extensor of the great toe)
Invertors
  Tibialis anterior
  Tibialis posterior
  Flexor digitorum longus (flexor of the lesser toes)
  Flexor hallucis longus (flexor of the great toe)

## Ankle and Foot Muscles by Compartment

Anterior compartment
  Tibialis anterior
  Extensor hallucis longus
  Extensor digitorum longus
  Peroneus (fibularis) tertius
Lateral compartment
  Peroneus (fibularis) longus
  Peroneus (fibularis) brevis
Deep posterior compartment
  Flexor digitorum longus
  Flexor hallucis longus
  Tibialis posterior
  Popliteus
Superficial posterior compartment
  Gastrocnemius (medial head)
  Gastrocnemius (lateral head)
  Soleus
  Plantaris

While viewing the muscles in Figs. 10.8 and 10.9, correlate them with Table 10.2.

Chapter
**10**

## TABLE 10.2 • Agonist muscles of the ankle and foot joints

| | Muscle | Origin | Insertion | Action | Plane of motion | Palpation | Innervation |
|---|---|---|---|---|---|---|---|
| **Superficial posterior compartment** | Gastrocnemius | Medial head: posterior surface of the medial femoral condyle; Lateral head: posterior surface of the lateral femoral condyle | Posterior surface of the calcaneus (Achilles tendon) | Plantar flexion of the ankle | Sagittal | Upper half of the posterior aspect of the lower leg | Tibial nerve (S1, S2) |
| | | | | Flexion of the knee | | | |
| | Soleus | Posterior surface of the proximal fibula and proximal two-third of the posterior tibial surface | Posterior surface of the calcaneus (Achilles tendon) | Plantar flexion of the ankle | Sagittal | Posteriorly under the gastrocnemius muscle on the medial and lateral sides of the lower leg, particularly while prone with knee flexed approximately 90 degrees and actively plantarflexing ankle | Tibial nerve (S1, S2) |
| **Deep posterior compartment** | Tibialis posterior | Posterior surface of the upper half of the interosseus membrane and the adjacent surfaces of the tibia and fibula | Inferior surfaces of the navicular, cuneiform, and cuboid bones and bases of the 2nd, 3rd, and 4th metatarsal bones | Inversion of the foot | Frontal | The tendon may be palpated both proximally and distally immediately behind medial malleolus with inversion and plantar flexion and is better distinguished from flexor digitorum longus and flexor hallucis longus if toes can be maintained in slight extension | Tibial nerve (L5, S1) |
| | | | | Plantar flexion of the ankle | Sagittal | | |
| | Flexor digitorum longus | Middle one-third of the posterior surface of the tibia | Base of the distal phalanx of each of the four lesser toes | Flexion of the four lesser toes at the metatarsophalangeal and the proximal and distal interphalangeal joints | Sagittal | The tendon may be palpated immediately posterior to the medial malleolus and tibialis posterior and immediately anterior to the flexor hallucis longus with flexion of the lesser toes while maintaining great toe extension, ankle dorsiflexion, and foot eversion | Tibial nerve (L5, S1) |
| | | | | Plantar flexion of the ankle | Sagittal | | |
| | | | | Inversion of the foot | Frontal | | |
| | Flexor hallucis longus | Middle two-third of the posterior surface of the fibula | Base of the distal phalanx of the great toe; plantar surface | Flexion of the great toe at the metatarsophalangeal and interphalangeal joints | Sagittal | Most posterior of three tendons immediately behind medial malleolus; between medial soleus and tibia with active great toe flexion while maintaining extension of four lesser toes, ankle dorsiflexion, and foot eversion | Tibial nerve (L5, S1, S2) |
| | | | | Plantar flexion of the ankle | Sagittal | | |
| | | | | Inversion of the foot | Frontal | | |

Chapter
**10**

TABLE 10.2 (continued) • **Agonist muscles of the ankle and foot joints**

| | Muscle | Origin | Insertion | Action | Plane of motion | Palpation | Innervation |
|---|---|---|---|---|---|---|---|
| Lateral compartment | Peroneus (fibularis) longus | Head and upper two-third of the lateral surface of the fibula | Undersurfaces of the medial cuneiform and 1st metatarsal bone | Eversion of the foot | Frontal | Upper lateral side of tibia just distal to fibular head and down to immediately posterior to lateral malleolus; just posterolateral from tibialis anterior and extensor digitorum longus with active eversion | Superficial peroneal nerve (L4, L5, S1) |
| | | | | Plantar flexion of the ankle | Sagittal | | |
| | Peroneus (fibularis) brevis | Mid to lower two-third of the lateral surface of the fibula | Tuberosity of the 5th metatarsal bone | Eversion of the foot | Frontal | Tendon of muscle at proximal end of 5th metatarsal; just proximal and posterior to lateral malleolus; immediately deep anteriorly and posteriorly to peroneus longus with active eversion | Superficial peroneal nerve (L4, L5, S1) |
| | | | | Plantar flexion of the ankle | Sagittal | | |
| Anterior compartment | Peroneus (fibularis) tertius | Distal one-third of the anterior fibula | Superior aspect of the base of the 5th metatarsal | Dorsiflexion of the ankle | Sagittal | Just medial to distal fibula; lateral to extensor digitorum longus tendon on anterolateral aspect of foot, down to medial side of base of 5th metatarsal with dorsiflexion and eversion | Deep peroneal nerve (L4, L5, S1) |
| | | | | Eversion of the foot | Frontal | | |
| | Extensor digitorum longus | Lateral condyle of the tibia, head of the fibula, and upper two-third of the anterior surface of the fibula | Tops of the middle and distal phalanges of the four lesser toes | Extension of the four lesser toes at the metatarsophalangeal and the proximal and distal interphalangeal joints | Sagittal | Second muscle to lateral side of anterior tibial border; upper lateral side of tibia between tibialis anterior medially and fibula laterally; divides into four tendons just distal to anterior ankle with active toe extension | Deep peroneal nerve (L4, L5, S1) |
| | | | | Dorsiflexion of the ankle | | | |
| | | | | Eversion of the foot | Frontal | | |
| | Extensor hallucis longus | Middle two-third of the medial surface of the anterior fibula | Base of the distal phalanx of the great toe | Extension of great toe at the metatarsophalangeal and interphalangeal joints | Sagittal | From dorsal aspect of great toe to just lateral to tibialis anterior and medial to extensor digitorum longus at anterior ankle joint | Deep peroneal nerve (L4, L5, S1) |
| | | | | Dorsiflexion of the ankle | | | |
| | | | | Weak inversion of the foot | Frontal | | |
| | Tibialis anterior | Upper two-third of the lateral surface of the tibia | Inner surface of the medial cuneiform and the base of the 1st metatarsal bone | Dorsiflexion of the ankle | Sagittal | First muscle to the lateral side of the anterior tibial border, particularly palpable with fully active ankle dorsiflexion; most prominent tendon crossing the ankle anteromedially | Deep peroneal nerve (L4, L5, S1) |
| | | | | Inversion of the foot | Frontal | | |

**Note:** The plantaris is not included because its contribution to knee flexion and plantar flexion is relatively minimal, but it is found in the posterior compartment with the gastrocnemius and has the same function.

Chapter
**10**

# Nerves

As described in Chapter 8, the sciatic nerve originates from the sacral plexus and becomes the tibial nerve and peroneal nerve. The tibial division of the sciatic nerve (Fig. 8.22) continues down to the posterior aspect of the lower leg to innervate the gastrocnemius, soleus, tibialis posterior, flexor digitorum longus, and flexor hallucis longus. Just before reaching the ankle, the tibial nerve branches to become the medial and lateral plantar nerves, which innervate the intrinsic muscles of the foot. The medial plantar nerve innervates the abductor hallucis, flexor hallucis brevis, first lumbrical, and flexor digitorum brevis. The lateral plantar nerve supplies the adductor hallucis, quadratus plantae, lumbricals (2, 3, and 4), dorsal interossei, plantar interossei, abductor digiti minimi, and flexor digiti minimi.

The peroneal, or fibular, nerve (Fig. 10.10) divides just below the head of the fibula to become the superficial and deep peroneal nerves. The superficial branch innervates the peroneus longus and peroneus brevis, while the deep branch innervates the tibialis anterior, extensor digitorum longus, extensor hallucis longus, peroneus tertius, and extensor digitorum brevis. See Fig. 10.11 for a listing of the major ankle and foot muscular innervations.

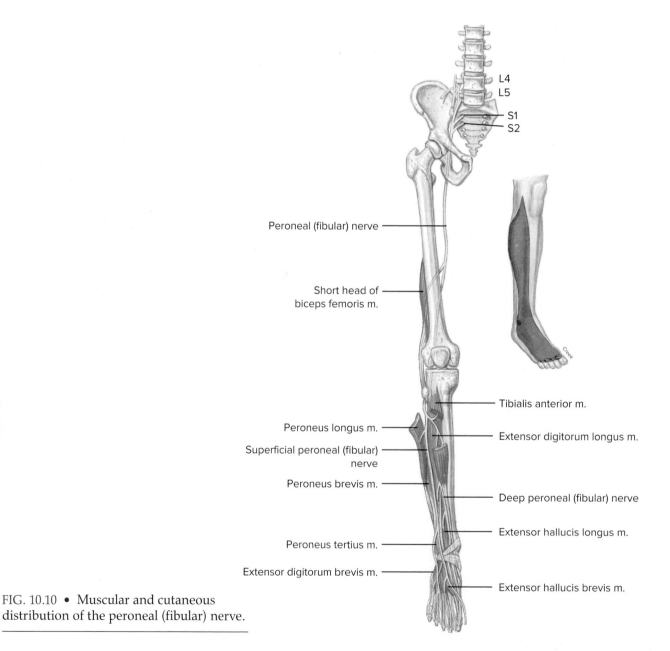

FIG. 10.10 • Muscular and cutaneous distribution of the peroneal (fibular) nerve.

Chapter 10

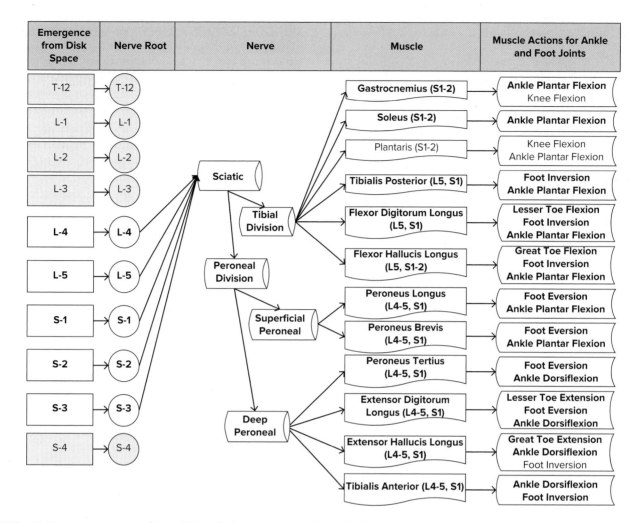

FIG. 10.11 • Innervation for ankle and foot joint muscles and their actions. The neuromuscular structures associated with the primary movers for the ankle and foot joints and their actions are bolded.

# Gastrocnemius Muscle  FIG. 10.12

(gas-trok-ne´mi-us)

## Origin

Medial head: posterior surface of the medial femoral condyle

Lateral head: posterior surface of the lateral femoral condyle

## Insertion

Posterior surface of the calcaneus (Achilles tendon)

## Action

Plantar flexion of the ankle

Flexion of the knee

## Palpation

Easiest muscle in the lower extremity to palpate; upper one-half of posterior aspect of lower leg

## Innervation

Tibial nerve (S1, S2)

## Application, strengthening, and flexibility

The gastrocnemius and soleus together are known as the triceps surae with triceps referring to the heads of the medial and lateral gastrocnemius and the soleus and surae referring to the calf. Because the gastrocnemius is a biarticular muscle, it is more effective as a knee flexor if the ankle is dorsiflexed and more effective as a plantar flexor of the foot if the knee is held in extension. This is observed when one sits too close to the steering wheel in driving a car, which significantly shortens the entire muscle, reducing its effectiveness. When the knees are bent, the muscle becomes an ineffective plantar flexor, and it is more difficult to depress the brakes. Running, jumping, hopping, and skipping exercises all depend significantly on the gastrocnemius and soleus to propel the body upward and forward. Heel-raising exercises with the knees in full extension and the toes resting on a block of wood are an excellent way to strengthen the muscle through the full range of motion. Holding a barbell on the shoulders can increase the resistance. See Appendix 2 for more commonly used exercises for the gastrocnemius and other muscles in this chapter.

The gastrocnemius may be stretched by standing and placing both palms on a wall about 3 feet away and leaning into the wall. The feet should be pointed straight ahead, and the heels should remain on the floor. The knees should remain fully extended throughout the exercise to accentuate the stretch on the gastrocnemius.

O, Posterior surfaces of two condyles of femur

Medial head

Lateral head

Knee flexion

Gastrocnemius m.

I, Posterior surface of calcaneus

Plantar flexion

FIG. 10.12 • Gastrocnemius muscle, posterior view. O, Origin; I, Insertion.

## Soleus muscle   FIG. 10.13

(so´le-us)

### Origin

Posterior surface of the proximal fibula and proximal two-thirds of the posterior tibial surface

### Insertion

Posterior surface of the calcaneus (Achilles tendon)

### Action

Plantar flexion of the ankle

### Palpation

Posteriorly under the gastrocnemius muscle on the medial and lateral sides of the lower leg, particularly prone with knee flexed approximately 90 degrees and actively plantarflexing ankle

### Innervation

Tibial nerve (S1, S2)

### Application, strengthening, and flexibility

The soleus muscle is one of the most important plantar flexors of the ankle. Some anatomists believe that it is nearly as important in this movement as the gastrocnemius. This is especially true when the knee is flexed. When one rises up on the toes, the soleus muscle can plainly be seen on the outside of the lower leg if one has exercised the legs extensively, as in running and walking.

The soleus muscle is used whenever the ankle plantar flexes. Any movement with body weight on the foot with the knee flexed or extended calls it into action. When the knee is flexed slightly, the effect of the gastrocnemius is reduced, placing more work on the soleus. Running, jumping, hopping, skipping, and dancing on the toes are all exercises that depend heavily on the soleus. It may be strengthened through any plantar flexion exercise against resistance, particularly if the knee is flexed slightly to deemphasize the gastrocnemius. Heel-raising exercises as described for the gastrocnemius, except with the knees flexed slightly, are one way to isolate this muscle for strengthening. Resistance may be increased by holding a barbell on the shoulders.

The soleus is stretched in the same manner as the gastrocnemius, except that the knees must be flexed slightly, which releases the stretch on the gastrocnemius and places it on the soleus. Again, it is important to attempt to keep the heels on the floor.

Ankle plantar flexion

O, Posterior surface of proximal fibula and proximal two-thirds of posterior tibial surface

Soleus m.

I, Posterior surface of calcaneus

Plantar flexion

FIG. 10.13 • Soleus muscle, posterior view. O, Origin; I, Insertion.

Chapter
**10**

Transverse tarsal and subtalar eversion

Ankle plantar flexion

# Peroneus (fibularis) longus muscle FIG. 10.14

(per-o-ne´us lon´gus)

## Origin

Head and upper two-thirds of the lateral surface of the fibula

## Insertion

Undersurfaces of the medial cuneiform and first metatarsal bones

## Action

Eversion of the foot
Plantar flexion of the ankle

## Palpation

Upper lateral side of the tibia; just distal to fibular head and down to immediately posterior to lateral malleolus; just posterolateral from the tibialis anterior and extensor digitorum longus with active eversion

## Innervation

Superficial peroneal nerve (L4, L5, S1)

## Application, strengthening, and flexibility

The peroneus longus muscle passes posteroinferiorly to the lateral malleolus and under the foot from the outside to under the inner surface. Because of its line of pull, it is a strong evertor and assists in plantar flexion.

When the peroneus longus muscle is used effectively with the other ankle flexors, it helps bind the transverse arch as it contracts. Developed without the other plantar flexors, it would produce a weak, everted foot. In running, jumping, hopping, and skipping, the foot should be placed so that it is pointing forward to ensure proper development of the group. Walking barefoot or in stocking feet on the inside of the foot (everted position) is a good exercise for this muscle.

Eversion exercises to strengthen this muscle may be performed by turning the sole of the foot outward while resistance is applied in the opposite direction.

The peroneus longus may be stretched by passively taking the foot into extreme inversion and dorsiflexion while the knee is flexed.

FIG. 10.14 • Peroneus longus muscle, lateral plantar views, right leg and foot. O, Origin; I, Insertion.

Chapter
10

# Peroneus (fibularis) brevis muscle   FIG. 10.15

(per-o-ne´us bre´vis)

## Origin

Mid to lower two-thirds of the lateral surface of the fibula

## Insertion

Tuberosity of the fifth metatarsal

## Action

Eversion of the foot
Plantar flexion of the ankle

## Palpation

Tendon of the muscle at the proximal end of the fifth metatarsal just proximal and posterior to the lateral malleolus; immediately deep anteriorly and posteriorly to the peroneus longus with active eversion

## Innervation

Superficial peroneal nerve (L4, L5, S1)

## Application, strengthening, and flexibility

The peroneus brevis muscle passes posteroinferiorly to the lateral malleolus to pull on the base of the fifth metatarsal. It is a primary evertor of the foot and assists in plantar flexion. In addition, it aids in maintaining the lateral longitudinal arch as it depresses the foot.

The peroneus brevis muscle is exercised with other plantar flexors in the powerful movements of running, jumping, hopping, and skipping. It may be strengthened in a fashion similar to that for the peroneus longus by performing eversion exercises, such as turning the sole of the foot outward against resistance.

The peroneus brevis is stretched in the same manner as the peroneus longus.

Transverse tarsal and subtalar eversion

Ankle plantar flexion

Peroneus brevis m.

Plantar flexion

Foot eversion

O, Mid to lower two-thirds of lateral surface of fibula

I, Tuberosity of fifth metatarsal

Chapter **10**

FIG. 10.15 • Peroneus brevis muscle, lateral and plantar views, right leg and foot. O, Origin; I, Insertion.

# Peroneus (fibularis) tertius muscle  FIG. 10.16

(per-o-ne´us ter´shi-us)

## Origin

Distal third of the anterior fibula

## Insertion

Superior aspect of the base of the fifth metatarsal

## Action

Eversion of the foot
Dorsiflexion of the ankle

## Palpation

Just medial to distal fibula; lateral to the extensor digitorum longus tendon on the anterolateral aspect of the foot, down to the medial side of the base of the fifth metatarsal with dorsiflexion and eversion

## Innervation

Deep peroneal nerve (L4, L5, S1)

## Application, strengthening, and flexibility

The peroneus tertius, absent in some humans, is a small muscle that assists in dorsiflexion and eversion. Some authorities refer to it as the fifth tendon of the extensor digitorum longus. It may be strengthened by pulling the foot up toward the shin against a weight or resistance. Everting the foot against resistance, such as weighted eversion towel drags, can also be used for strength development.

The peroneus tertius may be stretched by passively taking the foot into extreme inversion and plantar flexion.

Peroneus tertius m.

O, Distal third of anterior fibula

I, Superior aspect of base of fifth metatarsal

Ankle dorsiflexion

Foot eversion

FIG. 10.16 • Peroneus tertius muscle, anterior view, right leg and foot. O, Origin; I, Insertion.

# Extensor digitorum longus muscle   FIG. 10.17

(eks-ten´sor dij-i-to´rum lon´gus)

## Origin

Lateral condyle of the tibia, head of the fibula, and upper two-thirds of the anterior surface of the fibula

## Insertion

Tops of the middle and distal phalanxes of the four lesser toes

## Action

Extension of the four lesser toes at the metatarsophalangeal and the proximal and distal interphalangeal joints
Dorsiflexion of the ankle
Eversion of the foot

## Palpation

Second muscle to the lateral side of the anterior tibial border; upper lateral side of the tibia between the tibialis anterior medially and the fibula laterally; divides into four tendons just distal to anterior ankle with active toe extension

## Innervation

Deep peroneal nerve (L4, L5, S1)

## Application, strengthening, and flexibility

Strength is necessary in the extensor digitorum longus muscle to maintain balance between the plantar and dorsal flexors.

Action that involves dorsal flexion of the ankle and extension of the toes against resistance strengthens both the extensor digitorum longus and the extensor hallucis longus muscles. This may be accomplished by manually applying a downward force on the toes while attempting to extend them up.

The extensor digitorum longus may be stretched by passively taking the four lesser toes into full flexion while the foot is inverted and plantarflexed.

2nd–5th MTP, PIP, and DIP extension

Ankle dorsal flexion

Transverse tarsal and subtalar eversion

O, Lateral condyle of tibia

Extensor digitorum longus m.

O, Head of fibula, upper two-thirds of anterior surface of fibula

Ankle dorsiflexion

Toe extension

Foot eversion

I, Tops of middle and distal phalanxes of four lesser toes

FIG. 10.17 • Extensor digitorum longus muscle, anterior view, right leg and foot. O, Origin; I, Insertion.

Chapter
**10**

**Great toe MTP and IP extension**

**Ankle dorsal flexion**

**Transverse tarsal and subtalar inversion**

# Extensor hallucis longus muscle

FIG. 10.18

(eks-ten´sor hal-u´sis lon´gus)

## Origin

Middle two-thirds of the medial surface of the anterior fibula

## Insertion

Top of the base of the distal phalanx of the great toe

## Action

Dorsiflexion of the ankle

Extension of the great toe at the metatarsophalangeal and interphalangeal joints

Weak inversion of the foot

## Palpation

From the dorsal aspect of the great toe to just lateral to the tibialis anterior and medial to the extensor digitorum longus at the anterior ankle joint

## Innervation

Deep peroneal nerve (L4, L5, S1)

## Application, strengthening, and flexibility

The four dorsiflexors of the foot—tibialis anterior, extensor digitorum longus, extensor hallucis longus, and peroneus tertius—may be exercised by attempting to walk on the heels with the ankle flexed dorsally and toes extended. Extension of the great toe, as well as ankle dorsiflexion against resistance, will provide strengthening for this muscle.

The extensor hallucis longus may be stretched by passively taking the great toe into full flexion while the foot is everted and plantarflexed.

Extensor hallucis longus m.

Ankle dorsiflexion

Toe extension

O, Middle two-thirds of medial surface of anterior fibula

I, Top of base of distal phalanx of great toe

FIG. 10.18 • Extensor hallucis longus muscle, anterior view, right leg and foot. O, Origin; I, Insertion.

# Tibialis anterior muscle   FIG. 10.19

(tib-i-a´lis an-te´ri-or)

## Origin

Upper two-thirds of the lateral surface of the tibia

## Insertion

Inner surface of the medial cuneiform and the base of the first metatarsal bone

## Action

Dorsiflexion of the ankle
Inversion of the foot

## Palpation

First muscle to the lateral side of the anterior tibial border, particularly palpable with fully active ankle dorsiflexion; most prominent tendon crossing the ankle anteromedially

## Innervation

Deep peroneal nerve (L4, L5, S1)

## Application, strengthening, and flexibility

By its insertion, the tibialis anterior muscle is in a fine position to hold up the inner margin of the foot. However, as it contracts concentrically, it dorsiflexes the ankle and is used as an antagonist to the plantar flexors of the ankle. The tibialis anterior is forced to contract strongly when a person ice skates or walks on the outside of the foot. It strongly supports the medial longitudinal arch in inversion.

Turning the sole of the foot to the inside against resistance to perform inversion exercises is one way to strengthen this muscle. Dorsal flexion exercises against resistance may also be used for this purpose. Walking barefoot or in stocking feet on the outside of the foot (inversion) is an excellent exercise for the tibialis anterior muscle.

The tibialis anterior may be stretched by passively taking the foot into extreme eversion and plantar flexion.

Ankle dorsal flexion

Transverse tarsal and subtalar inversion

**Medial view**

Tibialis anterior m.

O, Upper two-thirds of lateral surface of tibia

I, Inner surface of medial cuneiform, base of first metatarsal bone

Ankle dorsiflexion

Foot inversion

FIG. 10.19 • Tibialis anterior muscle, anterior view, right leg and foot. O, Origin; I, Insertion.

Chapter
**10**

# Tibialis posterior muscle  FIG. 10.20

(tib-i-a′lis pos-te′ri-or)

### Origin

Posterior surface of the upper half of the interosseus membrane and adjacent surfaces of the tibia and fibula

### Insertion

Inferior surfaces of the navicular, cuneiform, and cuboid bones and bases of the second, third, and fourth metatarsal bones

### Action

Plantar flexion of the ankle
Inversion of the foot

### Palpation

The tendon may be palpated both proximally and distally immediately behind the medial malleolus with inversion and plantar flexion and is better distinguished from the flexor digitorum longus and flexor hallucis longus if the toes can be maintained in slight extension

### Innervation

Tibial nerve (L5, S1)

## Application, strengthening, and flexibility

Passing down the back of the leg, under the medial malleolus, then forward to the navicular and medial cuneiform bones, the tibialis posterior muscle pulls down from the underside and, when contracted concentrically, inverts and plantar flexes the foot. As a result, it is in position to support the medial longitudinal arch. **Shin splints** is a slang term frequently used to describe an often chronic condition in which the tibialis posterior, tibialis anterior, and extensor digitorum longus muscles are inflamed. This inflammation is usually a tendinitis of one or more of these structures but may be a result of stress fracture, periostitis, tibial stress syndrome, or compartment syndrome. Sprints and long-distance running are common causes, particularly if the athlete has not developed appropriate strength, flexibility, and endurance in the leg musculature.

Use of the tibialis posterior muscle in plantar flexion and inversion gives support to the longitudinal arch of the foot. This muscle is generally strengthened by performing heel raises, as described for the gastrocnemius and soleus, as well as inversion exercises against resistance.

The tibialis posterior may be stretched by passively taking the foot into extreme eversion and dorsiflexion while the knee and toes are passively flexed.

I, Inferior surfaces of the navicular, cuneiform, and cuboid bones and bases of the second, third, and fourth metatarsal bones

O, Posterior surface of upper half of interosseous membrane, adjacent surfaces of tibia and fibula

Foot inversion

Plantar flexion of ankle

FIG. 10.20 • Tibialis posterior muscle, posterior and plantar views. O, Origin; I, Insertion.

# Flexor digitorum longus muscle   FIG. 10.21

(fleks´or dij-i-to´rum lon´gus)

## Origin

Middle third of the posterior surface of the tibia

## Insertion

Base of the distal phalanx of each of the four lesser toes

## Action

Flexion of the four lesser toes at the metatarsophalan-
geal and the proximal and distal interphalangeal joints
Inversion of the foot
Plantar flexion of the ankle

## Palpation

The tendon may be palpated immediately posterior
to the medial malleolus and tibialis posterior and
immediately anterior to the flexor hallucis longus
with flexion of the four lesser toes while maintaining great toe extension, ankle dorsiflexion, and foot
eversion

## Innervation

Tibial nerve (L5, S1)

## Application, strengthening, and flexibility

Passing down the back of the lower leg under the
medial malleolus and then forward, the flexor
digitorum longus muscle draws the four lesser
toes down into flexion toward the heel as it plantar flexes the ankle. It is very important in helping
other foot muscles maintain the longitudinal arch.
Some of the weak foot and ankle conditions result
from ineffective use of the flexor digitorum longus.
Walking barefoot with the toes curled downward
toward the heel and with the foot inverted will
exercise this muscle. It may be strengthened by
performing towel grabs against resistance in which
the heel rests on the floor while the toes extend
to grab a flat towel and then flex to pull the towel
under the foot. This may be repeated numerous
times, with a small weight placed on the opposite
end of the towel for added resistance.

The flexor digitorum longus may be stretched by
passively taking the four lesser toes into extreme
extension while the foot is everted and dorsiflexed.
The knee should be flexed.

2nd–5th
MTP, PIP,
and DIP
flexion

Transverse
tarsal and
subtalar
inversion

Ankle
plantar
flexion

FIG. 10.21 • Flexor
digitorum longus muscle,
posterior and plantar views.
O, Origin; I, Insertion.

## Flexor hallucis longus muscle  FIG. 10.22

(fleks´or hal-u´sis lon´gus)

**Great toe MTP and IP flexion**

### Origin

Middle two-thirds of the posterior surface of the fibula

**Transverse tarsal and subtalar inversion**

### Insertion

Base of the distal phalanx of the great toe, plantar surface

### Action

Flexion of the great toe at the metatarsophalangeal and interphalangeal joints
Inversion of the foot
Plantar flexion of the ankle

**Ankle plantar flexion**

### Palpation

Most posterior of the three tendons immediately behind the medial malleolus; between the medial soleus and the tibia with active great toe flexion while maintaining extension of the four lesser toes, ankle dorsiflexion, and foot eversion

### Innervation

Tibial nerve (L5, S1, S2)

### Application, strengthening, and flexibility

Pulling from the underside of the great toe, the flexor hallucis longus muscle may work independently of the flexor digitorum longus muscle or together with it. If these two muscles are poorly developed, they cramp easily when they are called on to do activities to which they are unaccustomed.

These muscles are used effectively in walking if the toes are used (as they should be) in maintaining balance as each step is taken. Walking "with the toes" rather than "over" them is an important action for them.

When the gastrocnemius, soleus, tibialis posterior, peroneus longus, peroneus brevis, flexor digitorum longus, flexor digitorum brevis, and flexor hallucis longus muscles are all used effectively in walking, the strength of the ankle is evident. If an ankle and a foot are weak, in most cases it is because of lack of use of all the muscles just mentioned. Running, walking, jumping, hopping, and skipping provide exercise for this muscle group. The flexor hallucis longus muscle may be specifically strengthened by performing towel grabs as described for the flexor digitorum longus muscle.

The flexor hallucis longus may be stretched by passively taking the great toe into extreme extension while the foot is everted and dorsiflexed. The knee should be flexed.

**Chapter 10**

O, Middle two-thirds of posterior surface of fibula

Flexor hallucis longus m.

I, Base of distal phalanx of great toe, plantar surface

Plantar flexion of ankle

Toe flexion

Foot inversion

FIG. 10.22 • Flexor hallucis longus muscle, medial view, right leg and foot. O, Origin; I, Insertion.

# Intrinsic muscles of the foot

FIGS. 10.23, 10.24

The intrinsic muscles of the foot have their origins and insertions on the bones within the foot (Figs. 10.23 and 10.24). One of these muscles, the extensor digitorum brevis, is found on the dorsum of the foot. This muscle includes a band that attaches to the base of the first proximal phalanx and is sometimes labeled the extensor hallucis brevis. The remainder of the muscles are found in a plantar compartment in four layers on the plantar surface of the foot, as follows:

**First (superficial) layer:** abductor hallucis, flexor digitorum brevis, abductor digiti minimi (quinti)

**Second layer:** quadratus plantae, lumbricals (four)

**Third layer:** flexor hallucis brevis, adductor hallucis, flexor digiti minimi (quinti) brevis

**Fourth (deep) layer:** dorsal interossei (four), plantar interossei (three)

The intrinsic foot muscles may be grouped by location as well as by the parts of the foot on which they act. The abductor hallucis, flexor hallucis brevis, and adductor hallucis all insert either medially or laterally on the proximal phalanx of the great toe. The abductor hallucis and flexor hallucis brevis are located somewhat medially, whereas the adductor hallucis is more centrally located beneath the metatarsals.

The quadratus plantae, four lumbricals, four dorsal interossei, three plantar interossei, flexor digitorum brevis, and extensor digitorum brevis are all located somewhat centrally. All are beneath the foot except the extensor digitorum brevis, which is the only intrinsic muscle in the foot located in the dorsal compartment. Although the entire extensor digitorum brevis has its origin on the anterior and lateral calcaneus, some anatomists refer to its first tendon as the extensor hallucis brevis in order to maintain consistency in naming according to function and location.

Located laterally beneath the foot are the abductor digiti minimi and the flexor digiti minimi brevis, which both insert on the lateral aspect of the base of the proximal phalanx of the fifth phalange.

Because of these two muscles' insertion and action on the fifth toe, the name "quinti" is sometimes used instead of "minimi."

Four muscles act on the great toe. The abductor hallucis is solely responsible for abduction of the great toe but assists the flexor hallucis brevis in flexing the great toe at the metatarsophalangeal joint. The adductor hallucis is the sole adductor of the great toe, while the extensor digitorum brevis is the only intrinsic extensor of the great toe at the metatarsophalangeal joint.

The four lumbricals are flexors of the second, third, fourth, and fifth phalanges at their metatarsophalangeal joints, while the quadratus plantae muscles are flexors of these phalanges at their distal interphalangeal joints. The three plantar interossei are adductors and flexors of the proximal phalanxes of the third, fourth, and fifth phalanges, while the four dorsal interossei are abductors and flexors of the second, third, and fourth phalanges, also at their metatarsophalangeal joints. The flexor digitorum brevis flexes the middle phalanxes of the second, third, fourth, and fifth phalanges. The extensor digitorum brevis, as previously mentioned, is an extensor of the great toe but also extends the second, third, and fourth phalanges at their metatarsophalangeal joints.

There are two muscles that act solely on the fifth toe. The proximal phalanx of the fifth phalange is abducted by the abductor digiti minimi and is flexed by the flexor digiti minimi brevis.

Refer to Table 10.3 for further details regarding the intrinsic muscles of the foot.

Muscles are developed and maintain their strength only when they are used. One factor in the great increase in weak foot conditions is the lack of exercise to develop these muscles. Walking is one of the best activities for maintaining and developing the many small muscles that help support the arch of the foot. Some authorities advocate walking without shoes or with shoes designed to enhance proper mechanics. Additionally, towel exercises such as those described for the flexor digitorum longus and flexor hallucis longus are helpful in strengthening the intrinsic muscles of the foot.

**Great toe MTP and IP flexion**

**2nd–5th MTP, PIP, and DIP flexion**

**A**

Fibrous digital sheaths

Lumbricals

Flexor digiti minimi brevis

Abductor digiti minimi

Plantar interosseous

Tendon of flexor hallucis longus

Flexor hallucis brevis

Flexor digitorum brevis

Abductor hallucis

Plantar aponeurosis (cut)

Calcaneal tuberosity

**B**

Tendons of flexor digitorum brevis (cut)

Tendon of flexor hallucis longus

Flexor hallucis brevis

Lumbricals

Tendon of flexor digitorum longus

Quadratus plantae

Flexor digitorum brevis (cut)

Abductor hallucis (cut)

Flexor digiti minimi brevis

Abductor digiti minimi

**C**

Tendon of flexor digitorum brevis (cut)

Plantar interossei

Quadratus plantae (cut)

Plantar ligaments

Tendon of lumbrical (cut)

Tendons of flexor digitorum longus (cut)

Transverse head and

Oblique head of adductor hallucis

Flexor hallucis brevis

Tendon of flexor hallucis longus (cut)

Tendon of flexor digitorum longus (cut)

**D**

Articular capsules

Sesamoid bones

Dorsal interossei

Plantar interossei

Tendon of peroneus longus

Tendon of tibialis posterior

Long plantar ligament

Opponens digiti minimi

Peroneus brevis tendon

FIG. 10.23 • The four musculotendinous layers of the plantar aspect of the right foot, detailing the intrinsic muscles. **A,** First (superficial) layer; **B,** Second layer; **C,** Third layer; **D,** Fourth (deep) layer.

Anterior tibial artery and
deep fibular nerve

Superior extensor
retinaculum

Peroneus tertius

Tendon of extensor
digitorum longus

Lateral malleolus

Inferior extensor
retinaculum

Extensor digitorum
brevis

Tendon of peroneus
brevis

Tuberosity of fifth
metatarsal bone

Tendon of peroneus tertius

Abductor digiti minimi

Tendons of extensor
digitorum brevis

Tendons of extensor
digitorum longus

Tendon of tibialis anterior

Medial malleolus

Anterior medial malleolar artery

Tendon of extensor hallucis longus

Dorsalis pedis artery

Deep fibular nerve

Extensor hallucis brevis

Arcuate artery

Abductor hallucis

First dorsal interosseous

Tendon of extensor hallucis
brevis

Extensor expansions

Dorsal digital arteries

Dorsal digital branches of
superficial fibular nerve

Creek

FIG. 10.24 • Anterior view of dorsum of right foot.

## TABLE 10.3 • **Intrinsic muscles of the foot**

| | Muscle | Origin | Insertion | Action | Palpation | Innervation |
|---|---|---|---|---|---|---|
| Superficial layer | Flexor digitorum brevis | Tuberosity of cal-caneus, plantar aponeurosis | Medial and lateral aspects of 2nd, 3rd, 4th, and 5th middle phalanxes | MP and PIP flexion of 2nd, 3rd, 4th, and 5th phalanges | Cannot be palpated | Medial plantar nerve (L4, L5) |
| | Abductor digiti minimi (quinti) | Tuberosity of cal-caneus, plantar aponeurosis | Lateral aspect of 5th proxi-mal phalanx | MP abduc-tion of 5th phalange | Cannot be palpated | Lateral plantar nerve (S1, S2) |
| | Abductor hallucis | Tuberosity of cal-caneus, flexor reti-naculum, plantar aponeurosis | Medial aspect of base of 1st proximal phalanx | MP flexion, abduction of 1st phalange | On the plantar aspect of the foot from medial tuber-cle of calcaneus to medial side of great toe proximal phalanx with great toe abduction | Medial plantar nerve (L4, L5) |

Chapter
**10**

321

TABLE 10.3 (Continued) • **Intrinsic muscles of the foot**

| | Muscle | Origin | Insertion | Action | Palpation | Innervation |
|---|---|---|---|---|---|---|
| **Second layer** | Quadratus plantae | Medial head: medial surface of calcaneus Lateral head: lateral border of inferior surface of calcaneus | Lateral margin of flexor digitorum longus tendon | DIP flexion of 2nd, 3rd, 4th, and 5th phalanges | Cannot be palpated | Lateral plantar nerve (S1, S2) |
| | Lumbricals (4) | Tendons of flexor digitorum longus | Dorsal surface of 2nd, 3rd, 4th, and 5th proximal phalanxes | MP flexion of 2nd, 3rd, 4th, and 5th phalanges | Cannot be palpated | 1st lumbricals: medial plantar nerve (L4, L5) 2nd, 3rd, 4th lumbricals: lateral plantar nerve (S1, S2) |
| **Third layer** | Adductor hallucis | Oblique head: 2nd, 3rd, and 4th metatarsals and sheath of peroneus longus tendon Transverse head: plantar metatarsophalangeal ligaments of 3rd, 4th, and 5th phalanges and transverse metatarsal ligaments | Lateral aspect of base of 1st proximal phalanx | MP adduction of 1st phalange | Cannot be palpated | Lateral plantar nerve (S1, S2) |
| | Flexor hallucis brevis | Cuboid, lateral cuneiform | Medial head: medial aspect of 1st proximal phalanx Lateral head: lateral aspect of 1st proximal phalanx | MP flexion of 1st phalange | Cannot be palpated | Medial plantar nerve (L4, L5, S1) |
| | Flexor digiti minimi (quinti) brevis | Base of 5th metatarsal, sheath of peroneus longus tendon | Lateral aspect of base of 5th proximal phalanx | MP flexion of 5th phalange | Cannot be palpated | Lateral plantar nerve (S2, S3) |
| **Fourth layer** | Plantar interossei (3) | Bases and medial shafts of 3rd, 4th, and 5th metatarsals | Medial aspects of bases of 3rd, 4th, and 5th proximal phalanxes | MP adduction and flexion of 3rd, 4th, and 5th phalanges | Cannot be palpated | Lateral plantar nerve (S1, S2) |

## TABLE 10.3 (Continued) • Intrinsic muscles of the foot

| | Muscle | Origin | Insertion | Action | Palpation | Innervation |
|---|---|---|---|---|---|---|
| **Fourth layer** | Dorsal interossei (4) | Two heads on shafts of adjacent metatarsals | 1st interosseus: medial aspect of 2nd proximal phalanx<br>2nd, 3rd, and 4th interossei: lateral aspects of 2nd, 3rd, and 4th proximal phalanxes | MP abduction and flexion of 2nd, 3rd, and 4th phalanges | Cannot be palpated | Lateral plantar nerve (S1, S2) |
| | Extensor digitorum brevis (including extensor hallucis brevis) | Anterior and lateral calcaneus, lateral talocalcaneal ligament, inferior extensor retinaculum | Base of proximal phalanx of 1st phalange, lateral sides of extensor digitorum longus tendons of 2nd, 3rd, and 4th phalanges | Assists in MP extension of 1st phalange and extension of middle three phalanges | As a mass anterior to and slightly below lateral malleolus on dorsum of foot | Deep peroneal nerve (L5, S1) |

## REVIEW EXERCISES

1. List the planes in which each of the following movements occurs. List the axis of rotation for each movement in each plane.
   a. Plantar flexion
   b. Dorsiflexion
   c. Inversion
   d. Eversion
   e. Flexion of the toes
   f. Extension of the toes
   g. Abduction of the metatarsophalangeal joints
   h. Adduction of the metatarsophalangeal joints
2. Why are "low arches" and "flat feet" not synonymous terms?
3. Discuss the value of proper footwear in various sports and activities.
4. What are orthotics and how do they function?
5. Research common foot and ankle disorders, such as flat feet, lateral ankle sprains, high ankle sprains, bunions, plantar fasciitis, and hammertoes. Report your findings in class.
6. Research the anatomical factors related to the prevalence of inversion versus eversion ankle sprains and report your findings in class.
7. Report orally or in writing on magazine articles that rate running and walking shoes.

8. Muscle analysis chart • Ankle, transverse tarsal and subtalar joints, and toes

| Complete the chart by listing the muscles primarily involved in each movement. | |
|---|---|
| **Ankle** | |
| Dorsiflexion | Plantar flexion |
| **Transverse tarsal and subtalar joints** | |
| Eversion | Inversion |
| **Toes** | |
| Flexion | Extension |

9. Antagonistic muscle action chart • Ankle, transverse tarsal and subtalar joints, and toes

Complete the chart by listing the muscle(s) or parts of muscles that are antagonist in their actions to the muscles in the left column.

| Agonist | Antagonist |
|---|---|
| Gastrocnemius | |
| Soleus | |
| Tibialis posterior | |
| Flexor digitorum longus | |
| Flexor hallucis longus | |
| Peroneus longus/Peroneus brevis | |
| Peroneus tertius | |
| Extensor digitorum longus | |
| Extensor hallucis longus | |

# LABORATORY EXERCISES

1. Locate the following bony landmarks of the ankle and foot on a human skeleton and on a subject:
   a. Lateral malleolus
   b. Medial malleolus
   c. Calcaneus
   d. Navicular
   e. Three cuneiform bones
   f. Cuboid
   g. Metatarsal bones
   h. Phalanges and individual phalanxes

2. How and where can the following muscles be palpated on a human subject?
   a. Tibialis anterior
   b. Extensor digitorum longus
   c. Peroneus longus
   d. Peroneus brevis
   e. Soleus
   f. Gastrocnemius
   g. Extensor hallucis longus
   h. Flexor digitorum longus
   i. Flexor hallucis longus
   j. Tibialis posterior

3. Demonstrate and palpate the following movements:
   a. Plantar flexion of ankle
   b. Dorsiflexion of ankle
   c. Inversion of the foot
   d. Eversion of the foot
   e. Flexion of the toes
   f. Extension of the toes

4. Have a laboratory partner rise up on the toes (heel raise) with the knees fully extended and then repeat with the knees flexed approximately 20 degrees. Which exercise position appears to be more difficult to maintain for an extended period of time and why? What are the implications for strengthening these muscles? For stretching these muscles?

5. Ankle and foot joint exercise movement analysis chart

After analyzing each exercise in the chart, break each into two primary movement phases, such as a lifting phase and a lowering phase. For each phase, determine what ankle and foot joint movements occur, and then list the ankle and foot joint muscles primarily responsible for causing/controlling those movements or maintaining the position. Beside each muscle in each movement, indicate the type of contraction as follows: I—isometric; C—concentric; E—eccentric.

| Exercise | Initial movement (lifting) phase | | Secondary movement (lowering) phase | |
| --- | --- | --- | --- | --- |
| | Movement(s) | Agonist(s)—(contraction type) | Movement(s) | Agonist(s)—(contraction type) |
| Push-up | | | | |
| Squat | | | | |
| Dead lift | | | | |
| Hip sled | | | | |
| Forward lunge | | | | |
| Rowing exercise | | | | |
| Stair machine | | | | |

## 6. Ankle and foot joint sport skill analysis chart

Analyze each skill in the chart, and list the movements of the right and left ankle and foot joints in each phase of the skill. You may prefer to list the initial positions that the ankle and foot joints are in for the stance phase. After each movement, list the ankle and foot joint muscle(s) primarily responsible for causing/controlling the movement or maintaining the position. Beside each muscle in each movement, indicate the type of contraction as follows: I—isometric; C—concentric; E—eccentric. It may be desirable to review the concepts for analysis in Chapter 12 for the various phases. Assume right hand/leg dominant where applicable.

| Exercise | | Stance phase | Preparatory phase | Movement phase | Follow-through phase |
| --- | --- | --- | --- | --- | --- |
| Baseball pitch | (R) | | | | |
| | (L) | | | | |
| Football punt | (R) | | | | |
| | (L) | | | | |
| Walking | (R) | | | | |
| | (L) | | | | |
| Softball pitch | (R) | | | | |
| | (L) | | | | |
| Soccer pass | (R) | | | | |
| | (L) | | | | |
| Batting | (R) | | | | |
| | (L) | | | | |
| Bowling | (R) | | | | |
| | (L) | | | | |
| Basketball jump shot | (R) | | | | |
| | (L) | | | | |

# References

Astrom M, Arvidson T: Alignment and joint motion in the normal foot, *Journal of Orthopaedic and Sports Physical Therapy* 22:5, November 1995.

Booher JM, Thibodeau GA: *Athletic injury assessment,* ed 4, New York, 2000, McGraw-Hill.

Cleland JA, Koppenhaver S, Su J: *Netter's orthopaedic clinical examination: an evidence-based approach*, ed 3, Philadelphia, 2016, Elsevier.

Field D: *Anatomy: palpation and surface markings,* ed 3, Oxford, 2001, Butterworth-Heinemann.

Gench BE, Hinson MM, Harvey PT: *Anatomical kinesiology,* Dubuque, IA, 1995, Eddie Bowers.

Hamilton N, Weimer W, Luttgens K: *Kinesiology: scientific basis of human motion,* ed 12, New York, 2012, McGraw-Hill.

Hislop HJ, Montgomery J: *Daniels and Worthingham's muscle testing: techniques of manual examination,* ed 8, Philadelphia, 2007, Saunders.

Lindsay DT: *Functional human anatomy,* St. Louis, 1996, Mosby.

Magee DJ: *Orthopedic physical assessment,* ed 6, St. Louis, 2014, Elsevier Saunders.

Muscolino JE: *The muscular system manual: the skeletal muscles of the human body,* ed 3, St. Louis, 2010, Elsevier Mosby.

Norkus SA, Floyd RT: The anatomy and mechanisms of syndesmotic ankle sprains, *Journal of Athletic Training* 36(1): 68–73, January–March 2001.

Oatis CA: *Kinesiology: the mechanics and pathomechanics of human movement,* ed 2, Philadelphia, 2008, Lippincott Williams & Wilkins.

Prentice WE: *Principles of athletic training: a competency based approach,* ed 15, New York, 2014, McGraw-Hill.

Putz R, Pabst R: *Sobotta-Atlas of human anatomy one volume edition,* ed 14, Germany, 2008, Elsevier.

Rockar PA: The subtalar joint: anatomy and joint motion, *Journal of Orthopaedic and Sports Physical Therapy* 21:6, June 1995.

Saladin KS: *Anatomy & physiology: the unity of form and function,* ed 5, New York, 2010, McGraw-Hill.

Sammarco GJ: Foot and ankle injuries in sports, *American Journal of Sports Medicine* 14:6, November–December 1986.

Schünke M, Schulte E, Schumacher U: *Thieme atlas of anatomy: general anatomy and musculoskeletal system,* ed 2, New York, 2014, Thieme Medical.

Seeley RR, Stephens TD, Tate P: *Anatomy & physiology,* ed 8, New York, 2008, McGraw-Hill.

Sieg KW, Adams SP: *Illustrated essentials of musculoskeletal anatomy,* ed 4, Gainesville, FL, 2002, Megabooks.

Stone RJ, Stone JA: *Atlas of the skeletal muscles,* ed 6, New York, 2009, McGraw-Hill.

Thibodeau GA, Patton KT: *Anatomy & physiology,* ed 9, St. Louis, 1993, Mosby.

Van De Graaff KM: *Human anatomy,* ed 6, Dubuque, IA, 2002, McGraw-Hill.

# Worksheet Exercises

For in- or out-of-class assignments, or for testing, utilize this tear-out worksheet.

## Worksheet 1

Using crayons or colored markers, draw and label on the worksheet the following muscles. Indicate the origin and insertion of each muscle with an "O" and an "I," respectively, and draw in the origin and insertion on the anterior or posterior view as applicable.

a. Tibialis anterior

b. Extensor digitorum longus

c. Peroneus longus

d. Peroneus brevis

e. Peroneus tertius

f. Soleus

g. Gastrocnemius

h. Extensor hallucis longus

i. Tibialis posterior

j. Flexor digitorum longus

k. Flexor hallucis longus

# Worksheet Exercises

For in- or out-of-class assignments, or for testing, utilize this tear-out worksheet.

## Worksheet 2

Label and indicate with arrows the following movements of the talocrural, transverse tarsal, and subtalar joints. For each motion, complete the sentence by supplying the plane in which it occurs and the axis of rotation as well as the muscles causing the motion.

a.  Dorsiflexion occurs in the _____ plane about the _____ axis and is accomplished by concentric contractions of the _____ _____ muscles.

b.  Plantar flexion occurs in the _____ plane about the _____ axis and is accomplished by concentric contractions of the _____ _____ muscles.

c.  Eversion occurs in the _____ plane about the _____ axis and is accomplished by concentric contractions of the _____ _____ muscles.

d.  Inversion occurs in the _____ plane about the _____ axis and is accomplished by concentric contractions of the _____ _____ muscles.

A

B

C

D

# THE TRUNK AND SPINAL COLUMN

## Objectives

- To identify and differentiate the different types of vertebrae in the spinal column

- To label on a skeletal chart the types of vertebrae and their important features

- To appreciate the joint structure and role of the ligaments in providing stability to the spinal column

- To draw and label on a skeletal chart the larger muscles of the trunk and the spinal column

- To demonstrate and palpate on a human subject the movements of the spine and trunk and list their respective planes of motion and axes of rotation

- To palpate on a human subject the larger muscles of the trunk and spinal column

- To list and organize the muscles that produce the primary movements of the trunk and spinal column and their antagonists

- To learn and understand the innervation of the major trunk and spinal column muscles

- To determine, through analysis, the trunk and spinal column movements and muscles involved in selected skills and exercises

The trunk and spinal column present problems in kinesiology that are not found in the study of other parts of the body. The vertebral column is quite elaborate, consisting of 24 articulating vertebrae with an additional 9 nonmovable vertebrae. These vertebrae contain the spinal column, with its 31 pairs of spinal nerves. Most would agree that it is one of the more complex parts of the human body.

The anterior portion of the trunk contains the abdominal muscles, which are somewhat different from most other muscles in that some sections are linked by fascia and tendinous bands and thus do not attach from bone to bone. In addition, there are many small intrinsic muscles acting on the head, vertebral column, and thorax that assist in spinal stabilization or respiration, depending on their location. These muscles are generally too deep to palpate and consequently will not be given the full attention that the larger superficial muscles receive in this chapter.

## Bones

### Vertebral column

The intricate and complex bony structure of the vertebral column consists of 24 articulating vertebrae (freely movable) and 9 that are fused (Fig. 11.1). The column is further divided into the 7 cervical (neck) vertebrae, 12 thoracic (chest) vertebrae, and 5 lumbar (lower back) vertebrae. The sacrum (posterior pelvic girdle) and the coccyx (tailone) consist of 5 and 4 fused vertebrae, respectively.

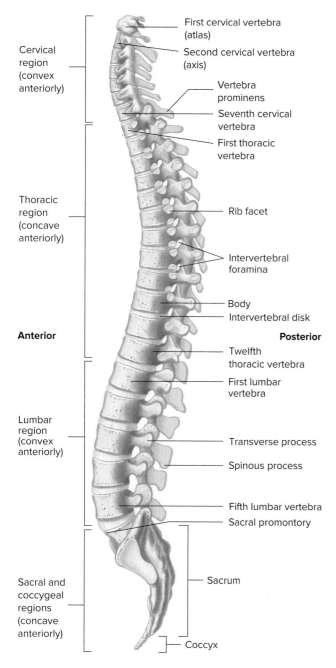

Cervical region (convex anteriorly)

First cervical vertebra (atlas)

Second cervical vertebra (axis)

Vertebra prominens

Seventh cervical vertebra

First thoracic vertebra

Thoracic region (concave anteriorly)

Rib facet

Intervertebral foramina

Body

Intervertebral disk

**Anterior**

**Posterior**

Twelfth thoracic vertebra

First lumbar vertebra

Lumbar region (convex anteriorly)

Transverse process

Spinous process

Fifth lumbar vertebra

Sacral promontory

Sacral and coccygeal regions (concave anteriorly)

Sacrum

Coccyx

FIG. 11.1 • Complete vertebral column, left lateral view.

The bones in each region of the spine have slightly different sizes and shapes to allow for various functions (Fig. 11.2). The vertebrae increase in size from the cervical region to the lumbar region, primarily because they have to support more weight in the lower back than in the neck. The first two cervical vertebrae are known as the atlas and the axis, respectively, and are unique in that their shapes allow for extensive rotary movement of the head to the sides, as well as movement forward and backward. Vertebrae C3 through L5 have similar architecture: each has a

bony block anteriorly, known as the body, a vertebral foramen centrally for the spinal cord to pass through, a transverse process projecting laterally to each side, and a spinous process projecting posteriorly that is easily palpable. These processes are key attachment points for many of the spinal column muscles.

The spine has three normal curves within its movable vertebrae. The primary spinal curve prior to birth and briefly afterward is kyphotic, or C-shaped. As muscle development occurs and activity increases, the secondary curves, which are lordotic, develop in the cervical and lumbar regions. The thoracic curve is concave anteriorly and convex posteriorly, whereas the cervical and lumbar curves are convex anteriorly and concave posteriorly. Finally, the sacral curve, including the coccyx, is concave anteriorly and convex posteriorly. The normal curves of the spine enable it to absorb blows and shocks.

Undesirable deviations from the normal curvatures occur due to a number of factors including genetics, pathological conditions, and posture. Increased posterior concavity of the lumbar and cervical curves is known as **lordosis**, and increased anterior concavity of the normal thoracic curve is known as **kyphosis**. The lumbar spine may have a reduction of its normal lordotic curve, resulting in a flat-back appearance referred to as **lumbar kyphosis**. **Scoliosis** consists of lateral curvatures or sideward deviations of the spine.

## Thorax

Skeletal foundation of the thorax is formed by 12 pairs of ribs (Fig. 11.3). Seven pairs are true ribs (vertebrosternal), in that they attach directly to the sternum via a separate costal cartilage. Five pairs are considered false ribs. Of these, three pairs (vertebrochondral) attach indirectly to the sternum via a shared costal cartilage and two pairs (vertebral) are floating ribs, with their ends free. The manubrium, the body of the sternum, and the xiphoid process are the other bones of the thorax. All the ribs are attached posteriorly to the thoracic vertebrae.

Key bony landmarks for locating the muscles of the neck include the mastoid process, transverse and spinous processes of the cervical spine, spinous processes of the upper four thoracic vertebrae, manubrium of the sternum, and medial clavicle. The spinous and transverse processes of the thoracic spine and the posterior ribs are key areas of attachment for the posterior muscles of the spine. The anterior trunk muscles have attachments on the borders of the lower eight ribs, the costal cartilages of the ribs, the iliac crest, and the pubic crest. The transverse processes of the upper

FIG. 11.2 • Vertebral column. **A,** Atlas (first cervical vertebra), superior view; **B,** Axis (second cervical vertebra), superior view; **C,** Typical cervical vertebra, superior view; **D,** Typical cervical vertebra, lateral view; **E,** Typical thoracic vertebra, superior view; **F,** Typical thoracic vertebra, lateral view; **G,** Third lumbar vertebra, superior view; **H,** Third lumbar vertebra, lateral view.

Chapter
**11**

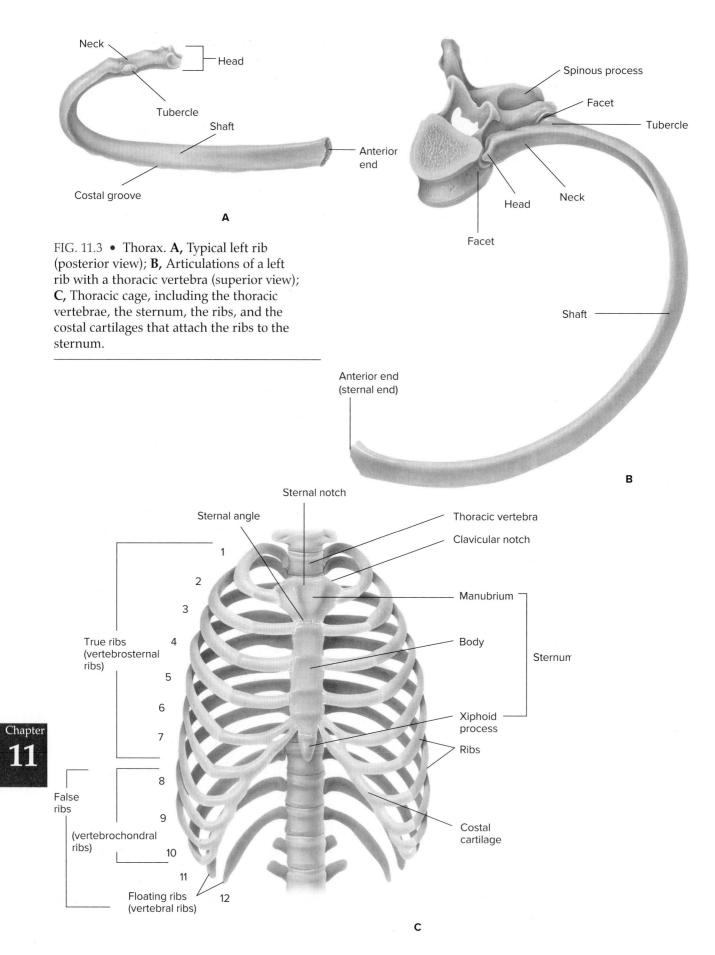

Neck

Head

Tubercle

Shaft

Anterior end

Costal groove

**A**

FIG. 11.3 • Thorax. **A,** Typical left rib (posterior view); **B,** Articulations of a left rib with a thoracic vertebra (superior view); **C,** Thoracic cage, including the thoracic vertebrae, the sternum, the ribs, and the costal cartilages that attach the ribs to the sternum.

Spinous process

Facet

Tubercle

Head

Neck

Facet

Shaft

Anterior end
(sternal end)

**B**

Sternal notch

Sternal angle

Thoracic vertebra

Clavicular notch

1
2
3

True ribs
(vertebrosternal ribs)

4

Manubrium

Body

Sternum

5
6
7

False
ribs

8

Xiphoid process

Ribs

9

(vertebrochondral ribs)

10

11

Floating ribs
(vertebral ribs)

12

Costal cartilage

**C**

four lumbar vertebrae also serve as points of insertion for the quadratus lumborum, along with the lower border of the twelfth rib.

## Joints

The first joints in the axial skeleton are the atlanto-occipital joints, formed by the occipital condyles of the skull sitting on the articular fossae of the first vertebra, which allow approximately 20 degrees of capital flexion and approximately 10 degrees of capital extension or flexion and extension of the head on the neck. This pair of joints is also referred to as the craniovertebral joints and consists of a right and left condyloid joint which also allows approximately 5 degrees of lateral flexion. Although these are separate articulations, their movements are often grouped with those of the cervical spine. The atlas (C1) in turn sits on the axis (C2) to form the atlantoaxial joint (Fig. 11.4, *A*). Except for the atlantoaxial joint, there is not a great deal of movement possible between any two vertebrae. However, the cumulative effect of combining the movement from several adjacent vertebrae allows for substantial movements

within a given area. Most of the rotation within the cervical region occurs in the atlantoaxial joint, which is classified as a trochoid or pivot-type joint. The remainder of the vertebral articulations are classified as arthrodial or gliding-type joints because of their limited, gliding movements that occur in the facet joints on each side. These synovial joints are sometimes referred to as zygapophyseal or apophyseal joints. Movement also occurs between C2 through S1 at the intervertebral disks which are classified as amphiarthrodial symphysis joints.

As shown in Fig 11.4, *B* and *D,* the anterior longitudinal ligament runs the entire length of the spine from the base of the skull to the sacrum and attaches to the anterior surface of each vertebral

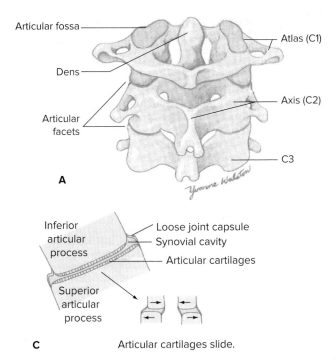

FIG. 11.4 • Articular facets of the vertebrae. **A,** The facets of the superior and inferior articular processes articulate between adjacent cervical vertebrae (posterior view); **B,** Ligaments limit motion between vertebrae, shown in sagittal section through three lumbar vertebrae; **C,** Articular cartilages slide back and forth on each other, a motion allowed by the loose articular capsule; **D,** Intervertebral articulations (right posterolateral view). Vertebrae articulate with adjacent vertebrae at both their superior and inferior articular processes. Intervertebral disks separate the bodies of adjacent vertebrae.

body. The posterior longitudinal ligament is located inside the vertebral canal, attaching on the posterior vertebral bodies, and runs from the axis to the sacrum. The ligamentum flavum binds the laminae of adjacent vertebrae together. The interspinous ligaments connect the spinous processes, and the intertransverse ligaments connect the transverse processes. The ligamentum nuchae connects the tips of the cervical spinous processes from the occipital protuberance to all seven cervical vertebrae, and the supraspinous ligament connects the tips of the spinous processes of the remaining vertebrae.

TABLE 11.1 • **Major ligaments of the trunk and spinal column**

| Joint/structure | Ligaments | Attachments | Functional application and joint stability |
|---|---|---|---|
| Intervertebral intersegmental | Anterior longitudinal ligament | The anterior surface of each vertebral body and annulus fibrosis from the occipital bone to the pelvic surface of the sacrum | A primary spine stabilizer running the entire length of the spine, about one inch wide; limits hyperextension of the vertebral column |
| | Posterior longitudinal ligament | The posterior vertebral bodies and the annulus fibrosis from the axis to the sacrum | A primary spine stabilizer inside the vertebral canal, about one inch wide; limits hyperflexion of the vertebral column; prevents posterior protrusion of the nucleus pulposus of the intervertebral disk |
| | Ligamentum nuchae (nuchal ligament) | External occipital protuberance to the posterior tubercle of C1 and posterior border of the foramen magnum of each cervical vertebrae down to the seventh cervical spinous process | Forms a septum that divides the posterior aspect of the neck in the midline to provide an attachment for several muscles, supports the head and neck when flexion puts tension on it; limits cervical hyperflexion |
| | Supraspinous ligament | Tips of the spinous processes of the vertebrae from C7 to the sacrum | Considered by some to be part of the ligamentum nuchae in the cervical spine; thicker and broader in the lumbar than in the thoracic region; limits hyperflexion |
| Intervertebral intrasegmental | Ligamentum flavum | Superiorly to the anterior surfaces of the inferior border of the lamina and inferiorly to the posterior surfaces of the superior border of the next lamina | The strongest spinal ligament consists mostly of yellow elastin tissue; protects the spinal cord and nerves; located just anterior to the facet joint capsules from the intervertebral foramina laterally to interspinous ligament posteriorly between the laminae of each vertebrae from the base of the skull to the pelvis; lengthened by flexion; its elastic fibers assist in returning to the normal curvature |
| | Interspinous ligaments | From the root to the apex of each spinous process joining the ligament flavum anteriorly and the supraspinous ligament posteriorly | Thinnest in the cervical region and considered by some to be part of the ligamentum nuchae in the cervical spine, narrow in the thoracic region, but broader and thicker in the lumbar region; connect the spinous processes and are taut in flexion |
| | Intertransverse ligaments | From the inferior aspect of the transverse process down to the superior aspect of the adjacent transverse process | Often blend with the intertransverse muscles; limit lateral flexion |

TABLE 11.1 (continued) • **Major ligaments of the trunk and spinal column**

| Joint/ structure | Ligaments | Attachments | Functional application and joint stability |
|---|---|---|---|
| Costovertebral joints | Superior costotransverse ligament (costovertebral) | From the neck of the rib to the transverse process above | Provide stability to the costotransverse joint and limit gliding movement of the joint |
| | Lateral costotransverse ligament (costovertebral) | From the posterior surface of the transverse process tip to the non-articular portion of the corresponding rib tubercle | |
| | Costotransverse ligament (middle costovertebral) | From the neck of the rib to anterior portion of the adjacent transverse process | |
| | Radiate ligament | Anterior aspect of the head of the rib to the lateral bodies of the two adjacent vertebrae | Consists of three flat fasciculi (superior, middle, and inferior); stabilizes the costovertebral joints |

Gliding movement occurs between the superior and inferior articular processes that form the facet joints of the vertebrae, as depicted in Figs. 11.2 and 11.4, *C* and *D*. Located in between and adhering to the articular cartilage of the vertebral bodies are the intervertebral disks (Fig. 11.4, *B*). These disks are composed of an outer rim of dense fibrocartilage known as the annulus fibrosus and a central gelatinous, pulpy substance known as the nucleus pulposus (Fig. 11.4, *D*). This arrangement of compressed elastic material allows compression in all directions, along with torsion. With age, injury, or improper use of the spine, the intervertebral disks become less resilient, resulting in a weakened annulus fibrosus. Substantial weakening combined with compression can result in the nucleus protruding through the annulus, known as a herniated nucleus pulposus. Commonly referred to as a herniated or "slipped" disk, this protrusion puts pressure on the spinal nerve root, causing a variety of symptoms, including radiating pain, tingling, numbness, and weakness in the dermatomes and myotomes of the extremity supplied by the spinal nerve (Fig. 11.5).

A substantial number of low back problems are caused by improper use of the back over time. These improper mechanics often result in acute strains and muscle spasm of the lumbar extensors and chronic mechanical changes leading to disk herniation. Most problems occur from using the relatively small back muscles to lift objects from a lumbar spine flexed position instead of keeping the lumbar spine in a neutral position while squatting and using the larger, more powerful muscles of the lower extremity. Additionally, our lifestyles chronically place us in lumbar flexion, which over time leads to a gradual loss of lumbar lordosis. This flat-back syndrome results in increased pressure on the lumbar disk and intermittent or chronic low back pain.

Additionally, a number of cervical issues occur throughout life due to poor posture in the lumbar and thoracic area leading to a forward head posture and cervical hyperextension. Inappropriate pillow height, sleeping habits, telephone and computer usage may all be contributing factors. Most of the spinal column movement occurs in the cervical and lumbar regions. There is, of course, some thoracic movement, but it is slight in comparison with that of the neck and low back. In discussing movements of the head, it must be remembered that this movement occurs between the occiput of the cranium and the first cervical vertebra, as well as within the cervical vertebrae. With the understanding that these motions usually occur together, for simplification purposes this text refers to all movements of the head and neck as cervical movements. Similarly, in discussing trunk movements, lumbar motion terminology is used to describe the combined motion that occurs in both the thoracic and the lumbar regions. A closer investigation of specific motion between any two vertebrae is beyond the scope of this text.

Chapter

**11**

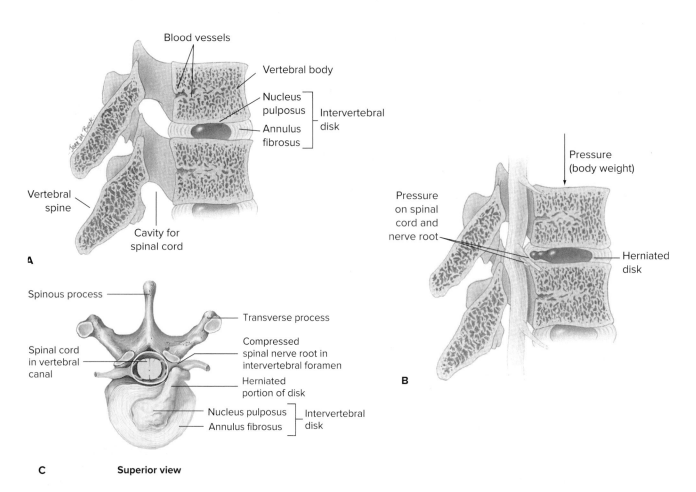

FIG. 11.5 • Intervertebral disks. **A,** Sagittal section of normal disks; **B,** Sagittal section of herniated disks; **C,** Herniated disk, superior view.

FIG. 11.6 • Active ROM of the cervical spine. **A,** Flexion and extension can be estimated in degrees or indicated by the distance the chin lacks from touching the chest; **B,** Rotation can be estimated in degrees or percentages of motion compared in each direction; **C,** Lateral flexion can be estimated in degrees or indicated by the distance the ear lacks from reaching the shoulders.

The cervical region (Fig. 11.6) can flex 45 degrees and extend 45 degrees. The cervical area laterally flexes 45 degrees and can rotate approximately 60 degrees. The lumbar spine, accounting for most of the trunk movement (Fig. 11.7), flexes approximately 80 degrees and extends 20 to 30 degrees. Lumbar lateral flexion to each side is usually within 35 degrees, and approximately 45 degrees of rotation occurs to the left and right.

FIG. 11.7 • Active ROM of the thoracic and lumbar spine. **A,** Forward flexion. Motion can be estimated in degrees or by measurement from fingertips to leg or to floor; **B** and **C,** Using a tape measure to compare the increased length of the lumbar and thoracic spine from the anatomical position to the fully flexed position (not shown, the reverse movement may be done to assess the amount of extension); **D,** Extension (hyperextension) with the subject standing; **E,** Extension (hyperextension) with the subject lying prone; **F,** Lateral bending; **G,** Rotation of the spine.

Cervical
flexion

Cervical
extension

Cervical
lateral
flexion

Cervical
rotation
unilaterally

## Movements FIG. 11.8

Spinal movements are often preceded by the name given to the region of movement. For example, flexion of the trunk at the lumbar spine is known as lumbar flexion, and extension of the neck is often referred to as cervical extension. Though usually included with cervical flexion and extension, isolated movements of the head or the neck at the atlantoccipital joint are technically known as capital flexion and extension. Additionally, as discussed in Chapter 8, the pelvic girdle rotates as a unit due to movement occurring in the hip joints and the lumbar spine. Refer to Table 8.2.

**Spinal flexion:** anterior movement of the spine in the sagittal plane; in the cervical region, the head moves toward the chest; in the lumbar region, the thorax moves toward the pelvis

**Spinal extension:** return from flexion; posterior movement of the spine in the sagittal plane; in the cervical spine, the head moves away from the chest; in the lumbar spine, the thorax moves away from the pelvis; sometimes referred to as hyperextension

**Lateral flexion (left or right):** sometimes referred to as side bending; the head moves laterally toward the shoulder, and the thorax moves laterally toward the pelvis; both in the frontal plane

**Spinal rotation (left or right):** rotary movement of the spine in the transverse plane; the chin rotates from neutral toward the shoulder, and the thorax rotates toward one iliac crest

**Reduction:** return movement from lateral flexion to neutral in the frontal plane

**Cervical flexion
(combined with capital flexion)**

A

**Cervical extension
(combined with capital extension)
(hyperextension)**

B

**Capital flexion**

C

Courtesy of Britt Jones (all)

**Capital extension**

D

**Cervical lateral flexion
to the right**

E

**Cervical rotation
to the right**

F

FIG. 11.8 • Movements of the spine.

Chapter
**11**

338

Courtesy of Britt Jones (all)

| **Lumbar flexion** | **Lumbar extension (hyperextension)** | **Lumbar lateral flexion to the right** | **Lumbar rotation to the right** |
| G | H | I | J |

Lumbar flexion

Lumbar extension

Lumbar lateral flexion

Lumbar rotation unilaterally

FIG. 11.8 (continued) • Movements of the spine.

## Trunk and spinal column muscles

The largest muscle group in this area is the erector spinae (sacrospinalis), which extends on each side of the spinal column from the pelvic region to the cranium. It is divided into three muscles: the spinalis, the longissimus, and the iliocostalis. From the medial to the lateral side, it has attachments in the lumbar, thoracic, cervical, and capital regions. Thus, the erector spinae group is actually made up of nine muscle segments which extend, lateral flex, and rotate the spine. The spinalis is medial and closest to the vertebrae and has, from inferior to superior, the thoracis, cervicis, and capitis segments. The intermediate layer is the longissimus with the thoracis, cervicis, and capitis segments. Most lateral is the iliocostalis with lumborum, thoracis, and cervicis segments. Additionally, the sternocleidomastoid and splenius muscles are large muscles involved in cervical and head movements. Large abdominal muscles involved in lumbar movements include the rectus abdominis, external oblique abdominal, and internal oblique abdominal. The quadratus lumborum is involved in the lumbar spine movements of lateral flexion and lumbar extension. See Table 11.2 for further details. Also, the psoas major and minor, as discussed in Chapter 8, are involved in ipsilateral lateral flexion and in flexion of the lumbar spine.

Numerous small muscles are found in the spinal column region. Many of them originate on one vertebra and insert on the next. They are important in the functioning of the spine, but detailed knowledge of these muscles is of limited value to most people who use this text. Consequently, discussion will concentrate on the larger muscles primarily involved in trunk and spinal column movements (see Table 11.2) and will only briefly address the smaller muscles.

So that the muscles of the trunk and spinal column may be better understood, they can be grouped according to both location and function. It should be noted that some muscles have multiple segments. As a result, one segment of a particular muscle may be located and perform movement in one region, while another segment of the same muscle may be located in another region and perform movements there. Many of the muscles of the trunk and spinal column function in moving the spine as well as in aiding respiration. All the muscles of the thorax are primarily involved in respiration. The abdominal wall muscles are different from most other muscles in that they do not go bone to bone but they instead attach from bone to an aponeurosis (fascia) around the rectus abdominis area. They are the external oblique abdominal, internal oblique abdominal, and transversus abdominis.

Chapter **11**

339

## TABLE 11.2 • Agonist muscles of the spine

| | Muscle | Origin | Insertion | Action | Plane of motion | Palpation | Innervation |
|---|---|---|---|---|---|---|---|
| **Anterior neck** | Sternoclei-domastoid (both sides) | Manubrium of sternum, anterior superior surface of medial clavicle | Mastoid process | Extension of head at atlanto-occipital joint | Sagittal | Anterolateral neck, diagonally between the origin and insertion, particularly with rotation to contralateral side | Spinal accessory (Cr11, C2, C3) |
| | | | | Flexion of the cervical spine | Sagittal | | |
| | | | | Each side: rotation to contralateral side | Transverse | | |
| | | | | Each side: lateral flexion to ipsilateral side | Frontal | | |
| **Posterior neck** | Splenius cervicis | Spinous processes of the 3rd through 6th thoracic vertebrae | Transverse processes of the first three cervical vertebrae | Both sides: extension of the cervical spine | Sagittal | Palpate in lower posterior cervical spine just medial to inferior levator scapulae with resisted ipsilateral rotation | Posterior lateral branches of cervical nerves four through eight (C4–C8) |
| | | | | Each side: rotation to ipsilateral side | Transverse | | |
| | | | | Each side: lateral flexion to ipsilateral side | Frontal | | |
| | Splenius capitis | Lower half of the ligamentum nuchae and the spinous processes of the 7th cervical and upper three or four thoracic vertebrae | Mastoid process and occipital bone | Both sides: extension of the head and cervical spine | Sagittal | Deep to trapezius inferiorly and sternocleidomastoid superiorly; with subject seated, palpate in posterior triangle of neck between upper trapezius and sternocleidomastoid with resisted rotation to ipsilateral side | Posterior lateral branches of cervical nerves four through eight (C4–C8) |
| | | | | Each side: rotation to ipsilateral side | Transverse | | |
| | | | | Each side: lateral flexion to ipsilateral side | Frontal | | |
| **Posterior spine** | Erector spinae: Iliocostalis (cervicis, thoracis, lumborum) | Medial iliac crest, thoracolumbar aponeurosis from sacrum, posterior ribs 3–12 | Posterior ribs 1–12, cervical 4–7 transverse processes | Extension of the spine | Sagittal | Deep and difficult to distinguish from other muscles in the cervical thoracic regions; with subject prone, palpate immediately lateral to spinous processes in lumbar region with active extension | Posterior branches of the spinal nerves |
| | | | | Anterior pelvic rotation | Sagittal | | |
| | | | | Lateral flexion of the spine | Frontal | | |
| | | | | Lateral pelvic rotation to contralateral side | Frontal | | |
| | | | | Ipsilateral rotation of the spine and head | Transverse | | |
| | Erector spinae: Longissimus (capitis, cervicis, thoracis) | Medial iliac crest, thoracolumbar aponeurosis from sacrum, lumbar 1–5 transverse processes, and thoracic 1–5 transverse processes, cervical 5–7 articular processes | Cervical 2–6 spinous processes, thoracic 1–12 transverse processes, lower 9 ribs, mastoid process | Extension of the spine and head | Sagittal | | Posterior branches of the spinal nerves |
| | | | | Anterior pelvic rotation | Sagittal | | |
| | | | | Lateral flexion of the spine and head | Frontal | | |
| | | | | Lateral pelvic rotation to contralateral side | Frontal | | |
| | | | | Ipsilateral rotation of the spine and head | Transverse | | |
| | Erector spinae: Spinalis (capitis, cervicis, thoracis) | Ligamentum nuchae, 7th cervical spinous process, thoracic 11 and 12 spinous processes, and lumbar 1 and 2 spinous processes | Second cervical spinous process, thoracic 5–12 spinous processes, occipital bone | Extension of the spine | Sagittal | | Posterior branches of the spinal nerves |
| | | | | Anterior pelvic rotation | Sagittal | | |
| | | | | Lateral flexion of the spine and head | Frontal | | |
| | | | | Lateral pelvic rotation to contralateral side | Frontal | | |
| | | | | Ipsilateral rotation of the spine and head | Transverse | | |

TABLE 11.2 (continued) • **Agonist muscles of the spine**

| | Muscle | Origin | Insertion | Action | Plane of motion | Palpation | Innervation |
|---|---|---|---|---|---|---|---|
| Lateral lumbar | Quadratus lumborum | Posterior inner lip of the iliac crest | Approximately one-half the length of the lower border of the 12th rib and the transverse process of the upper four lumbar vertebrae | Lateral flexion to the ipsilateral side | Frontal | With subject prone, just superior to iliac crest and lateral to lumbar erector spinae with isometric lateral flexion | Branches of T12, L1–L4 nerves |
| | | | | Lateral pelvic rotation to contralateral side | | | |
| | | | | Lumbar extension | Sagittal | | |
| | | | | Anterior pelvic rotation | | | |
| | | | | Stabilizes the pelvis and lumbar spine | All planes | | |
| Anterior trunk | Rectus abdominis | Crest of pubis | Cartilage of 5th, 6th, and 7th ribs and xiphoid process | Both sides: lumbar flexion | Sagittal | Anteromedial surface of abdomen, between rib cage and pubic bone with isometric trunk flexion | Intercostal nerves (T 7–T11), subcostal nerve (T12) |
| | | | | Both sides: posterior pelvic rotation | | | |
| | | | | Each side: weak lateral flexion to ipsilateral side | Frontal | | |
| | External oblique abdominal | Borders of lower eight ribs at side of chest dovetailing with serratus anterior | Anterior half of crest of ilium, inguinal ligament, crest of pubis, and fascia of rectus abdominis at lower front | Both sides: lumbar flexion | Sagittal | With subject supine, palpate lateral to the rectus abdominis between iliac crest and lower ribs with active rotation to the contralateral side | Intercostal nerves (T 8–T11), subcostal nerve (T12), iliohypogastric nerve (T12, L1), and ilioinguinal nerve (L1) |
| | | | | Both sides: posterior pelvic rotation | | | |
| | | | | Each side: lumbar lateral flexion to ipsilateral side | Frontal | | |
| | | | | Each side: lateral pelvic rotation to contralateral side | | | |
| | | | | Each side: lumbar rotation to contralateral side | Transverse | | |
| | Internal oblique abdominal | Upper half of inguinal ligament, anterior two-third of crest of ilium, and lumbar fascia | Costal cartilage of 8th, 9th, and 10th ribs and linea alba | Both sides: lumbar flexion | Sagittal | With subject supine, palpate anterolateral abdomen between iliac crest and lower ribs with active rotation to the ipsilateral side | Intercostal nerves (T 8–T11), subcostal nerve (T12), iliohypogastric nerve (T12, L1), and ilioinguinal nerve (L1) |
| | | | | Both sides: posterior pelvic rotation | | | |
| | | | | Each side: lumbar lateral flexion to ipsilateral side | Frontal | | |
| | | | | Each side: lateral pelvic rotation to contralateral side | | | |
| | | | | Each side: lumbar rotation to ipsilateral side | Transverse | | |
| | Transversus abdominis | Lateral one-third of inguinal ligament, inner rim of iliac crest, inner surface of costal cartilages of lower six ribs, lumbar fascia | Crest of the pubis and the iliopectineal line, abdominal aponeurosis to the linea alba | Forced expiration by pulling the abdominal wall inward | Transverse | With subject supine, anterolateral abdomen between iliac crest and lower ribs during forceful exhalation; very difficult to distinguish from abdominal obliques | Intercostal nerves (T 8–T11), subcostal nerve (T12), iliohypogastric nerve (T12, L1), and ilioinguinal nerve (L1) |

Chapter **11**

## Muscles that move the head

Anterior
>   Rectus capitis anterior
>   Longus capitis

Posterior
>   Longissimus capitis
>   Obliquus capitis superior
>   Obliquus capitis inferior
>   Rectus capitis posterior—major and minor
>   Trapezius, superior fibers
>   Splenius capitis
>   Semispinalis capitis

Lateral
>   Rectus capitis lateralis
>   Sternocleidomastoid

## Muscles of the vertebral column

Superficial
>   Erector spinae (sacrospinalis)
>   Spinalis—capitis, cervicis, thoracis
>   Longissimus—capitis, cervicis, thoracis
>   Iliocostalis—cervicis, thoracis, lumborum
>   Splenius cervicis
>   Quadratus lumborum

Deep
>   Longus colli—superior oblique, inferior oblique, vertical
>   Interspinales—entire spinal column
>   Intertransversales—entire spinal column
>   Multifidus—entire spinal column
>   Psoas minor
>   Rotatores—entire spinal column
>   Semispinalis—cervicis, thoracis
>   Psoas major and minor (see Chapter 8)

## Muscles of the thorax

Diaphragm
Intercostalis—external, internal
Levator costarum
Subcostales
Scalenus—anterior, medius, posterior
Serratus posterior—superior, inferior
Transversus thoracis

## Muscles of the abdominal wall

Rectus abdominis
External oblique abdominal (obliquus externus abdominis)
Internal oblique abdominal (obliquus internus abdominis)
Transverse abdominis (transversus abdominis)

## Core Training

In recent years, a great deal of attention has focused on core training aimed specifically at muscles in and around the abdominal area. Discussions in Chapters 4 and 5 pointed out the importance of the scapula muscles in providing dynamic stability for proper upper extremity function. The same importance and concepts apply to the abdominal core for total body functioning. All too often, there is significant focus on major muscle groups that are strengthened through typical exercises like bench press and squats, with minimal attention given to the link between the upper and lower body—the low back and abdominal core. It is beyond the scope of this text to provide a detailed description of all exercises that may be used to address this area, but it is important to consider core training in designing programs to improve performance and prevent injury.

In addressing the core, some attention is given to the inner core as well as the outer core. The inner core consists of deeper muscles that should be activated as the first step in stabilizing the trunk and pelvis. These muscles consist of the diaphragm, transversus abdominis, lumbar multifidus, and the muscles of the pelvic floor—those that attach to the bony ring of the pelvis. Activating these muscles requires a level of focus and concentration. The outer core consists of the rectus abdominis, external obliques, internal obliques, and erector spinae. These muscles are exercised in a variety of means—including, but not limited to, sit-ups, V-sit-ups, crunches, curl-ups, abdominal twists, prone extensions, superman exercises, and so on.

## Nerves

Cranial nerve 11 and the spinal nerves of C2 and C3 innervate the sternocleidomastoid. The posterior lateral branches of C4 through C8 innervate the splenius muscles. The entire erector spinae group is supplied by the posterior branches of the spinal nerves, whereas the intercostal nerves of T7 through T12 innervate the rectus abdominis. Both the internal and external oblique abdominal muscles receive innervation from the intercostal nerves (T8–T12), the iliohypogastric nerve (T12, L1), and the ilioinguinal nerve (L1). The same innervation is provided to the transverse abdominis, except that innervation begins with the T7 intercostal nerve. Branches from T12 and L1 supply the quadratus lumborum. Review Figs. 2.6, 4.8, 4.9, and 8.19. See Fig. 11.9 for a full listing of the major trunk and spine muscular innervations.

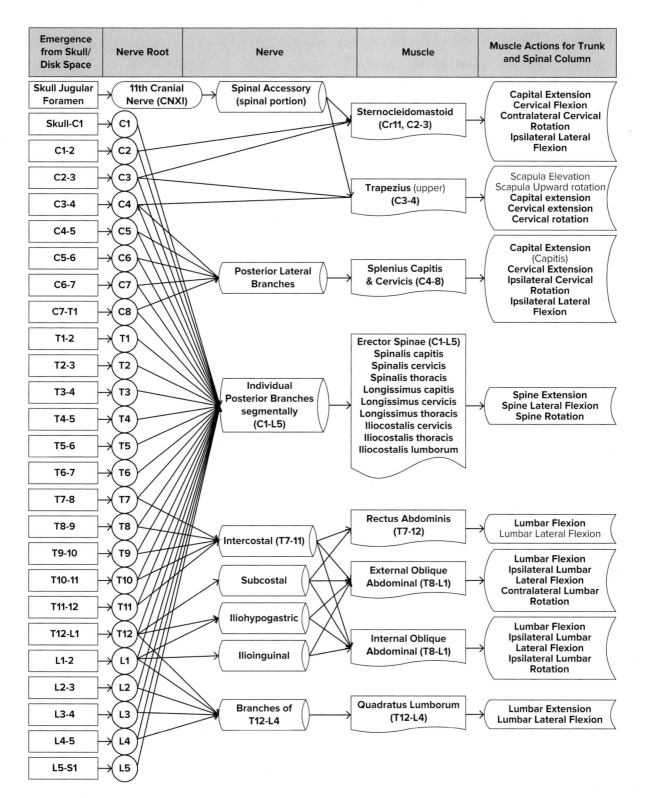

FIG. 11.9 • Innervation for the trunk and spinal column muscles and their actions. The neuromuscular structures associated with the primary movers for the trunk and spinal column and their actions are bolded.

# Muscles that move the head

All muscles featured here originate on the cervical vertebrae and insert on the occipital bone of the skull, as implied by their "capitis" name (Figs. 11.10 and 11.11; Table 11.3). Three muscles make up the

FIG. 11.10 • Anterior muscles of the neck.

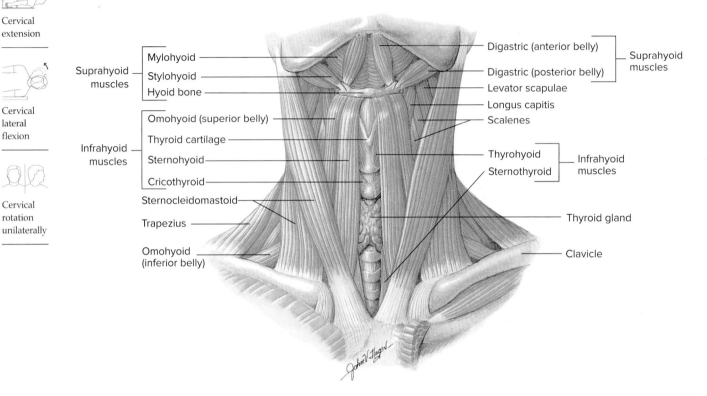

FIG. 11.11 • Deep muscles of the posterior neck and upper back regions.

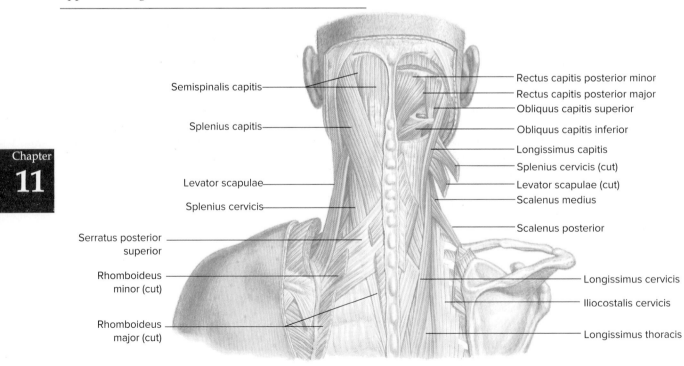

TABLE 11.3 · **Muscles that move the head**

| | Muscle | Origin | Insertion | Action | Innervation |
|---|---|---|---|---|---|
| **Anterior** | Longus capitis | Transverse processes of C3–C6 | Basilar part of occipital bone | Flexion of head and cervical spine | C1–3 |
| | Rectus capitis anterior | Anterior surface of lateral mass of atlas | Basilar part of occipital bone anterior to foramen magnum | Flexion of head and stabilization of atlantooccipital joint | C1–3 |
| | Rectus capitis lateralis | Superior surface of transverse processes of atlas | Jugular process of occipital bone | Lateral flexion of head and stabilization of atlantooccipital joint | C1–3 |
| **Posterior** | Rectus capitis posterior (major) | Spinous process of axis | Lateral portion of inferior nuchal line of occipital bone | Extension and rotation of head to ipsilateral side | Posterior rami of C1 |
| | Rectus capitis posterior (minor) | Spinous tubercle of posterior arch of atlas | Medial portion of inferior nuchal line of occipital bone | Extension of head | Posterior rami of C1 |
| | Obliquus capitis superior | Transverse process of atlas | Occipital bone between inferior and superior nuchal line | Extension and lateral flexion of head | Posterior rami of C1 |
| | Obliquus capitis inferior | Spinous process of axis | Transverse process of atlas | Rotation of atlas | Posterior rami of C1 |
| | Semispinalis capitis | Transverse processes of C4–T7 | Occipital bone, between superior & inferior nuchal lines | Extension and contralateral rotation of head | Posterior primary rami of C3–C5 |

anterior vertebral muscles—the longus capitis, the rectus capitis anterior, and the rectus capitis lateralis. All are flexors of the head and upper cervical spine. The rectus capitis lateralis laterally flexes the head, in addition to assisting the rectus capitis anterior in stabilizing the atlantooccipital joint.

The rectus capitis posterior major and minor, obliquus capitis superior and inferior, and semispinalis capitis are located posteriorly. All are extensors of the head, except for the obliquus capitis inferior, which rotates the atlas. The obliquus capitis superior assists the rectus capitis lateralis in lateral flexion of the head. In addition to extension, the rectus capitis posterior major is responsible for rotation of the head to the ipsilateral side. It is assisted by the semispinalis capitis, which rotates the head to the contralateral side. The splenius capitis and the sternocleidomastoid (see Table 11.2) are much larger and more powerful in moving the head and cervical spine; they are covered in detail on the following pages. The remaining muscles that act on the cervical spine are addressed with the muscles of the vertebral column.

Chapter
**11**

# Sternocleidomastoid muscle FIG. 11.12

(ster´no-kli-do-mas´toyd)

Cervical
flexion

Cervical
lateral
flexion

Cervical
rotation
contrala-
terally

## Origin

Manubrium of the sternum
Anterior superior surface of the medial clavicle

## Insertion

Mastoid process

## Action

Extension of the head at the atlantooccipital joint
Flexion of the cervical spine
Right side: rotation to the left and lateral flexion to
   the right
Left side: rotation to the right and lateral flexion to
   the left

## Palpation

Anterolateral side of the neck, diagonally between
   the origin and insertion, particularly with rotation to
   contralateral side and slight cervical flexion

## Innervation

Spinal accessory nerve (Cr11, C2, C3)

## Application, strengthening, and flexibility

The sternocleidomastoid is primarily responsible for flexion and rotation of the head and neck. One side of this muscle may be easily visualized and palpated when the head is rotated to the opposite side.

The sternocleidomastoid is easily worked for strength development by placing the hands on the forehead to apply force posteriorly while using these muscles to pull the head forward into flexion. The hand may also be used on one side of the jaw to apply rotary force in the opposite direction while the sternocleidomastoid is contracting concentrically to rotate the head in the direction of the hand.

Cervical hyperextension in combination with capital flexion provides some bilateral stretching of the sternocleidomastoid. Each side may be stretched individually. The right side is stretched by moving into left lateral flexion and right cervical rotation combined with extension. The opposite movements in extension stretch the left side.

Chapter
11

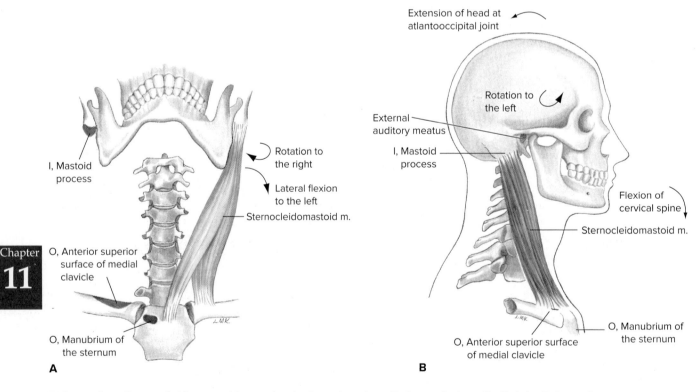

FIG. 11.12 • Sternocleidomastoid muscle. **A,** Anterior view; **B,** Lateral view. O, Origin; I, Insertion.

# Splenius muscles (cervicis, capitis)

FIG. 11.13

(sple´ni-us) (ser´vi-sis) (kap´i-tis)

## Origin

Splenius cervicis: spinous processes of the third through sixth thoracic vertebrae

Splenius capitis: lower half of the ligamentum nuchae and spinous processes of the seventh cervical and upper three or four thoracic vertebrae

## Insertion

Splenius cervicis: transverse processes of the first three cervical vertebrae

Splenius capitis: mastoid process and occipital bone

## Action

Both sides: extension of the head (splenius capitis) and neck (splenius cervicis and capitis)

Right side: rotation and lateral flexion to the right

Left side: rotation and lateral flexion to the left

## Palpation

Splenius cervicis: palpate in lower posterior cervical spine just medial to inferior levator scapulae with resisted ipsilateral rotation

Splenius capitis: deep to the trapezius inferiorly and the sternocleidomastoid superiorly; with subject seated, palpate in the posterior triangle of the neck between the upper trapezius and the sternocleidomastoid with resisted rotation to ipsilateral side

## Innervation

Posterior lateral branches of cervical nerves four through eight (C4–C8)

## Application, strengthening, and flexibility

Any movement of the head and neck into extension, particularly extension and rotation, would bring the splenius muscle strongly into play, together with the erector spinae and the upper trapezius muscles. Tone in the splenius muscle tends to hold the head and neck in proper posture position.

A good exercise for the splenius muscle is to lace the fingers behind the head with it in flexion and then slowly contract the posterior head and neck muscles to move the head and neck into full extension. This exercise may also be performed by using a towel or a partner for resistance.

The entire splenius may be stretched with maximal flexion of the head and cervical spine. The right side can be stretched through combined movements of left rotation, left lateral flexion, and flexion. The same movements to the right apply when stretching the left side.

Cervical extension

Cervical rotation ipsilaterally

Cervical lateral flexion

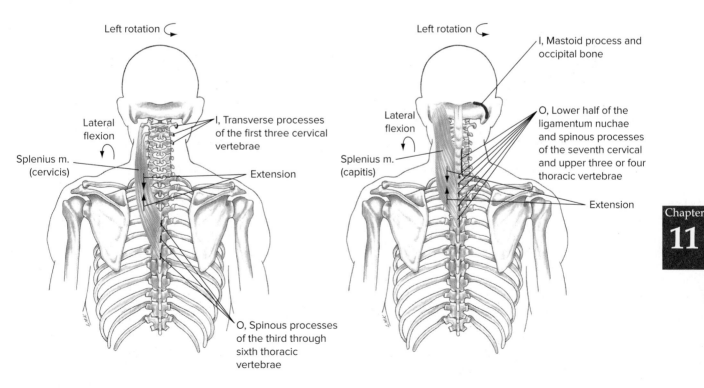

FIG. 11.13 • Splenius muscles (cervicis on the left, capitis on the right), posterior view. O, Origin; I, Insertion.

# Muscles of the vertebral column

Cervical
flexion

Cervical
extension

Cervical
lateral
flexion

Cervical
rotation
ipsilaterally

Lumbar
extension

Lumbar
lateral
flexion

Lumbar
rotation
ipsilaterally

In the cervical area, the longus colli muscles are located anteriorly and flex the cervical and upper thoracic vertebrae. Posteriorly, the erector spinae group, the transversospinalis group, the interspinalintertransverse group, and the splenius all run vertically parallel to the spinal column (Figs. 11.14 and 11.15; Table 11.4). This location enables them to extend the spine as well as assist in rotation and lateral flexion. The splenius and erector spinae group are addressed in detail elsewhere in this chapter. The transversospinalis group consists of the semispinalis, multifidus, and rotatores muscles. These muscles all originate on the transverse processes of their respective vertebrae and generally run posteriorly to attach to the spinous processes on the vertebrae just above their vertebrae of origin. All are extensors of the spine and contract to rotate their respective vertebrae to the contralateral side. The interspinal-intertransverse group lies deep to the rotatores and consists of the interspinalis and the intertransversarii muscles. As a group, they laterally flex and extend but do not rotate the vertebrae. The interspinalis muscles are extensors that connect from the spinous process of one vertebra to the spinous process of the adjacent vertebra. The intertransversarii muscles flex the vertebral column laterally by connecting to the transverse processes of adjacent vertebrae.

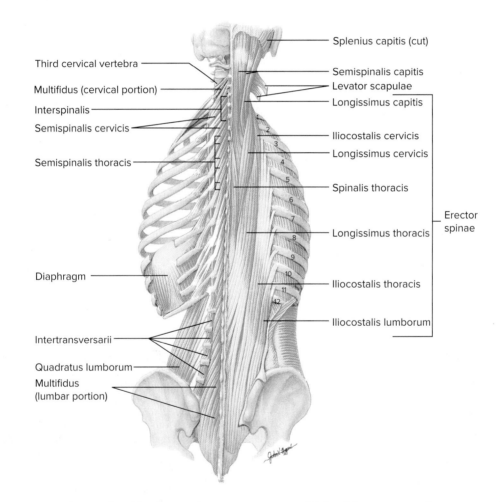

FIG. 11.14 • Deep and superficial back muscles, posterior view. **Right,** The erector spinae group of muscles is demonstrated. **Left,** Those muscles have been removed to reveal the deeper back muscles.

## TABLE 11.4 • **Muscles of the vertebral column**

| | Muscle | Origin | Insertion | Action | Innervation |
|---|---|---|---|---|---|
| **Anterior** | Longus colli (superior oblique) | Transverse processes of C3–C5 | Anterior arch of atlas | Flexion of cervical spine | C2–7 |
| | Longus colli (inferior oblique) | Bodies of T1–T3 | Transverse processes of C5 and C6 | Flexion of cervical spine | C2–7 |
| | Longus colli (vertical) | Bodies of C5–C7 and T1–T3 | Anterior surface of bodies of C2–C4 | Flexion of cervical spine | C2–7 |
| **Lateral** | Intertransversarii | Tubercles of transverse processes of each vertebra | Tubercles of transverse processes of next vertebra | Lateral flexion of spinal column | Anterior primary ramus of spinal nerves |
| **Posterolateral** | Multifidus | Sacrum, iliac spine, transverse processes of lumbar, thoracic, and lower four cervical vertebrae | Spinous processes of 2nd, 3rd, or 4th vertebra above origin | Extension and contralateral rotation of spinal column | Posterior primary ramus of spinal nerves |
| | Rotatores | Transverse process of each vertebra | Base of spinous process of next vertebra above | Extension and contralateral rotation of spinal column | Posterior primary ramus of spinal nerves |
| | Semispinalis cervicis | Transverse processes of T1–T5 or T6 | Spinous processes from C2–C5 | Extension and contralateral rotation of spinal column | All divisions, posterior primary ramus of spinal nerves |
| | Semispinalis thoracis | Transverse processes of T6–10 | Spinous processes of C6, C7, and T1–T4 | Extension and contralateral rotation of spinal column | Posterior primary ramus of spinal nerves |
| **Posterior** | Interspinalis | Spinous process of each vertebra | Spinous process of next vertebra | Extension of spinal column | Posterior primary ramus of spinal nerves |

**Right posterolateral view**

Rotatores
Interspinalis
Spinous process
Multifidus
Intertransversarii
Transverse process

FIG. 11.15 • Deep muscles associated with the vertebrae posterolateral view from the right.

## Muscles of the thorax

The thoracic muscles are involved almost entirely in respiration (Fig. 11.16). During quiet rest, the diaphragm is responsible for breathing movements. As it contracts and flattens, the thoracic volume is increased and air is inspired to equalize the pressure. When larger amounts of air are needed, such as during exercise, the other thoracic muscles take on a more significant role in inspiration. The scalene muscles elevate the first two ribs to increase the thoracic volume. Further expansion of the chest is accomplished by the external intercostals. Additional muscles of inspiration are the levator costarum and the serratus posterior (Fig. 11.17). Forced expiration occurs with contraction of the internal intercostals, transversus thoracis, and subcostales. All these muscles are detailed in Table 11.5.

Chapter **11**

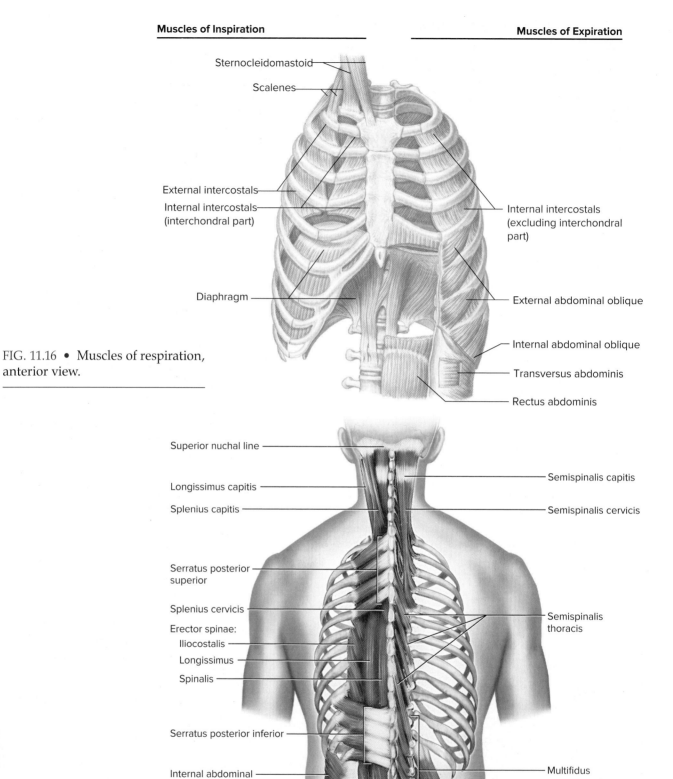

FIG. 11.16 • Muscles of respiration, anterior view.

FIG. 11.17 • Deeper muscles acting on the vertebral column and thorax.

## TABLE 11.5 • Muscles of the thorax

| Muscle | Origin | Insertion | Action | Innervation |
|---|---|---|---|---|
| Diaphragm | Circumference of thoracic inlet from xiphoid process, costal cartilages of ribs 6–12, and lumbar vertebrae | Central tendon of diaphragm | Depresses and draws central tendon forward in inhalation, reduces pressure in thoracic cavity, and increases pressure in abdominal cavity | Phrenic nerve (C3–C5) |
| Internal intercostals | Longitudinal ridge on inner surface of ribs and costal cartilages | Superior border of next rib below | Elevate costal cartilages of ribs 1–4 during inhalation, depress all ribs in exhalation | Intercostal branches of T1–T11 |
| External intercostals | Inferior border of ribs | Superior border of next rib below | Elevate ribs | Intercostal branches of T1–T11 |
| Levator costarum | Ends of transverse processes of C7, T2–T12 | Outer surface of angle of next rib below origin | Elevates ribs, lateral flexion of thoracic spine | Intercostal nerves |
| Subcostales | Inner surface of each rib near its angle | Medially on the inner surface of second or third rib below | Draws the ventral part of the ribs downward, decreasing the volume of the thoracic cavity | Intercostal nerves |
| Scalenus anterior | Transverse processes of C3–C6 | Inner border and upper surface of first rib | Elevates first rib, flexion, lateral flexion, and contralateral rotation of cervical spine | Ventral rami of C5, C6, sometimes C4 |
| Scalenus medius | Transverse processes of C2–C7 | Superior surface of first rib | Elevates first rib, flexion, lateral flexion, and contralateral rotation of cervical spine | Ventral rami of C3–C8 |
| Scalenus posterior | Transverse processes of C5–C7 | Outer surface of second rib | Elevates second rib, flexion, lateral flexion, and slight contralateral rotation of cervical spine | Ventral rami of C6–C8 |
| Serratus posterior (superior) | Ligamentum nuchae, spinous processes of C7, T1, and T2 or T3 | Superior borders lateral to angles of ribs 2–5 | Elevates upper ribs | Branches from anterior primary rami of T1–T4 |
| Serratus posterior (inferior) | Spinous processes of T10–T12 and L1–L3 | Inferior borders lateral to angles of ribs 9–12 | Counteracts inward pull of diaphragm by drawing last four ribs outward and downward | Branches from anterior primary rami of T9–T12 |
| Transversus thoracis | Inner surface of sternum and xiphoid process, sternal ends of costal cartilages of ribs 3–6 | Inner surfaces and inferior borders of costal cartilages 3–6 | Depresses ribs | Intercostal branches of T3–T6 |

Chapter

**11**

Lumbar
extension

Cervical
extension

Lumbar
lateral
flexion

Cervical
lateral
flexion

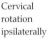

Lumbar
rotation
ipsilaterally

Cervical
rotation
ipsilaterally

# Erector spinae muscles*
# (sacrospinalis)   FIGS. 11.17, 11.18, 11.19

(e-rek´tor spi´ne) (sa´kro-spi-na´lis)

## Iliocostalis

(il´i-o-kos-ta´lis): lateral layer

## Longissimus

(lon-jis´i-mus): middle layer

## Spinalis

(spi-na´lis): medial layer

---

*This muscle group includes the iliocostalis, the longissimus dorsi, the spinalis dorsi, and divisions of these muscles in the lumbar, thoracic, and cervical sections of the spinal column.

## Origin

Iliocostalis: medial iliac crest, thoracolumbar aponeurosis from sacrum, posterior ribs 3–12

Longissimus: medial iliac crest, thoracolumbar aponeurosis from sacrum, lumbar 1–5 transverse processes and thoracic 1–5 transverse processes, cervical 5–7 articular processes

Spinalis: ligamentum nuchae, seventh cervical spinous process, thoracic 11 and 12 spinous processes, and lumbar 1 and 2 spinous processes

## Insertion

Iliocostalis: posterior ribs 1–12, cervical 4–7 transverse processes

Longissimus: cervical 2–6 spinous processes, thoracic 1–12 transverse processes, lower nine ribs, mastoid process

Spinalis: second cervical spinous process, thoracic 5–12 spinous processes, occipital bone

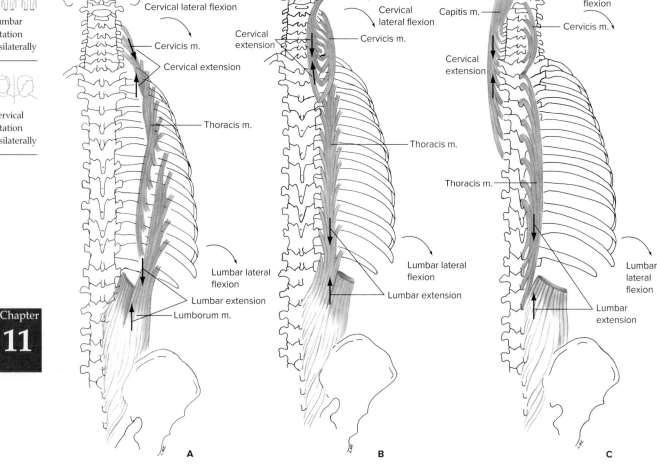

FIG. 11.18 • Erector spinae (sacrospinalis) muscle, posterior view. **A,** Iliocostalis lumborum, thoracis, and cervicis; **B,** Longissimus thoracis, cervicis, and capitis; **C,** Spinalis thoracis, cervicis, and capitis.

## Action

Extension, lateral flexion, and ipsilateral rotation of the spine and head

Anterior pelvic rotation

Lateral pelvic rotation to contralateral side

## Palpation

Deep and difficult to distinguish from other muscles in the cervical and thoracic regions; with subject prone, palpate immediately lateral to spinous processes in lumbar region with active extension, resistance against rotation may further assist in palpation for ipsilateral muscles

## Innervation

Posterior branches of the spinal nerves

## Application, strengthening, and flexibility

The erector spinae muscles function best when the pelvis is posteriorly rotated. This lowers the origin of the erector spinae and makes it more effective in keeping the spine straight. As the spine is held straight, the ribs are raised, thus fixing the chest high and consequently making the abdominal muscles more effective in holding the pelvis up in front and flattening the abdominal wall.

An exercise known as the dead lift, employing a barbell, uses the erector spinae in extending the spine. In this exercise, the subject bends over, keeping the arms and legs straight; picks up the barbell; and returns to a standing position. In performing this type of exercise, it is very important to always use correct technique to avoid back injuries. Voluntary static contraction of the erector spinae in the standing position can provide a mild exercise and improve body posture.

The erector spinae and the various divisions may be strengthened through numerous forms of back extension exercises. These are usually done in a prone or face-down position in which the spine is already in some state of flexion. The subject then uses these muscles to move part or all of the spine toward extension against gravity. A weight may be held in the hands behind the head to increase resistance. Seated trunk twists assist as well, especially with the iliocostalis and longissimus.

Maximal hyperflexion of the entire spine stretches the erector spinae muscle group. Stretch may be isolated to specific segments through specific movements. Maximal flexion of the head and cervical spine stretches the capitis and cervicis segments. Flexion combined with lateral flexion to one side accentuates the stretch on the contralateral side. Thoracic and lumbar flexion places the stretch primarily on the thoracis and lumborum segments.

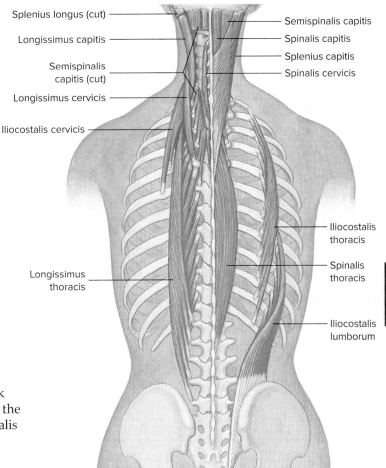

Splenius longus (cut) — Semispinalis capitis — Spinalis capitis — Splenius capitis — Spinalis cervicis — Longissimus capitis — Semispinalis capitis (cut) — Longissimus cervicis — Iliocostalis cervicis — Iliocostalis thoracis — Spinalis thoracis — Longissimus thoracis — Iliocostalis lumborum

Chapter **11**

FIG. 11.19 • Muscles of the back and the neck help move the head (posterior view) and hold the torso erect. The splenius capitis and semispinalis have been cut on the left to show underlying muscles.

# Quadratus lumborum muscle  FIG. 11.20

(kwad-ra´tus lum-bo´rum)

## Origin

Posterior inner lip of the iliac crest

## Insertion

Approximately one-half the length of the lower border of the twelfth rib and the transverse process of the upper four lumbar vertebrae

## Action

Lateral flexion to the ipsilateral side
Stabilizes the pelvis and lumbar spine
Extension of the lumbar spine
Anterior pelvic rotation
Lateral pelvic rotation to contralateral side

## Palpation

With subject prone, just superior to iliac crest and lateral to lumbar erector spinae with isometric lateral flexion

## Innervation

Branches of T12, L1 nerves

## Application, strengthening, and flexibility

The quadratus lumborum is important in lumbar lateral flexion and in elevating the pelvis on the same side in the standing position. Trunk rotation and lateral flexion movements against resistance are good exercises for development of this muscle. The position of the body relative to gravity may be changed to increase resistance on this and other trunk and abdominal muscles. Left lumbar lateral flexion while in lumbar flexion stretches the right quadratus lumborum, and vice versa.

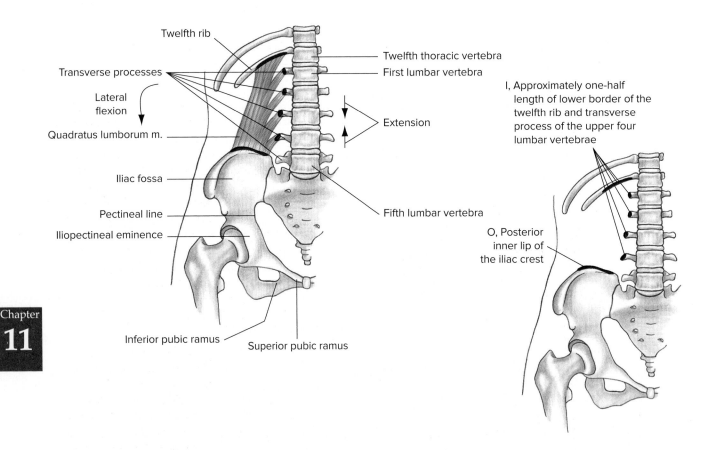

FIG. 11.20 • Quadratus lumborum muscle. O, Origin; I, Insertion.

Lumbar lateral flexion

Lumbar extension

Chapter 11

# Muscles of the abdominal wall

FIGS. 11.21, 11.22, 11.23

Lumbar flexion

Lumbar lateral flexion

Lumbar rotation unilaterally

**FIG. 11.21 • Muscles of the abdomen:** external oblique and rectus abdominis. The fibrous sheath around the rectus has been removed on the right side to show the muscle within.

**FIG. 11.22 • Muscles of the abdomen.** The external oblique has been removed on the right side to reveal the internal oblique. The external and internal obliques have been removed on the left side to reveal the transversus abdominis. The rectus abdominis has been cut to reveal the posterior rectus sheath.

**FIG. 11.23 • Abdominal wall above umbilicus.** The unique arrangement of the four abdominal muscles with their fascial attachment in and around the rectus abdominis muscle is shown. With no bones for attachments, these muscles can be adequately maintained through exercise.

Chapter

**11**

Lumbar
flexion

Lumbar
lateral
flexion

# Rectus abdominis muscle FIG. 11.24

(rek´tus ab-dom´i-nis)

### Origin

Crest of the pubis

### Insertion

Cartilage of the fifth, sixth, and seventh ribs and the
  xiphoid process

### Action

Both sides: lumbar flexion
Posterior pelvic rotation
Right side: weak lateral flexion to the right
Left side: weak lateral flexion to the left

### Palpation

Anteromedial surface of the abdomen, between the rib
  cage and the pubic bone with isometric trunk flexion

### Innervation

Intercostal nerves (T7–T12)

### Application, strengthening, and flexibility

The rectus abdominis muscle controls the tilt of
the pelvis and the consequent curvature of the

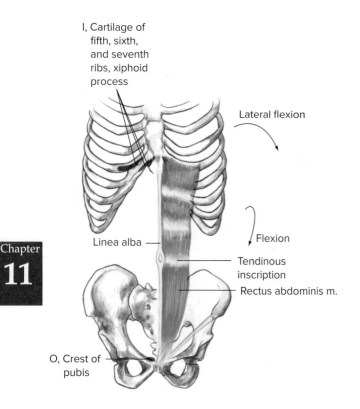

FIG. 11.24 • Rectus abdominis muscle, anterior view.
O, Origin; I, Insertion.

lower spine. By rotating the pelvis posteriorly, the
rectus abdominis flattens the lower back, making the erector spinae muscle more effective as an
extensor of the spine and the hip flexors (the iliopsoas muscle, particularly) more effective in raising the legs.

In a relatively lean person with well-developed
abdominals, three distinct sets of lines or depressions may be noted. Each represents an area of
tendinous connective tissue connecting or supporting the abdominal arrangement of muscles
in lieu of bony attachments. Running vertically
from the xiphoid process through the umbilicus to
the pubis is the **linea alba**. It divides each rectus
abdominis and serves as its medial border. Lateral
to each rectus abdominis is the **linea semilunaris**,
a crescent, or moon-shaped, line running vertically. This line represents the aponeurosis connecting the lateral border of the rectus abdominis
and the medial border of the external and internal
abdominal obliques. The **tendinous inscriptions**
are horizontal indentations that transect the rectus abdominis at three or more locations, giving
the muscle its segmented appearance. Refer to
Fig. 11.21.

There are several exercises for the abdominal
muscles, such as bent-knee sit-ups, crunches, and
isometric contractions. Bent-knee sit-ups with the
arms folded across the chest are considered by many
to be a safe and an efficient exercise. Crunches
are considered to be even more effective for isolating the work to the abdominals. Both of these
exercises shorten the iliopsoas muscle and other
hip flexors, thereby reducing their ability to generate force. Twisting to the left and right brings the
oblique muscles into more active contraction. In all
the above exercises, it is important to use proper
technique, which involves gradually moving to
the up position until the lumbar spine is actively
flexed maximally and then slowly returning to
the beginning position. Jerking movements using
momentum should be avoided. Movement continued beyond full lumbar flexion exercises only
the hip flexors, which is not usually an objective.
Even though all these exercises may be helpful in
strengthening the abdominals, careful analysis
should occur before deciding which are indicated
in the presence of various injuries and problems of
the lower back.

The rectus abdominis is stretched by simultaneously hyperextending both the lumbar and the
thoracic spine. Extending the hips assists in this
process by accentuating the anterior rotation of the
pelvis to hyperextend the lumbar spine.

# External oblique abdominal muscle  FIG. 11.25

(ek-stur´nel o-bleek´ ab-dom´i-nel)

## Origin

Borders of the lower eight ribs at the side of the chest, dovetailing with the serratus anterior muscle*

## Insertion

Anterior half of the crest of the ilium, the inguinal ligament, the crest of the pubis, and the fascia of the rectus abdominis muscle at the lower front

## Action

Both sides: lumbar flexion
Posterior pelvic rotation
Right side: lumbar lateral flexion to the right and rotation to the left, lateral pelvic rotation to the left
Left side: lumbar lateral flexion to the left and rotation to the right, lateral pelvic rotation to the right

*Sometimes the origin and insertion are reversed in anatomy books. This is the result of different interpretations of which bony structure is the more movable. The insertion is considered the most movable part of a muscle.

## Palpation

With subject supine, palpate lateral to the rectus abdominis between iliac crest and lower ribs with active rotation to the contralateral side

Lumbar flexion

## Innervation

Intercostal nerves (T8–T12), iliohypogastric nerve (T12, L1), and ilioinguinal nerve (L1)

Lumbar lateral flexion

## Application, strengthening, and flexibility

Working on each side of the abdomen, the external oblique abdominal muscles aid in rotating the trunk when working independently of each other. Working together, they aid the rectus abdominis muscle in its described action. The left external oblique abdominal muscle contracts strongly during sit-ups when the trunk rotates to the right, as in touching the left elbow to the right knee. Rotating to the left brings the right external oblique into action.

Each side of the external oblique must be stretched individually. The right side is stretched by moving into extreme left lateral flexion combined with extension, or by extreme lumbar rotation to the right combined with extension. The opposite movements combined with extension stretch the left side.

Lumbar rotation contralaterally

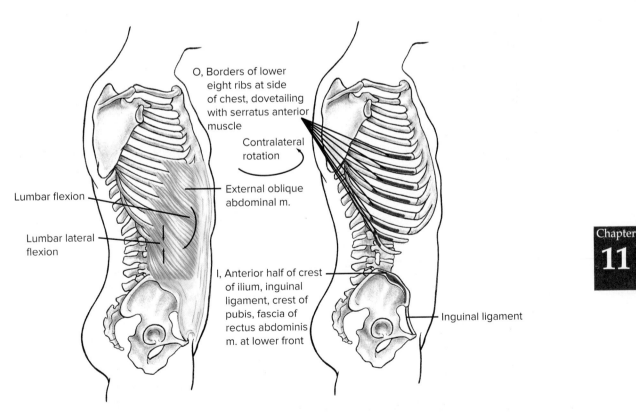

FIG. 11.25 • External oblique abdominal muscle, lateral view. O, Origin; I, Insertion.

Lumbar
flexion

Lumbar
lateral
flexion

Lumbar
rotation
ipsilaterally

# Internal oblique abdominal muscle  FIG. 11.26

(in-ter´nel o-bleek´ ab-dom´i-nel)

### Origin

Upper half of the inguinal ligament, anterior two-thirds of the crest of the ilium, and lumbar fascia

### Insertion

Costal cartilages of the eighth, ninth, and tenth ribs and the linea alba

### Action

Both sides: lumbar flexion
Posterior pelvic rotation
Right side: lumbar lateral flexion to the right and rotation to the right, lateral pelvic rotation to the left
Left side: lumbar lateral flexion to the left and rotation to the left, lateral pelvic rotation to the right

### Palpation

With subject supine, palpate anterolateral abdomen between iliac crest and lower ribs with active rotation to the ipsilateral side

### Innervation

Intercostal nerves (T8–T12), iliohypogastric nerve (T12, L1), and ilioinguinal nerve (L1)

### Application, strengthening, and flexibility

The internal oblique abdominal muscles run diagonally in the direction opposite that of the external obliques. The left internal oblique rotates to the left, and the right internal oblique rotates to the right.

In touching the left elbow to the right knee in crunches, the left external oblique and the right internal oblique abdominal muscles contract at the same time, assisting the rectus abdominis muscle in flexing the trunk to make completion of the movement possible. In rotary movements, the internal oblique and the external oblique on the opposite side always work together.

Like the external oblique, each side of the internal oblique must be stretched individually. The right side is stretched by moving into extreme left lateral flexion and extreme left lumbar rotation combined with extension. The same movements to the right combined with extension stretch the left side.

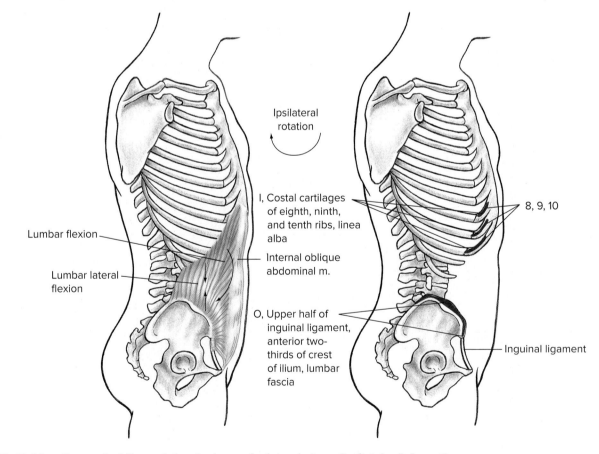

FIG. 11.26 • Internal oblique abdominal muscle, lateral view. O, Origin; I, Insertion.

## Transversus abdominis muscle  FIG. 11.27

(trans-vurs´us ab-dom´i-nis)

### Origin

Lateral third of the inguinal ligament, inner rim of the iliac crest, inner surface of the costal cartilages of the lower six ribs, and lumbar fascia

### Insertion

Crest of the pubis and the iliopectineal line
Abdominal aponeurosis to the linea alba

### Action

Forced expiration by pulling the abdominal wall inward

### Palpation

With subject supine, palpate anterolateral abdomen between iliac crest and lower ribs during forceful exhalation; very difficult to distinguish from the abdominal obliques

### Innervation

Intercostal nerves (T7–T12), iliohypogastric nerve (T12, L1), and ilioinguinal nerve (L1)

### Application, strengthening, and flexibility

The transversus abdominis is the chief muscle of forced expiration and is effective—together with the rectus abdominis, the external oblique abdominal, and the internal oblique abdominal muscles—in helping hold the abdomen flat. It, along with the other abdominal muscles, is considered by many to be key in providing and maintaining core stability. This abdominal flattening and forced expulsion of the abdominal contents are the only action of this muscle.

The transversus abdominis muscle is exercised effectively by attempting to draw the abdominal contents back toward the spine. This may be done isometrically in the supine position or while standing. A maximal inspiration held in the abdomen applies stretch.

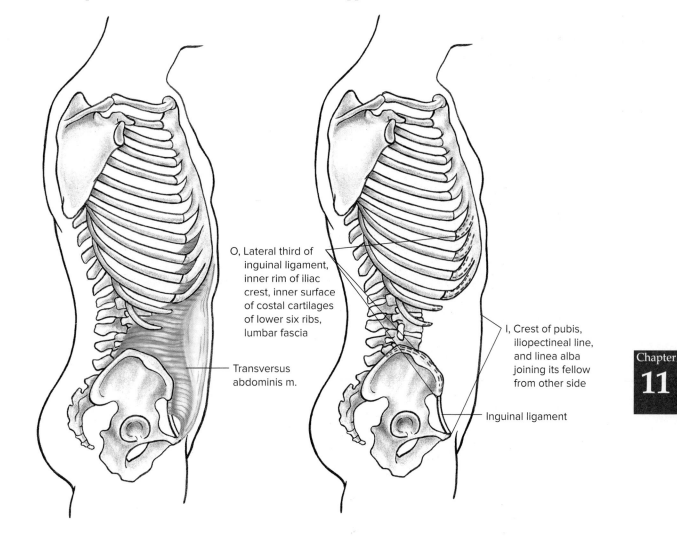

O, Lateral third of inguinal ligament, inner rim of iliac crest, inner surface of costal cartilages of lower six ribs, lumbar fascia

Transversus abdominis m.

I, Crest of pubis, iliopectineal line, and linea alba joining its fellow from other side

Inguinal ligament

Chapter 11

FIG. 11.27 • Transversus abdominis muscle, lateral view. O, Origin; I, Insertion.

# REVIEW EXERCISES

1. List the planes in which each of the following movements occurs. List the axis of rotation for each movement in each plane.

   a. Cervical flexion
   b. Cervical extension
   c. Cervical rotation
   d. Cervical lateral flexion
   e. Lumbar flexion
   f. Lumbar extension
   g. Lumbar rotation
   h. Lumbar lateral flexion
   i. Capital flexion
   j. Capital extension

2. Why is good abdominal muscular development so important? Why is this area so frequently neglected?

3. Why are weak abdominal muscles frequently blamed for lower back pain?

4. Prepare an oral or a written report on abdominal or back injuries found in the literature.

5. Research common spinal disorders such as brachial plexus neuropraxia, cervical radiculopathy, lumbosacral herniated nucleus pulposus, sciatica, spondylolysis, and spondylolisthesis. Report your findings in class.

6. **Muscle analysis chart** · Cervical and lumbar spine

| Complete the chart by listing the muscles primarily involved in each movement. | |
|---|---|
| Cervical spine | |
| Flexion | Extension |
| Lateral flexion right | Rotation right |
| Lateral flexion left | Rotation left |
| Lumbar spine | |
| Flexion | Extension |
| Lateral flexion right | Rotation right |
| Lateral flexion left | Rotation left |

## 7. **Antagonistic muscle action chart** • Cervical and lumbar spine

Complete the chart by listing the muscle(s) or parts of muscles that are antagonist in their actions to the muscles in the left column.

| Agonist | Antagonist |
|---|---|
| Splenius capitis | |
| Splenius cervicis | |
| Sternocleidomastoid | |
| Erector spinae | |
| Rectus abdominis | |
| External oblique abdominal | |
| Internal oblique abdominal | |
| Quadratus lumborum | |

## LABORATORY EXERCISES

1. Locate the following parts of the spine on a human skeleton and on a human subject:
   a. Cervical vertebrae
   b. Thoracic vertebrae
   c. Lumbar vertebrae
   d. Spinous processes
   e. Transverse processes
   f. Sacrum
   g. Manubrium
   h. Xiphoid process
   i. Sternum
   j. Rib cage (various ribs)

2. How and where can the following muscles be palpated on a human subject?
   a. Rectus abdominis
   b. External oblique abdominal
   c. Internal oblique abdominal
   d. Erector spinae
   e. Sternocleidomastoid
   f. Splenius cervicis
   g. Splenius capitis
   h. Quadratus lumborum

3. Contrast crunches, bent-knee sit-ups, V sit-ups, and straight-leg sit-ups. Does having a partner to hold the feet make a difference in the ability to do the bent-knee and straight-leg sit-ups? If so, why?

4. Have a laboratory partner stand and assume a position exhibiting good posture. What motions in each region of the spine does gravity attempt to produce? Which muscles are responsible for counteracting these motions against the pull of gravity?

5. Compare and contrast the spinal curves of a laboratory partner sitting erect with those of one sitting slouched in a chair. Which muscles are responsible for maintaining good sitting posture?

6. Which exercise is better for the development of the abdominal muscles—leg lifts or sit-ups? Analyze each exercise with regard to the activity of the abdominal muscles. Defend your answer.

Chapter
**11**

## 7. Trunk and spine exercise movement analysis chart

After analyzing each exercise in the chart, break each into two primary movement phases, such as a lifting phase and a lowering phase. For each phase, determine what trunk and spine movements occur, and then list the trunk and spine muscles primarily responsible for causing/controlling those movements or maintaining the position. Beside each muscle in each movement, indicate the type of contraction as follows: I—isometric; C—concentric; E—eccentric.

| Exercise | Initial movement phase | | Secondary movement phase | |
|---|---|---|---|---|
| | Movement(s) | Agonist(s)—(contraction type) | Movement(s) | Agonist(s)—(contraction type) |
| Push-up | | | | |
| Squat | | | | |
| Dead lift | | | | |
| Sit-up, bent knee | | | | |
| Prone extension | | | | |
| Rowing exercise | | | | |
| Leg raises | | | | |
| Stair machine | | | | |

## 8. Trunk and spine sport skill analysis chart

Analyze each skill in the chart, and list the movements of the trunk and spine in each phase of the skill. You may prefer to list the initial positions that the trunk and spine are in for the stance phase. After each movement, list the trunk and spine muscle(s) primarily responsible for causing/controlling the movement or maintaining the position. Beside each muscle in each movement, indicate the type of contraction as follows: I—isometric; C—concentric; E—eccentric. It may be desirable to review the concepts for analysis in Chapter 12 for the various phases. Assume right hand/leg dominant where applicable. Circle R or L to indicate the dominant extremity for the exercise, if appropriate.

| Exercise | | Stance phase | Preparatory phase | Movement phase | Follow-through phase |
|---|---|---|---|---|---|
| Baseball pitch | (R) | | | | |
| | (L) | | | | |
| Football punt | (R) | | | | |
| | (L) | | | | |
| Walking | (R) | | | | |
| | (L) | | | | |

## Trunk and spine sport skill analysis chart (continued)

| Exercise | | Stance phase | Preparatory phase | Movement phase | Follow-through phase |
|---|---|---|---|---|---|
| Softball pitch | (R) | | | | |
| | (L) | | | | |
| Soccer pass | (R) | | | | |
| | (L) | | | | |
| Batting | (R) | | | | |
| | (L) | | | | |
| Bowling | (R) | | | | |
| | (L) | | | | |
| Basketball jump shot | (R) | | | | |
| | (L) | | | | |

# References

Clarkson HM, Gilewich GB: *Musculoskeletal assessment: joint range of motion and manual muscle strength,* ed 2, Baltimore, 1999, Lippincott Williams & Wilkins.

Cleland JA, Koppenhaver S, Su J: *Netter's orthopaedic clinical examination: an evidence-based approach,* ed 3, Philadelphia, 2016, Elsevier.

Day AL: Observation on the treatment of lumbar disc disease in college football players, *American Journal of Sports Medicine* 15:275, January–February 1987.

Field D: *Anatomy: palpation and surface markings,* ed 3, Oxford, 2001, Butterworth-Heinemann.

Gench BE, Hinson MM, Harvey PT: *Anatomical kinesiology,* Dubuque, IA, 1995, Eddie Bowers.

Hamilton N, Weimer W, Luttgens K: *Kinesiology: scientific basis of human motion,* ed 12, New York, 2012, McGraw-Hill.

Hislop HJ, Montgomery J: *Daniels and Worthingham's muscle testing: techniques of manual examination,* ed 8, Philadelphia, 2007, Saunders.

Lindsay DT: *Functional human anatomy,* St. Louis, 1996, Mosby.

Magee DJ: *Orthopedic physical assessment,* ed 6, St. Louis, 2014, Elsevier Saunders.

Martens MA, et al: Adductor tendonitis and muscular abdominis tendopathy, *American Journal of Sports Medicine* 15:353, July–August 1987.

Marymont JV: Exercise-related stress reaction of the sacroiliac joint, an unusual cause of low back pain in athletes, *American Journal of Sports Medicine* 14:320, July–August 1986.

Muscolino JE: *The muscular system manual: the skeletal muscles of the human body,* ed 3, St. Louis, 2010, Elsevier Mosby.

National Strength and Conditioning Association; Baechle TR, Earle RW: *Essentials of strength training and conditioning,* ed 2, Champaign, IL, ed 3, 2008, Human Kinetics.

Oatis CA: *Kinesiology: the mechanics and pathomechanics of human movement,* ed 2, Philadelphia, 2008, Lippincott Williams & Wilkins.

Perry JF, Rohe DA, Garcia AO: *The kinesiology workbook,* Philadelphia, 1992, Davis.

Prentice WE: *Principles of athletic training: a competency-based approach,* ed 15, New York, 2014, McGraw-Hill.

Putz R, Pabst R: *Sobotta - Atlas of human anatomy one volume edition,* ed 14, Germany, 2008, Elsevier.

Rasch PJ: *Kinesiology and applied anatomy,* ed 7, Philadelphia, 1989, Lea & Febiger.

Saladin KS: *Anatomy & physiology: the unity of form and function,* ed 5, New York, 2010, McGraw-Hill.

Schünke M, Schulte E, Schumacher U: *Thieme atlas of anatomy: general anatomy and musculoskeletal system,* ed 2, New York, 2014, Thieme Medical.

Seeley RR, Stephens TD, Tate P: *Anatomy & physiology,* ed 8, New York, 2008, McGraw-Hill.

Sieg KW, Adams SP: *Illustrated essentials of musculoskeletal anatomy,* ed 4, Gainesville, FL, 2002, Megabooks.

Stone RJ, Stone JA: *Atlas of the skeletal muscles,* ed 6, New York, 2009, McGraw-Hill.

Thibodeau GA, Patton KT: *Anatomy & physiology,* ed 9, St. Louis, 1993, Mosby.

Van De Graaff KM: *Human anatomy,* ed 6, Dubuque, IA, 2002, McGraw-Hill.

Chapter

**11**

# Worksheet Exercises

For in- or out-of-class assignments, or for testing, utilize this tear-out worksheet.

## Worksheet 1

Using crayons or colored markers, draw and label on the worksheet the following muscles. Indicate the origin and insertion of each muscle with an "O" and an "I," respectively, and draw in the origin and insertion on the anterior view as applicable.

    a. Rectus abdominis

    b. External oblique abdominal

    c. Internal oblique abdominal

    d. Sternocleidomastoid

# Worksheet Exercises

For in- or out-of-class assignments, or for testing, utilize this tear-out worksheet.

## Worksheet 2

Using crayons or colored markers, draw and label on the worksheet the following muscles. Indicate the origin and insertion of each muscle with an "O" and an "I," respectively, and draw in the origin and insertion on the posterior view as applicable.

a. Erector spinae

b. Quadratus lumborum

c. Splenius—cervicis and capitis

# CHAPTER 12

# MUSCULAR ANALYSIS OF SELECTED EXERCISES AND RELATED CONCEPTS

## Objectives

- To begin analyzing sport skills in terms of phases and the various joint movements occurring in those phases

- To understand various conditioning principles and how to apply them to strengthening major muscle groups

- To analyze an exercise to determine the joint movements and positions, the specific muscles involved and their contraction types in accomplishing those movements, and/or maintaining those positions

- To learn, understand, and apply the concept of open versus closed kinetic chain

- To learn to group individual muscles into units that cause, control, or prevent certain joint movements

- To begin to think of exercises that increase the strength and endurance of individual muscle groups

- To learn to analyze and prescribe exercises to strengthen major muscle groups

Proper functioning of the upper and lower extremities as well as the trunk and spinal column is critical for most sports activities, as well as for many activities of daily living. Strength and endurance in the muscles of the human body are essential for improved appearance and posture, as well as for more efficient skill performance. Specific exercises and activities to condition the various body parts should be intelligently selected by becoming thoroughly familiar with the muscles involved.

Simple exercises may be used to begin teaching individuals how to group muscles to produce joint movement. Some of these simple introductory exercises are included in this chapter.

The early analysis of exercise makes the study of structural kinesiology more meaningful as we strive to better understand the importance of individual muscles and groups of muscles in bringing about joint movements in various exercises. This chapter contains analysis of exercises for the upper extremities, trunk, and lower extremities. Contrary to what many new to structural kinesiology may believe, muscular analysis of activities is not difficult once the basic concepts are understood.

## Upper extremity activities

Children seem to have an innate desire to climb, swing, and hang. Such movements use the muscles of the hands, wrists, elbows, and shoulder joints. But the opportunity to perform these types of activities is limited in our modern culture. Unless emphasis is placed on the development of this area of our bodies by physical education teachers in elementary schools, for both boys and girls, it will continue to be muscularly the weakest area of our bodies. Weakness in the upper extremities can impair skill development and performance in many common, enjoyable recreational activities, such as golf, tennis, softball, and racquetball. The same can be said for the trunk and lower extremities. As we age, muscle volume and joint function tend to decrease, primarily due to decreased activity. Working specifically to avoid this can prevent loss of function and related pathological conditions such as osteoarthritis. People enjoy what they can do well, and they can be taught to enjoy activities that will increase strength and endurance throughout the body. These and other such activities can be enjoyed throughout the entire adult life; therefore, adequate skill development built on an appropriate base of muscular strength and endurance is essential for enjoyment and prevention of injury.

Often, we perform typical strengthening exercises in the weight room, such as the bench press, overhead press, and biceps curl. These are good exercises, but they all concentrate primarily on the muscles of the anterior upper extremity. This can lead to an overdevelopment of these muscles with respect to the posterior muscles. As a result, individuals may become strong and tight anteriorly and weak and flexible posteriorly. It is for these reasons that one must be able to analyze specific strengthening exercises and determine the muscles involved so that overall muscular balance is addressed through appropriate exercise prescription.

More and more medical and allied health professionals have been emphasizing the development of muscle groups through resistance training and circuit-training activities. Athletes and non-athletes, both male and female, need overall muscular development. Even those who do not necessarily desire significant muscle mass are advised to develop and maintain muscle mass through resistance training. As we age, we normally tend to lose muscle mass, and as a result our metabolism decreases. This factor, combined with improper eating habits, results in unhealthful fat accumulation and excessive weight gain. Through increasing our muscle mass, we burn more calories and are less likely to gain excessive fat.

Sports participation does not ensure sufficient development of muscle groups. Also, more and more emphasis has been placed on mechanical kinesiology in physical education and athletic skill teaching. This is desirable and can help bring about more skillful performance. However, it is important to remember that mechanical principles will be of little or no value to performers without adequate strength and endurance of the muscular system, which is developed through planned exercises and activities. In the fitness and health revolution of recent decades, a much greater emphasis has been placed on exercises and activities that improve the physical fitness, strength, endurance, and flexibility of participants.

## Concepts for analysis

In analyzing activities, it is important to understand that muscles are usually grouped according to their concentric function and work in paired opposition to an antagonistic group. An example of this **aggregate** muscle grouping to perform a given joint action is seen with the elbow flexors all working together. In this example, the elbow flexors (biceps, brachii, brachialis, and brachioradialis) are concentrically contracting as an agonist group to achieve flexion. As they flex the elbow, each muscle contributes significantly to the task. They are working in opposition to their antagonists, the triceps brachii and anconeus. The triceps brachii and anconeus work together as an aggregate muscle group to cause elbow extension, but in this example they are cooperating in their lengthening to allow the flexors to perform their task. In this cooperative lengthening, the triceps and anconeus may or may not be under active tension. If there is no tension, then the lengthening is passive, caused totally by the elbow flexors. If there is active tension, then the elbow extensors are contracting eccentrically to control the amount and speed of lengthening.

An often confusing aspect is that, depending on the activity, these muscle groups can function to control the exact opposite actions by contracting eccentrically. That is, through eccentric contractions, the elbow flexors may control elbow extension, as in lowering the weight in a biceps curl, and the triceps brachii and anconeus may control elbow flexion, as in lowering the weight in a triceps extension (see Tables 12.3 and 12.4). Exercise professionals should be able to view an activity and not only determine which muscles are performing the movement but also know what type of contraction is occurring and what kinds of exercises are appropriate for developing the muscles. Chapter 2 provides a review of how muscles contract to work in groups to function in joint movement.

## Analysis of movement

In analyzing various exercises and sport skills, it is essential to break down all the movements into phases. The number of phases, usually three to five, will vary depending on the skill. Practically all sport skills will have at least a preparatory phase, a movement phase, and a follow-through phase. Many will also begin with a stance phase and end with a recovery phase. The names of the phases will vary from skill to skill to fit in with the terminology used in various sports, and they may also vary depending on the body part involved. In some cases, these major phases may be divided even further, as with baseball, in which the preparatory phase for the pitching arm is broken into early cocking and late cocking.

The **stance phase** allows the athlete to assume a comfortable and appropriately balanced body position from which to initiate the sport skill. The emphasis is on setting the various joint angles in their correct positions with respect to one another and to the sport surface. Generally, with respect to the subsequent phases, the stance phase is a relatively static phase involving fairly short ranges of motion. Due to the minimal amount of movement in this phase, the majority of the joint position maintenance throughout the body will be accomplished through isometric contractions.

The **preparatory phase**, often referred to as the cocking or wind-up phase, is used to lengthen the appropriate muscles so that they will be in position to generate more force and momentum as they concentrically contract in the next phase. It is the most critical phase in achieving the desired result of the activity and becomes more dynamic as the need for explosiveness increases. Generally, to lengthen the muscles needed in the next phase, concentric contractions occur in their antagonist muscles in this phase.

The **movement phase**, sometimes known as the acceleration, action, motion, or contact phase, is the action part of the skill. It is the phase in which the summation of force is generated directly to the ball, sport object, or opponent and is usually characterized by near-maximal concentric activity in the involved muscles.

The **follow-through phase** begins immediately after the climax of the movement phase, in order to bring about negative acceleration of the involved limb or body segment. In this phase, often referred to as the deceleration phase, the velocity of the body segment progressively decreases, usually over a wide range of motion. This velocity decrease is usually attributable to high eccentric activity in the muscles that were antagonist to the muscles used in the movement phase. Generally, the greater the acceleration in the movement phase, the greater the length and importance of the follow-though phase. Occasionally, some athletes may begin the follow-through phase too soon, thereby cutting short the movement phase and achieving a less-than-desirable result in the activity.

The **recovery phase** is used after follow-through to regain balance and positioning to be ready for the next sport demand. To a degree, the muscles used eccentrically in the follow-through phase to decelerate the body or body segment will be used concentrically in recovery to bring about the initial return to a functional position.

Skill analysis can be seen with the example of a baseball pitch in Fig. 12.1. The stance phase begins when the player assumes a position with the ball in the glove before receiving the signal from the catcher. The pitcher begins the preparatory phase by extending the throwing arm posteriorly and rotating the trunk to the right in conjunction with left hip

Stance        Preparatory

Movement

Follow-through

FIG. 12.1 • Skill analysis phases—baseball pitch. The stance phase is accomplished predominantly with isometric contractions. Movements in the preparatory phase are accomplished primarily through concentric contractions. The movement phase involves significant concentric activity. Eccentric activity is high during the follow-through phase.

flexion. The right shoulder girdle is fully retracted in combination with abduction and maximum external rotation of the glenohumeral joint to complete this phase. Immediately following, the movement phase begins with forward movement of the arm and continues until ball release. At ball release, the follow-through phase begins as the arm continues moving in the same direction established by the movement phase until the velocity decreases to the point that the arm can safely change movement direction. This deceleration of the body, and especially the arm, is accomplished by high amounts of eccentric activity. At this point, the recovery phase begins, enabling the player to reposition to field the batted ball. In this example, reference has been made primarily to the throwing arm, but there are many similarities in other overhand sport skills, such as the tennis serve, javelin throw, and volleyball serve. In actual practice, the movements of each joint in the body should be analyzed with respect to the various phases.

## The kinetic chain concept

As you have learned, our extremities consist of several bony segments linked by a series of joints. These bony segments and their linkage system of joints may be likened to a chain. Just as with a chain, any one link in the extremity may be moved individually without significantly affecting the other links if the chain is open or not attached at one end. However, if the chain is securely attached or closed, substantial movement of any one link cannot occur without substantial and subsequent movement of the other links.

In the body, an extremity may be seen as representing an **open kinetic chain** if the distal end of the extremity is not fixed to a relatively stable surface. This arrangement allows any one joint in the extremity to move or function separately without necessitating movement of other joints in the extremity. This does not mean that open kinetic chain activities have to involve only one joint but rather that motion at one joint does not require motion at other joints in the chain. Examples in the upper extremity of these single joint exercises include the shoulder shrug, deltoid raise (shoulder abduction), and biceps curl (Fig. 12.2, *A*). Lower extremity examples include seated hip flexion, knee extension, and ankle dorsiflexion exercises (Fig. 12.2, *C*). In all these examples, the core of the body and the proximal segment are stabilized while the distal segment is free to move in space through a single plane. These types of exercises are known as joint-isolation exercises and are beneficial in isolating a particular joint to concentrate on specific muscle groups. However, they are not very functional in that most physical activity, particularly for the lower extremity, requires multiple-joint activity involving numerous muscle groups simultaneously. Furthermore, since the joint is

Courtesy of Britt Jones

**A**

Courtesy of R.T. Floyd

**B**

FIG. 12.2 • Open versus closed kinetic chain activities. **A,** Open-chain activity for the upper extremity; **B,** Closed-chain activity for the upper extremity.

Courtesy of R.T. Floyd (left, right)

**C**

**D**

FIG. 12.2 (continued) • Open versus closed kinetic chain activities. **C,** Open-chain activity for the lower extremity; **D,** Closed-chain activity for the lower extremity.

stable proximally and loaded distally, shear forces are acting on the joint, with potential negative consequences.

If the distal end of the extremity is fixed, as in a pull-up, push-up, dip, squat, or dead lift, the extremity represents a **closed kinetic chain**. See Fig. 12.2, *B* and *D.* In this closed system, movement of one joint cannot occur without causing predictable movements of the other joints in the extremity. Closed-chain activities are very functional and involve the body moving in relation to the relatively fixed distal segment. The advantage of multiple-joint exercises is that several joints are involved and numerous muscle groups must participate in causing and controlling the multiple-plane movements, which strongly correlate to most physical activities. Additionally, the joint is more stable due to the joint compressive forces from weight bearing.

To state the differences another way, open-chain exercises involve the extremity being moved to or from the stabilized body, whereas closed-chain exercises involve the body being moved to or from the stabilized extremity. Table 12.1 provides a comparison of variables that differ between open- and closed-chain exercises, and Fig. 12.2 provides examples of each.

Not every exercise or activity can be classified totally as either an open- or closed-chain exercise. For example, walking and running are both open and closed due to their swing and stance phases,

respectively. Another case is bicycle riding, which is mixed in that the pelvis on the seat is the stablest segment, but the feet are attached to movable pedals.

Consideration of the open versus closed kinetic chain is important in determining both the muscles and their types of contractions when analyzing sports activities. Realizing the relative differences in demands on the musculoskeletal system through detailed analysis of skilled movements is critical for determining the most appropriate conditioning exercises to improve performance. Generally, closed kinetic chain exercises are more functional and applicable to the demands of sports and physical activity. Most sports involve closed-chain activities in the lower extremities and open-chain activities in the upper extremities. However, there are many exceptions, and closed-chain conditioning exercises may be beneficial for extremities primarily involved in open-chain sporting activities. Open-chain exercises are useful in developing a specific muscle group at a single joint.

## Conditioning considerations

It is not the intent of this book to thoroughly address conditioning principles, but a brief overview is provided to serve as a general reference and reminder of the importance of applying these concepts correctly when developing major muscle groups.

## TABLE 12.1 • Differences between open- and closed-chain exercises

| Variable | Open-chain exercise | Closed-chain exercise |
| --- | --- | --- |
| Distal end of extremity | Free in space and not fixed | Fixed to something |
| Movement pattern | Characterized by rotary stress in the joint (often nonfunctional) | Characterized by linear stress in the joint (functional) |
| Joint movements | Occur in isolation | Multiple occur simultaneously |
| Muscle recruitment | Isolated (minimal muscular co-contraction) | Multiple (significant muscular co-contraction) |
| Joint axis | Stable during movement patterns | Primarily transverse |
| Movement plane | Usually single | Multiple (triplanar) |
| Proximal segment of joint | Stable | Mobile |
| Distal segment of joint | Mobile | Mobile, except for most distal aspect |
| Motion occurs | Distal to instantaneous axis of rotation | Proximal and distal to instantaneous axis of rotation |
| Functionality | Often nonfunctional, especially lower extremity | Functional |
| Joint forces | Shear | Compressive |
| Joint stability | Decreased due to shear and distractive forces | Increased due to compressive forces |
| Stabilization | Artificial | Not artificial, rather realistic and functional |
| Loading | Artificial | Physiological, provides for normal proprioceptive and kinesthetic feedback |

Adapted from Ellenbecker TS, Davies GJ: *Closed kinetic chain exercise: a comprehensive guide to multiple-joint exercise*, Champaign, IL, 2001, Human Kinetics.

## Overload principle

A basic physiological principle of exercise is the overload principle. It states that, within appropriate parameters, a muscle or muscle group increases in strength in direct proportion to the overload placed on it. While it is beyond the scope of this text to fully explain specific applications of the overload principle for each component of physical fitness, some general concepts follow. To improve the strength and functioning of major muscles, this principle should be applied to every large muscle group in the body, progressively throughout each year, at all age levels. In actual practice, the amount of overload applied varies significantly based on several factors. For example, an untrained person beginning a strength-training program will usually make significant gains in the amount of weight he or she is able to lift in the first few weeks of the exercise program. Most of this increased ability is due to a refinement of neuromuscular function rather than to an actual increase in muscle tissue strength. Similarly, a well-trained person will see relatively minor improvements in

the amount of weight that can be lifted over a much longer period of time. Therefore, the amount and rate of progressive overload are extremely variable and must be adjusted to match the specific needs of the individual's exercise objectives.

Overload may be modified by changing any one or a combination of three exercise variables—**frequency**, **intensity**, and **duration**. Frequency usually refers to the number of times per week. Intensity is usually a certain percentage of the absolute maximum, and duration usually refers to the number of minutes per exercise bout. Increasing the speed of doing the exercise, the number of repetitions, the weight, and the bouts of exercise are all ways to modify these variables and apply the overload principle. All these factors are important in determining the total exercise volume.

Overload is not always progressively increased. In certain periods of conditioning, the overload should actually be prescriptively reduced or increased to improve the total results of the entire program. This intentional variance in a training program at regular intervals is known as **periodization** and is done to bring about optimal gains in physical performance. Part of the basis for periodization is so that the athlete will be at his or her peak level during the most competitive part of the season. To achieve this, a number of variables may be manipulated, including the number of sets per exercise or repetitions per set, types of exercises, number of exercises per training session, rest periods between sets and exercises, resistance used for a set, type of muscle contraction, and number of training sessions per day and per week.

## SAID principle

The **SAID** (**S**pecific **A**daptations to **I**mposed **D**emands) principle should be considered in all aspects of physiological conditioning and training. This principle, which states that the body will gradually, over time, adapt very specifically to the various stresses and overloads to which it is subjected, is applicable in every form of muscle training, as well as to the other systems of the body. For example, if an individual were to undergo several weeks of strength-training exercises for a particular joint through a limited range of motion, the specific muscles involved in performing the strengthening exercises would improve primarily in the ability to move against increased resistance through the specific range of motion used. There would be, in most cases, minimal strength gains significantly beyond the range of motion used in the training. Additionally, other components of physical fitness—such as flexibility, cardiorespiratory endurance, and muscular endurance—would be enhanced minimally, if at all. In other words, to achieve specific benefits, exercise programs must be specifically designed for the adaptation desired.

It should be recognized that this adaptation may be positive or negative, depending on whether the correct techniques are used and stressed in the design and administration of the conditioning program. Inappropriate or excessive demands placed on the body in too short a time span can result in injury. If the demands are too little or are administered too infrequently over too long a time period, less than desired improvement will occur. Conditioning programs and the exercises included in them should be analyzed to determine whether they are using the specific muscles for which they were intended in the correct manner.

## Specificity

Specificity of exercise strongly relates to the discussion of the SAID principle. The components of physical fitness—such as muscular strength, muscular endurance, and flexibility—are not general body characteristics but rather are specific to each body area and muscle group. Therefore, the specific needs of the individual must be addressed when designing an exercise program. Quite often, it will be necessary to analyze an individual's exercise and skill technique to design an exercise program to meet his or her specific needs. Potential exercises to be used in the conditioning program must be analyzed to determine their appropriateness for the individual's specific needs. The goals of the exercise program should be determined regarding specific areas of the body, preferred time to physically peak, and physical fitness needs such as strength, muscular endurance, power, agility, balance, coordination, flexibility, cardiorespiratory endurance, and body composition. After establishing goals, a regimen incorporating the overload variables of frequency, intensity, and duration may be prescribed to include the entire body or specific areas in such a way as to address the improvement of the preferred physical fitness components. Regular observation and follow-up exercise analysis are necessary to ensure proper adherence to correct technique.

## Muscular development

For years it was thought that a person developed adequate muscular strength, endurance, and flexibility through participation in sports activities. Now it is believed that a person needs to develop muscular strength, endurance, and flexibility in order to be able to participate safely and effectively in sports activities.

Adequate muscular strength, endurance, and flexibility of the entire body from head to toe should be developed through correct use of the proper exercise principles. Individuals responsible for this development need to prescribe appropriate exercises that meet these objectives.

In schools this development should start at an early age and continue throughout the school years. Results of fitness tests such as sit-ups, the standing long jump, and the mile run reveal the need for considerable improvement in this area in children in the United States. Adequate muscular strength and endurance are important in the adult years for the activities of daily living, as well as for job-related requirements and recreational needs. Many back problems and other physical ailments could be avoided through proper maintenance of the musculoskeletal system. Refer to Chapters 4 through 7 as needed.

Chapter

**12**

## Analysis of selected exercises

To assist in developing the practice of analyzing the muscles used for specific activities, the next several pages provide brief analyses of several common exercises. When these techniques are practiced extensively and mastered, the individual is ready to analyze and prescribe exercises and activities for the muscular strength and endurance needed in sports activities and for healthful living. All muscles listed in the analysis are contracting concentrically unless specifically noted to be contracting eccentrically or isometrically. It should be noted that these analyses focus only on the major joints involved in the movements and that a full analysis should include the activity in essentially all parts of the body. As an example of the importance of the total body, a baseball pitcher can be significantly impaired by functional deficits anywhere between the toes and fingertips.

To further assist in analyzing the muscles primarily involved in exercises, it would also be beneficial to utilize Appendix 4 and determine the muscles involved in the different phases using the Skill Analysis Worksheet found in the Worksheet Exercises at the end of this chapter. The worksheets together provide for analysis of up to six different phases.

**Note:** In many cases only the muscles that are considered the primary movers for a particular joint action are listed and the remaining muscles that contribute to a significantly lesser extent are not included in the analyses provided in this chapter.

## Valsalva maneuver

Many people bear down by holding their breath without thinking when attempting to lift something heavy. This bearing down, known as the Valsalva maneuver, is accomplished by exhaling against a closed epiglottis (the flap of cartilage behind the tongue that shuts the air passage when swallowing) and is thought by many to enhance lifting ability. It is mentioned here to caution against its use, because it causes a dramatic increase in blood pressure followed by an equally dramatic drop in blood pressure. Using the Valsalva maneuver can cause lightheadedness and fainting and can lead to complications in people with heart disease. Instead of using the Valsalva maneuver, people lifting should always be sure to use rhythmic and consistent breathing. It is usually advisable to exhale during the lifting or contracting phase and inhale during the lowering or recovery phase.

# Shoulder pull

## Description

In a standing or sitting position, the subject interlocks the fingers in front of the chest and then attempts to pull them apart (Fig. 12.3). This contraction is maintained for 5 to 20 seconds.

## Analysis

In this type of exercise, there is little or no movement of the contracting muscles. In certain isometric exercises, contraction of the antagonistic muscles is as strong as contraction of the muscles attempting to produce the force for movement. The muscle groups contracting to produce a movement are designated the **agonists**. In the exercise just described, the agonists in the right upper extremity are antagonistic to the agonists in the left upper extremity, and vice versa (Table 12.2). This exercise results in isometric contractions of the wrist and hand, elbow, shoulder joint, and shoulder girdle muscles. The strength of the contraction depends on the angle of pull and the

Courtesy of Britt Jones

FIG. 12.3 • Shoulder pull.

leverage of the joint involved. Thus, it is not the same at each point.

Isometric exercises vary in the number of muscles contracting, depending on the type of exercise and the joints at which movement is attempted. The shoulder pull exercise produces some contraction of agonist muscles at four sets of joints. See Tables 4.2, 5.3, 6.2, and 7.2.

## TABLE 12.2 • **Shoulder pull**

| Joint | \multicolumn{4}{c}{This entire exercise is designed so that all contractions presented are isometric} |
|---|---|---|---|---|
| | Action | Agonists | Action | Agonists |
| Wrist and hand | Flexion | Resisted by flexors of wrists and hand Agonists—wrist and MCP, PIP, PIP flexors | Flexion | Resisted by flexors of wrist and hand Antagonists—wrist and MCP, PIP, PIP flexors |
| Elbow joint | Extension | Resisted by flexors of elbow Agonists—triceps brachii, anconeus Antagonists—biceps brachii, brachialis, brachioradialis | Flexion | Resisted by extensors of elbow Agonists—biceps brachii, brachialis, brachioradialis Antagonists—triceps brachii, anconeus |
| Shoulder joint | Horizontal abduction | Resisted by horizontal abductors of contralateral shoulder joint Agonists—deltoid, infraspinatus, teres minor, latissimus dorsi Antagonists—contralateral deltoid, infraspinatus, teres minor, latissimus dorsi | Horizontal abduction | Resisted by horizontal abductors of contralateral shoulder joint Agonists—deltoid, infraspinatus, teres minor, latissimus dorsi Antagonists—contralateral deltoid, infraspinatus, teres minor, latissimus dorsi |
| Shoulder girdle | Adduction | Resisted by adductors of contralateral shoulder girdle Agonists—rhomboid and trapezius Antagonists—contralateral rhomboid and trapezius | Adduction | Resisted by adductors of contralateral shoulder girdle Agonists—rhomboid and trapezius Antagonists—contralateral rhomboid and trapezius |

Chapter
**12**

# Arm curl

## Description

With the subject in a standing position, the dumbbell is held in the hand with the palm to the front. The dumbbell is lifted until the elbow is completely flexed (Fig. 12.4). Then it is returned to the starting position.

## Analysis

This open kinetic chain exercise is divided into two phases for analysis: (1) lifting phase to flexed position and (2) lowering phase to extended position (Table 12.3). *Note:* An assumption is made that no movement occurs in the shoulder joint and shoulder girdle, although many of the muscles of both the shoulder and the shoulder girdle are isometrically acting as stabilizers. Review Tables 4.2, 5.3, 6.2, and 7.2.

Courtesy of Britt Jones (left, right)

FIG. 12.4 • Arm curl. **A,** Starting position in extension; **B,** Flexed position.

## TABLE 12.3 • Arm curl

| Joint | Lifting phase to flexed position | | Lowering phase to extended position | |
| | Action | Agonists | Action | Agonists |
|---|---|---|---|---|
| Wrist and hand | Flexion* | Wrist and MCP, PIP, PIP flexors (isometric contraction)<br>  Flexor carpi radialis<br>  Flexor carpi ulnaris<br>  Palmaris longus<br>  Flexor digitorum profundus<br>  Flexor digitorum superficialis<br>  Flexor pollicis longus | Flexion* | Wrist and MCP, PIP, PIP flexors (isometric contraction)<br>  Flexor carpi radialis<br>  Flexor carpi ulnaris<br>  Palmaris longus<br>  Flexor digitorum profundus<br>  Flexor digitorum superficialis<br>  Flexor pollicis longus |
| Elbow | Flexion | Elbow flexors<br>  Biceps brachii<br>  Brachialis<br>  Brachioradialis | Extension | Elbow flexors (eccentric contraction)<br>  Biceps brachii<br>  Brachialis<br>  Brachioradialis |

*The wrist is held in a position of slight extension by an isometric contraction of the wrist extensors (extensor carpi ulnaris, extensor carpi radialis longus, and extensor carpi radialis brevis) to facilitate greater active finger flexion in gripping the dumbbell.
(The flexors remain isometrically contracted throughout the entire exercise, to hold the dumbbell.)

Chapter
**12**

# Triceps extension

## Description

The subject may use the opposite hand to assist in maintaining the arm in a shoulder-flexed position. Then, grasping the dumbbell and beginning in full elbow flexion, the subject extends the elbow until the arm and forearm are straight. The shoulder joint and shoulder girdle are stabilized by the opposite hand. Consequently, no movement is assumed to occur in these areas (Fig. 12.5).

## Analysis

This open kinetic chain exercise is divided into two phases for analysis: (1) lifting phase to extended position and (2) lowering phase to flexed position (Table 12.4). *Note:* An assumption is made that no movement occurs in the shoulder joint and shoulder girdle, although it is critical that many of the shoulder and shoulder girdle muscles contract isometrically to stabilize this area so that the exercise may be performed correctly. Review Tables 4.2, 5.3, 6.2, and 7.2.

Courtesy of Britt Jones (left, right)

FIG. 12.5 • Triceps extension. **A,** Starting position in flexion; **B,** Extended position.

## TABLE 12.4 • Triceps extension

| Joint | Lifting phase to extended position | | Lowering phase to flexed position | |
| | Action | Agonists | Action | Agonists |
| --- | --- | --- | --- | --- |
| Wrist and hand | Flexion* | Wrist and MCP, PIP, PIP flexors (isometric contraction)<br>Flexor carpi radialis<br>Flexor carpi ulnaris<br>Palmaris longus<br>Flexor digitorum profundus<br>Flexor digitorum superficialis<br>Flexor pollicis longus | Flexion* | Wrist and MCP, PIP, PIP flexors (isometric contraction)<br>Flexor carpi radialis<br>Flexor carpi ulnaris<br>Palmaris longus<br>Flexor digitorum profundus<br>Flexor digitorum superficialis<br>Flexor pollicis longus |
| Elbow | Extension | Elbow extensors<br>Triceps brachii<br>Anconeus | Flexion | Elbow extensors (eccentric contraction)<br>Triceps brachii<br>Anconeus |

*The wrist is held in a position of slight extension by an isometric contraction of the wrist extensors (extensor carpi ulnaris, extensor carpi radialis longus, and extensor carpi radialis brevis) to facilitate greater active finger flexion in gripping the dumbbell.
(The flexors remain isometrically contracted throughout the entire exercise, to hold the dumbbell.)

Chapter
**12**

# Barbell press

## Description

This open kinetic chain exercise is sometimes referred to as the *overhead* or *military press*. The barbell is held in a position high in front of the chest, with palms facing forward, feet comfortably spread, and back and legs straight (Fig. 12.6, *A*). From this position, the barbell is pushed upward until fully overhead (Fig. 12.6, *B*), and then it is returned to the starting position. See Tables 4.2, 5.3, 6.2, and 7.2.

## Analysis

This exercise is separated into two phases for analysis: (1) lifting phase to full overhead position and (2) lowering phase to starting position (Table 12.5).

Courtesy of R.T. Floyd (left, right)

**A**  **B**

FIG. 12.6 • Barbell press. **A,** Starting position; **B,** Full overhead position.

TABLE 12.5 • **Barbell press**

| Joint | Lifting phase to full overhead position | | Lowering phase to starting position | |
| | Action | Agonists | Action | Agonists |
|---|---|---|---|---|
| Wrist and hand | Flexion* | Wrist and MCP, PIP, PIP flexors (isometric contraction)<br>Flexor carpi radialis<br>Flexor carpi ulnaris<br>Palmaris longus<br>Flexor digitorum profundus<br>Flexor digitorum superficialis<br>Flexor pollicis longus | Flexion* | Wrist and MCP, PIP, PIP flexors (isometric contraction)<br>Flexor carpi radialis<br>Flexor carpi ulnaris<br>Palmaris longus<br>Flexor digitorum profundus<br>Flexor digitorum superficialis<br>Flexor pollicis longus |
| Elbow joint | Extension | Elbow extensors<br>Triceps brachii<br>Anconeus | Flexion | Elbow extensors (eccentric contraction)<br>Triceps brachii<br>Anconeus |
| Shoulder | Flexion | Shoulder joint flexors<br>Pectoralis major (clavicular head or upper fibers)<br>Anterior deltoid<br>Coracobrachialis<br>Biceps brachii | Extension | Shoulder joint flexors (eccentric contraction)<br>Pectoralis major (clavicular head or upper fibers)<br>Anterior deltoid<br>Coracobrachialis<br>Biceps brachii |
| Shoulder girdle | Upward rotation and elevation | Shoulder girdle upward rotators and elevators<br>Trapezius<br>Levator scapulae<br>Serratus anterior | Downward rotation and depression | Shoulder girdle upward rotators and elevators (eccentric contraction)<br>Trapezius<br>Levator scapulae<br>Serratus anterior |

*The wrist is held in a position of slight extension by an isometric contraction of the wrist extensors (extensor carpi ulnaris, extensor carpi radialis longus, and extensor carpi radialis brevis) to facilitate greater active finger flexion in gripping the bar.

Chapter
**12**

# Chest press (bench press)

## Description

The subject lies on the exercise bench in the supine position, grasps the barbell, and presses the weight upward through the full range of arm and shoulder movement (Fig. 12.7). Then the weight is lowered to the starting position. Refer to Tables 4.2, 5.3, 6.2, and 7.2.

## Analysis

This open kinetic chain exercise can be divided into two phases for analysis: (1) lifting phase to up position and (2) lowering phase to starting position (Table 12.6).

FIG. 12.7 • Chest press (bench press). **A,** Starting position; **B,** Up position.

## TABLE 12.6 • **Chest press (bench press)**

| Joint | Lifting phase to up position | | Lowering phase to starting position | |
| | Action | Agonists | Action | Agonists |
|---|---|---|---|---|
| Wrist and hand | Flexion* | Wrist and MCP, PIP, PIP flexors (isometric contraction)<br>Flexor carpi radialis<br>Flexor carpi ulnaris<br>Palmaris longus<br>Flexor digitorum profundus<br>Flexor digitorum superficialis<br>Flexor pollicis longus | Flexion* | Wrist and MCP, PIP, PIP flexors (isometric contraction)<br>Flexor carpi radialis<br>Flexor carpi ulnaris<br>Palmaris longus<br>Flexor digitorum profundus<br>Flexor digitorum superficialis<br>Flexor pollicis longus |
| Elbow | Extension | Elbow extensors<br>Triceps brachii<br>Anconeus | Flexion | Elbow extensors (eccentric contraction)<br>Triceps brachii<br>Anconeus |
| Shoulder | Flexion and horizontal adduction | Shoulder flexors and horizontal adductors<br>Pectoralis major<br>Anterior deltoid<br>Coracobrachialis<br>Biceps brachii | Extension and horizontal abduction | Shoulder joint flexors and horizontal adductors (eccentric contraction)<br>Pectoralis major<br>Anterior deltoid<br>Coracobrachialis<br>Biceps brachii |
| Shoulder girdle | Abduction | Shoulder girdle abductors<br>Serratus anterior<br>Pectoralis minor | Adduction | Shoulder girdle abductors (eccentric contraction)<br>Serratus anterior<br>Pectoralis minor |

*The wrist is held in a position of slight extension by an isometric contraction of the wrist extensors (extensor carpi ulnaris, extensor carpi radialis longus, and extensor carpi radialis brevis) to facilitate greater active finger flexion in gripping the bar.

Courtesy of Lisa Floyd (left, right)

Chapter
**12**

# Chin-up (pull-up)

## Description

The subject grasps a horizontal bar or ladder with the palms away from the face (Fig. 12.8, *A*). From a hanging position on the bar, the subject pulls up until the chin is over the bar (Fig. 12.8, *B*) and then returns to the starting position (Fig. 12.8, *C*). The width of the grip on the chin-up bar affects the shoulder actions to a degree. A narrow grip will allow for more glenohumeral extension and flexion, whereas a wider grip, as shown in Fig. 12.8, requires more adduction and abduction, respectively. For a full review of the muscles involved in the chin-up, see Tables 4.2, 5.3, 6.2, and 7.2.

## Analysis

This closed kinetic chain exercise is separated into two phases for analysis: (1) pulling-up phase to chinning position and (2) lowering phase to starting position (Table 12.7).

FIG. 12.8 • Pull-up.
**A,** Straight-arm hang;
**B,** Chin over bar;
**C,** Bent-arm hang on way up or down.

Courtesy of R.T. Floyd (all)

## TABLE 12.7 · Chin-up (pull-up)

| Joint | Pulling-up phase to chinning position | | Lowering phase to starting position | |
|---|---|---|---|---|
| | Action | Agonists | Action | Agonists |
| Wrist and hand | Flexion* | Wrist and MCP, PIP, PIP flexors (isometric contraction) <br>     Flexor carpi radialis <br>     Flexor carpi ulnaris <br>     Palmaris longus <br>     Flexor digitorum profundus <br>     Flexor digitorum superficialis <br>     Flexor pollicis longus | Flexion | Wrist and MCP, PIP, PIP flexors (isometric contraction) <br>     Flexor carpi radialis <br>     Flexor carpi ulnaris <br>     Palmaris longus <br>     Flexor digitorum profundus <br>     Flexor digitorum superficialis <br>     Flexor pollicis longus |
| Elbow | Flexion | Elbow flexors <br>     Biceps brachii <br>     Brachialis <br>     Brachioradialis | Extension | Elbow flexors (eccentric contraction) <br>     Biceps brachii <br>     Brachialis <br>     Brachioradialis |
| Shoulder | Adduction | Shoulder joint adductors <br>     Pectoralis major <br>     Posterior deltoid <br>     Latissimus dorsi <br>     Teres major <br>     Subscapularis | Abduction | Shoulder joint adductors (eccentric contraction) <br>     Pectoralis major <br>     Posterior deltoid <br>     Latissimus dorsi <br>     Teres major <br>     Subscapularis |
| Shoulder girdle | Adduction, depression, and downward rotation | Shoulder girdle adductors, depressors, and downward rotators <br>     Trapezius (lower and middle) <br>     Pectoralis minor <br>     Rhomboids | Elevation, abduction, and upward rotation | Shoulder girdle adductors, depressors, and downward rotators (eccentric contraction) <br>     Trapezius (lower and middle) <br>     Pectoralis minor <br>     Rhomboids |

*The wrist is maintained in slight extension by an isometric contraction of the wrist extensors (extensor carpi ulnaris, extensor carpi radialis longus, and extensor carpi radialis brevis) to facilitate greater active finger flexion in gripping the bar.

# Latissimus pull (lat pull)

## Description

From a sitting position, the subject reaches up and grasps a horizontal bar (Fig. 12.9, *A*). The bar is pulled down to a position below the chin (Fig. 12.9, *B*). Then it is returned slowly to the starting position. The width of the grip on the horizontal bar affects the shoulder actions to a degree. A narrow grip will allow for more glenohumeral extension and flexion, whereas the more common wider grip, as shown in Fig. 12.9, requires more adduction and abduction, respectively. Tables 4.2, 5.3, 6.2, and 7.2 provide a thorough listing of the muscles utilized in this exercise.

## Analysis

This open kinetic chain exercise is separated into two phases for analysis: (1) pull-down phase to below the chin position and (2) return phase to starting position (Table 12.8).

**A**

Courtesy of R.T. Floyd (all)

**B**

FIG. 12.9 • Latissimus pull (lat pull). **A,** Starting position; **B,** Downward position.

## TABLE 12.8 • **Latissimus pull (lat pull)**

| Joint | Pull-down phase to below the chin position | | Return phase to starting position | |
|---|---|---|---|---|
| | Action | Agonists | Action | Agonists |
| Wrist and hand | Flexion* | Wrist and MCP, PIP, PIP flexors (isometric contraction)<br>　Flexor carpi radialis<br>　Flexor carpi ulnaris<br>　Palmaris longus<br>　Flexor digitorum profundus<br>　Flexor digitorum superficialis<br>　Flexor pollicis longus | Flexion | Wrist and MCP, PIP, PIP flexors (isometric contraction)<br>　Flexor carpi radialis<br>　Flexor carpi ulnaris<br>　Palmaris longus<br>　Flexor digitorum profundus<br>　Flexor digitorum superficialis<br>　Flexor pollicis longus |
| Elbow | Flexion | Elbow flexors<br>　Biceps brachii<br>　Brachialis<br>　Brachioradialis | Extension | Elbow flexors (eccentric contraction)<br>　Biceps brachii<br>　Brachialis<br>　Brachioradialis |
| Shoulder | Adduction | Shoulder joint adductors<br>　Pectoralis major<br>　Latissimus dorsi<br>　Teres major<br>　Subscapularis | Abduction | Shoulder joint adductors (eccentric contraction)<br>　Pectoralis major<br>　Latissimus dorsi<br>　Teres major<br>　Subscapularis |
| Shoulder girdle | Adduction, depression, and downward rotation | Shoulder girdle adductors, depressors, and downward rotators<br>　Trapezius (lower and middle)<br>　Pectoralis minor<br>　Rhomboids | Abduction, elevation, and upward rotation | Shoulder girdle adductors, depressors, and downward rotators (eccentric contraction)<br>　Trapezius (lower and middle)<br>　Pectoralis minor<br>　Rhomboids |

*The wrist is maintained in slight extension by an isometric contraction of the wrist extensors (extensor carpi ulnaris, extensor carpi radialis longus, and extensor carpi radialis brevis) to facilitate greater active finger flexion in gripping the bar.

Chapter
**12**

## Push-up

### Description

The subject lies on the floor in a prone position with the legs together, the palms touching the floor, and the hands pointed forward and approximately under the shoulders (Fig. 12.10, *A*). Keeping the back and legs straight, the subject pushes up to the up position and then returns to the starting position (Fig. 12.10, *B*).

The push-up is a total body exercise in that the muscles of the cervical and lumbar spine, hips, knees, ankles, and feet are active isometrically to stabilize the respective areas. Table 12.9 includes only the muscles of the upper extremity in the analysis. Tables 4.2, 5.3, 6.2, and 7.2 provide additional coverage of the muscles involved in the upper extremity portion of the push-up.

### Analysis

This closed kinetic chain exercise is separated into two phases for analysis: (1) pushing phase to up position and (2) lowering phase to starting position (Table 12.9).

Chin-ups and push-ups are excellent exercises for the shoulder area, shoulder girdle, shoulder joint, elbow joint, and wrist and hand (see Figs. 12.8 and 12.10). The use of free weights, machines, and other conditioning exercises helps develop strength and endurance for this part of the body.

FIG. 12.10 • Push-up. **A,** Starting position; **B,** Up position.

### TABLE 12.9 • **Push-up**

| Joint | Pushing phase to up position | | Lowering phase to starting position | |
|---|---|---|---|---|
| | Action | Agonists | Action | Agonists |
| Wrist and hand | Flexion | Wrist and MCP, PIP, PIP flexors (isometric contraction)<br>Flexor carpi radialis<br>Flexor carpi ulnaris<br>Palmaris longus<br>Flexor digitorum profundus<br>Flexor digitorum superficialis<br>Flexor pollicis longus | Flexion | Wrist and MCP, PIP, PIP flexors (isometric contraction)<br>Flexor carpi radialis<br>Flexor carpi ulnaris<br>Palmaris longus<br>Flexor digitorum profundus<br>Flexor digitorum superficialis<br>Flexor pollicis longus |
| Elbow | Extension | Elbow extensors<br>Triceps brachii<br>Anconeus | Flexion | Elbow extensors (eccentric contraction)<br>Triceps brachii<br>Anconeus |
| Shoulder | Horizontal adduction | Shoulder joint horizontal adductors<br>Pectoralis major<br>Anterior deltoid<br>Biceps brachii<br>Coracobrachialis | Horizontal abduction | Shoulder joint horizontal adductors (eccentric contraction)<br>Pectoralis major<br>Anterior deltoid<br>Biceps brachii<br>Coracobrachialis |
| Shoulder girdle | Abduction | Shoulder girdle abductors<br>Serratus anterior<br>Pectoralis minor | Adduction | Shoulder girdle abductors (eccentric contraction)<br>Serratus anterior<br>Pectoralis minor |

# Dumbbell bent-over row

## Description

This open kinetic chain exercise may also be performed in the prone position and is therefore sometimes referred to as a prone row. The subject is kneeling on a bench or lying prone on a table so that the involved arm is free from contact with the floor (Fig. 12.11, *A*). When kneeling, the subject uses the contralateral arm to support the body. The dumbbell is held in the hand with the arm and shoulder hanging straight to the floor. From this position, the subject adducts the shoulder girdle and horizontally abducts the shoulder joint (Fig. 12.11, *B*). Then the dumbbell is lowered slowly to the starting position. Tables 4.2, 5.3, 6.2, and 7.2 provide more details on the muscles used in this exercise.

## Analysis

This exercise is separated into two phases for analysis: (1) pull-up phase to horizontal abducted position and (2) lowering phase to starting position (Table 12.10).

FIG. 12.11 • Dumbbell bent-over row. **A,** Starting position; **B,** Up position.

Courtesy of Britt Jones (left, right)

A        B

## TABLE 12.10 • **Dumbbell bent-over row**

| Joint | Pull-up phase to horizontal abducted position | | Lowering phase to starting position | |
| | Action | Agonists | Action | Agonists |
|---|---|---|---|---|
| Hand | Flexion* | MCP, PIP, PIP flexors (isometric contraction)<br>  Flexor digitorum profundus<br>  Flexor digitorum superficialis<br>  Flexor pollicis longus | Flexion | MCP, PIP, PIP flexors (isometric contraction)<br>  Flexor digitorum profundus<br>  Flexor digitorum superficialis<br>  Flexor pollicis longus |
| Elbow | Flexion | Passive flexion as the arm becomes parallel to the floor due to gravity | Extension | Passive extension as the arm becomes perpendicular to the floor due to gravity |
| Shoulder | Horizontal abduction | Shoulder joint horizontal abductors<br>  Posterior deltoid<br>  Infraspinatus<br>  Teres minor<br>  Latissimus dorsi | Horizontal adduction | Shoulder joint horizontal abductors (eccentric contraction)<br>  Posterior deltoid<br>  Infraspinatus<br>  Teres minor<br>  Latissimus dorsi |
| Shoulder girdle | Adduction | Shoulder girdle adductors<br>  Trapezius (lower and middle)<br>  Rhomboids | Abduction | Shoulder girdle adductors (eccentric contraction)<br>  Trapezius (lower and middle)<br>  Rhomboids |

*The wrist is maintained in slight extension by an isometric contraction of the wrist extensors (extensor carpi ulnaris, extensor carpi radialis longus, and extensor carpi radialis brevis) to facilitate greater active finger flexion in gripping the dumbbell.

# Abdominal curl-up  FIG. 12.12

## Description

The participant lies on the back, forearms crossed and lying across the chest, with the knees flexed approximately 90 degrees and the feet about hip width apart. The hips and knees are flexed in this manner to reduce the length of the hip flexors, thereby reducing their contribution to the curl-up.

The participant performs trunk flexion up to a curl-up position, rotates the trunk to the right and points the left elbow toward the anterior right pelvis (anterior superior iliac spine), and then returns to the starting position. On the next repetition, the participant should rotate to the left instead of to the right for balanced muscular development.

## Analysis

This open kinetic chain exercise is divided into four phases for analysis: (1) trunk flexion phase to curl-up position, (2) rotating to right/left phase, (3) return phase to curl-up position, and (4) return phase to starting position (Table 12.11).

FIG. 12.12 • Abdominal curl-up. **A,** Beginning relaxed position; **B,** Trunk flexion to curl-up position; **C,** Trunk flexion and right rotation curl-up position.

Courtesy of Britt Jones (all)

## TABLE 12.11 • Abdominal curl-up

| Joint | Trunk flexion phase to curl-up position | | Rotating to right/left phase | | Return phase to curl-up position | | Return phase to starting position | |
|---|---|---|---|---|---|---|---|---|
| | Action | Agonists | Action | Agonists | Action | Agonists | Action | Agonists |
| Cervical spine | Flexion | Cervical spine flexors Sternocleidomastoid | Maintenance of cervical flexion | Cervical spine flexors (isometric contraction) Sternocleidomastoid | Maintenance of cervical flexion | Cervical spine flexors (isometric contraction) Sternocleidomastoid | Extension | Cervical spine flexors (eccentric contraction) Sternocleidomastoid |
| Trunk | Flexion | Trunk flexors Rectus abdominis External oblique abdominal Internal oblique abdominal | Right lumbar rotation | Right lumbar rotators (R) Rectus abdominis (L) External oblique abdominal (R) Internal oblique abdominal (R) Erector spinae | Left lumbar rotation to neutral position | Right lumbar rotators (eccentric contraction) (R) Rectus abdominis (L) External oblique abdominal (R) Internal oblique abdominal (R) Erector spinae | Extension | Trunk flexors (eccentric contraction) Rectus abdominis External oblique abdominal Internal oblique abdominal |
| Hip | Flexion | Hip flexors Iliopsoas Rectus femoris Pectineus | Maintenance of hip flexion | Hip flexors (isometric contraction) Iliopsoas Rectus femoris Pectineus | Maintenance of hip flexion | Hip flexors (isometric contraction) Iliopsoas Rectus femoris Pectineus | Extension | Hip flexors (eccentric contraction) Iliopsoas Rectus femoris Pectineus |

## Alternating prone extensions  FIG. 12.13

### Description

The participant lies in a prone position, face-down, with the shoulders fully flexed in a relaxed position lying in front of the body. The head, upper trunk, right upper extremity, and left lower extremity are raised from the floor. The knees are kept in full extension. Then the participant returns to the starting position. On the next repetition, the head, upper trunk, left upper extremity, and right lower extremity are raised from the floor.

### Analysis

This open kinetic chain exercise is separated into two phases for analysis: (1) lifting phase to raise the right upper extremity off the surface and raise the left lower extremity off the floor and (2) lowering phase to relaxed position (Table 12.12).

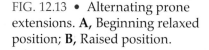

FIG. 12.13 • Alternating prone extensions. **A,** Beginning relaxed position; **B,** Raised position.

Courtesy of Britt Jones (left, right)

### TABLE 12.12 • Alternating prone extensions (

| Joint | Lifting phase to raise upper and lower extre... | | | ...laxed position | |
|---|---|---|---|---|---|
| | Action | Agonists | | | |
| Shoulder | Flexion | Shoulder joint flexors<br>Pectoralis major (clav...<br>upper fibers)<br>Deltoid<br>Coracobrachialis<br>Biceps brachii | | ...(eccentric<br>...(clavicular head or<br>...ialis<br>...hii | |
| Shoulder girdle | Adduction | Shoulder girdle add...<br>Trapezius<br>Rhomboids | | ...rdle adductors (eccentric<br>...n)<br>...zius<br>...nboids | |
| Trunk | Extension | Trunk extensors<br>Erector spinae<br>Splenius<br>Quadratus lumborum | Flexion...<br>neutral relax...<br>position) | ...k and cervical spine extensors<br>...entric contraction)<br>Erector spinae<br>Splenius<br>Quadratus lumborum | |
| Hip | Extension | Hip extensors<br>Gluteus maximus<br>Semitendinosus<br>Semimembranosus<br>Biceps femoris | Flexion (return to<br>neutral relaxed<br>position) | Hip extensors (eccentric contraction)<br>Gluteus maximus<br>Semitendinosus<br>Semimembranosus<br>Biceps femoris | |

# Squat   FIG. 12.14

## Description

The participant places a barbell on the shoulders behind the neck and grasps it with the palms-forward position of the hands. The participant squats down, flexing at the hips while keeping the spine in normal alignment, until the thighs are parallel to the floor. The participant then returns to the starting position. This exercise is commonly performed improperly by allowing the knees to move forward beyond the plane of the feet, which greatly increases the risk of injury. Care should be taken to ensure that the shins remain as vertical as possible during this exercise.

The feet should be parallel, with slight external rotation of the lower extremity. The knees should point over the ankles and feet without going in front of, between, or outside of the vertical plane of the feet. This exercise may also be done improperly by allowing lumbar flexion to occur, which contributes to low back problems. It is essential that the lumbar extensors be used more as isometric stabilizers of the low back to maintain the lordotic position while the hip and knee extensors along with the ankle plantar flexors perform the lift in this exercise.

## Analysis

This closed kinetic chain exercise is separated into two phases for analysis: (1) lowering phase to the squatted position and (2) lifting phase to the starting position (Table 12.13). *Note:* It is assumed that no movement will take place in the shoulder joint, shoulder girdle, wrists, hands, back, or feet, although isometric muscle activity is required in these areas to maintain proper positioning.

Courtesy of R.T. Floyd (left, right)

A          B

FIG. 12.14 • Squat. **A,** Starting position; **B,** Squatted position.

## TABLE 12.13 • **Squat**

| Joint | Lowering phase to squatted position | | Lifting phase to starting position | |
|-------|-------------------|-------------------|-------------------|-------------------|
| | Action | Agonists | Action | Agonists |
| Trunk | Maintenance of extension | Trunk extensors (isometric contraction) Erector spinae (sacrospinalis) Quadratus lumborum | Maintenance of extension | Trunk extensors (isometric contraction) Erector spinae (sacrospinalis) Quadratus lumborum |
| Hip | Flexion | Hip extensors (eccentric contraction) Gluteus maximus Semimembranosus Semitendinosus Biceps femoris | Extension | Hip extensors Gluteus maximus Semimembranosus Semitendinosus Biceps femoris |
| Knee | Flexion | Knee extensors (eccentric contraction) Rectus femoris Vastus medialis Vastus intermedius Vastus lateralis | Extension | Knee extensors Rectus femoris Vastus medialis Vastus intermedius Vastus lateralis |
| Ankle | Dorsiflexion | Plantar flexors (eccentric contraction) Gastrocnemius Soleus | Plantar flexion | Plantar flexors Gastrocnemius Soleus |

# Dead lift FIG. 12.15

## Description

The participant begins in hip/knee flexed and ankle dorsiflexed position, keeping the arms, legs, and back straight, and grasps the barbell on the floor. Then a movement to the standing position is made by extending the hips. This exercise, when done improperly by allowing lumbar flexion, may contribute to low back problems. Similar to the squat, it is essential that the lumbar extensors be used more as isometric stabilizers of the low back while the hip extensors perform the majority of the lift in this exercise.

## Analysis

This closed kinetic chain exercise is divided into two phases for analysis: (1) lifting phase to the hip extended/knee extended position and (2) lowering phase to the hip flexed/knee flexed starting position (Table 12.14).

Courtesy of R.T. Floyd (left, right)

FIG. 12.15 • Dead lift. **A,** Beginning hip flexed/knee flexed position; **B,** Ending hip extended/knee extended position.

A                    B

## TABLE 12.14 · **Dead lift**

| Joint | Lifting phase to hip/knee extended position | | Lowering phase to hip/knee flexed position | |
|---|---|---|---|---|
| | Action | Agonists | Action | Agonists |
| Wrist and hand | Flexion* | Wrist and hand flexors (isometric contraction)<br>  Flexor carpi radialis<br>  Flexor carpi ulnaris<br>  Palmaris longus<br>  Flexor digitorum profundus<br>  Flexor digitorum superficialis<br>  Flexor pollicis longus | Flexion | Wrist and hand flexors (isometric contraction)<br>  Flexor carpi radialis<br>  Flexor carpi ulnaris<br>  Palmaris longus<br>  Flexor digitorum profundus<br>  Flexor digitorum superficialis<br>  Flexor pollicis longus |
| Trunk | Maintenance of extension | Trunk extensors (isometric contraction)<br>  Erector spinae (sacrospinalis)<br>  Quadratus lumborum | Maintenance of extension | Trunk extensors (isometric contraction)<br>  Erector spinae (sacrospinalis)<br>  Quadratus lumborum |
| Hip | Extension | Hip extensors<br>  Gluteus maximus<br>  Semimembranosus<br>  Semitendinosus<br>  Biceps femoris | Flexion | Hip extensors (eccentric contraction)<br>  Gluteus maximus<br>  Semimembranosus<br>  Semitendinosus<br>  Biceps femoris |
| Knee | Extension | Knee extensors (quadriceps)<br>  Rectus femoris<br>  Vastus medialis<br>  Vastus intermedius<br>  Vastus lateralis | Flexion | Knee extensors (quadriceps) (eccentric contraction)<br>  Rectus femoris<br>  Vastus medialis<br>  Vastus intermedius<br>  Vastus lateralis |
| Ankle | Plantar flexion | Plantar flexors<br>  Gastrocnemius<br>  Soleus | Dorsiflexion | Plantar flexors (eccentric contraction)<br>  Gastrocnemius<br>  Soleus |

**Note:** Slight movement of the shoulder joint and shoulder girdle is not being analyzed.

*The wrist is in a position of slight extension to facilitate greater active finger flexion in gripping the dumbbell by an isometric contraction of the wrist extensors (extensor carpi ulnaris, extensor carpi radialis longus, and extensor carpi radialis brevis).

Chapter
**12**

## Isometric exercises

An exercise technique called isometrics is a type of muscular activity in which there is contraction of muscle groups with little or no muscle shortening. Though not as productive as isotonics in terms of overall strength gains, isometrics are an effective way to build and maintain muscular strength in a limited range of motion.

A few selected isometric exercises are analyzed muscularly to show how they are designed to develop specific muscle groups. Although there are varying approaches to isometrics, most authorities agree that isometric contractions should be held approximately 7 to 10 seconds for a training effect.

## Abdominal contraction   FIG. 12.6

### Description

The participant contracts the muscles in the anterior abdominal region as strongly as possible, with no movement of the trunk or hips. This exercise can be performed in the sitting, standing, or supine position. The longer the contraction in seconds, the more valuable the exercise will be, to a degree.

### Analysis

Abdomen
Contraction
Rectus abdominis
External oblique abdominal
Internal oblique abdominal
Transversus abdominis

Courtesy of Britt Jones (left, right)

**A**   **B**

FIG. 12.16 • Abdominal contraction. **A,** Beginning position; **B,** Contracted position.

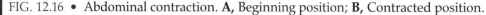

# Rowing exercise  FIG. 12.17

## Description

The participant sits on a movable seat with the knees and hips flexed close to the chest. The arms are reaching forward to grasp a horizontal bar. The legs are extended forcibly as the arms are pulled toward the chest. Then the legs and arms are returned to the starting position.

## Analysis

This closed kinetic chain exercise is divided into two movements for analysis: (1) arm pull to chest/leg push to extend knees and hip phase and (2) return phase to the starting position (Table 12.15).

FIG. 12.17 • Rowing exercise machine. **A,** Starting position; **B,** Movement.

## TABLE 12.15 • **Rowing exercise**

| Joint | Arm pull/leg push phase | | Return phase to starting position | |
|---|---|---|---|---|
| | Action | Agonists | Action | Agonists |
| Foot and ankle | Plantar flexion | Ankle plantar flexors<br>   Gastrocnemius<br>   Soleus | Dorsiflexion | Ankle dorsiflexors<br>   Tibialis anterior<br>   Extensor hallucis longus<br>   Extensor digitorum longus<br>   Peroneus tertius |
| Knee | Extension | Quadriceps (knee extensors)<br>   Rectus femoris<br>   Vastus medialis<br>   Vastus intermedius<br>   Vastus lateralis | Flexion | Knee flexors (hamstrings)<br>   Biceps femoris<br>   Semitendinosus<br>   Semimembranosus |
| Hip | Extension | Hip extensors<br>   Gluteus maximus<br>   Biceps femoris<br>   Semimembranosus<br>   Semitendinosus | Flexion | Hip flexors<br>   Iliopsoas<br>   Rectus femoris<br>   Pectineus |
| Trunk | Extension | Trunk extensors<br>   Erector spinae | Flexion | Trunk flexors<br>   Rectus abdominis<br>   Internal oblique abdominal<br>   External oblique abdominal |

Chapter

**12**

TABLE 12.15 (continued) • **Rowing exercise**

| Joint | Arm pull/leg push phase | | Return phase to starting position | |
|---|---|---|---|---|
| | Action | Agonists | Action | Agonists |
| Shoulder girdle | Adduction, downward rotation, and depression | Shoulder girdle adductors, downward rotators, and depressors<br>Trapezius (lower)<br>Rhomboid<br>Pectoralis minor | Abduction, upward rotation, and elevation | Shoulder girdle adductors, downward rotators, and depressors (eccentric contraction)<br>Trapezius (lower)<br>Rhomboid<br>Pectoralis minor |
| Shoulder joint | Extension | Shoulder joint extensors<br>Latissimus dorsi<br>Teres major<br>Posterior deltoid<br>Teres minor<br>Infraspinatus | Flexion | Shoulder joint extensors (eccentric contraction)<br>Latissimus dorsi<br>Teres major<br>Posterior deltoid<br>Teres minor<br>Infraspinatus |
| Elbow joint | Flexion | Elbow joint flexors<br>Biceps brachii<br>Brachialis<br>Brachioradialis | Extension | Elbow joint flexors (eccentric contraction)<br>Biceps brachii<br>Brachialis<br>Brachioradialis |
| Wrist and hand | Flexion | Wrist and hand flexors (isometric contraction)<br>Flexor carpi radialis<br>Flexor carpi ulnaris<br>Palmaris longus<br>Flexor digitorum profundus<br>Flexor digitorum superficialis<br>Flexor pollicis longus | Flexion | Wrist and hand flexors (isometric contraction)<br>Flexor carpi radialis<br>Flexor carpi ulnaris<br>Palmaris longus<br>Flexor digitorum profundus<br>Flexor digitorum superficialis<br>Flexor pollicis longus |

# REVIEW EXERCISES

1. Select, describe, and completely analyze five conditioning exercises.
2. Collect, analyze, and evaluate exercises that are found in newspapers, in magazines, and on the Internet or are observed on television.
3. Prepare a set of exercises that will ensure development of all large muscle groups in the body.
4. Select exercises from exercise books for analysis.
5. Bring to class other typical exercises for members to analyze.
6. Analyze the conditioning exercises given by your physical education teachers, coaches, and athletic trainers.
7. Consider a sport (basketball or any other sport) and develop exercises applying the overload principle that would develop all the large muscle groups used in that sport.
8. Discuss how you would teach and train boys and girls who cannot perform one chin-up to learn to do chin-ups. Additionally, discuss how you would teach and train subjects who cannot perform one push-up to do push-ups.

9. Should boys and girls attempt to do chin-ups and push-ups to see whether they have adequate strength in the shoulder area?
10. What, if any, benefit would result from doing fingertip push-ups as opposed to push-ups with the hands flat on the floor?
11. Prepare a list of exercises not found in this chapter to develop the upper extremity, lower extremity, and spinal muscles. Separate the list into open versus closed kinetic chain exercises.
12. Consider the change in muscle activation when performing the same movements with the body in different positions relative to gravity. For example, lumbar extension in standing versus supine versus prone.

# LABORATORY EXERCISES

1. Observe children using playground equipment. Analyze muscularly the activities they are performing.
2. Visit the facility on your campus where the free weights and specific or multifunction exercise machines are located. Analyze exercises that can be done with each machine. Compare and contrast similar exercises using different exercise

machines and free weights. *Note:* Manufacturers of all types of exercise apparatus have a complete list of exercises that can be performed with their machines. Secure a copy of recommended exercises and muscularly analyze each exercise.

3. Test yourself doing chin-ups and push-ups to determine your strength and muscular endurance in the shoulder area.

4. Stand slightly farther than arm's length from a wall with your arms by your side and hands facing forward at shoulder-level height. Perform each of the following movements fully before proceeding to the next. When finished, you should be reaching with the palm of your hand straight in front of your shoulder to attempt contact with the wall. Your elbow should be fully extended with your glenohumeral joint flexed 90 degrees.

- Glenohumeral flexion to 90 degrees
- Full elbow extension
- Wrist extension to 70 degrees
- Full shoulder girdle protraction

Analyze the movements and muscles responsible for each movement at the shoulder girdle, glenohumeral joint, elbow, and wrist. Include the type of contraction for each muscle for each movement.

5. Face a wall and stand about 6 inches from it. Place both hands on the wall at shoulder level and put your nose and chest against the wall. Keeping your palms in place on the wall, slowly push your body from the wall as in a push-up until your chest is as far away from the wall as possible without removing your palms from the wall surface. Analyze the movements and muscles responsible for each movement at the shoulder girdle, glenohumeral joint, elbow, and wrist. Include the type of contraction for each muscle for each movement.

6. What is the difference between the two exercises in Questions 4 and 5? Can you perform the movements in Question 5 one step at a time, as you did in Question 4?

7. Lie supine on a table with the knees flexed and hips flexed 90 degrees and the ankles in the neutral 90-degree position. Extend each joint until your knee is fully extended, your hip is flexed only 10 degrees, and your ankle is plantarflexed 10 degrees by performing each of the following movements in the order given:

- Full knee extension
- Hip extension to within 10 degrees of neutral
- Plantar flexion to 10 degrees

Analyze the movements and the muscles responsible for each movement at the hip, knee, and ankle.

8. Stand with your back and buttocks against a smooth wall, and place your feet shoulder width apart with approximately 12 inches between your heels and the wall. Maintain your feet in position, with hips and knees each flexed approximately 90 degrees so that your thighs are parallel to the floor. Keeping your feet in place, slowly slide your back and buttocks up the wall until your buttocks are as far away from the floor as possible without moving your feet. Analyze the movements and the muscles responsible for each movement at the hip, knee, and ankle.

9. What is the difference between the two exercises in Questions 7 and 8? Can you perform the movement in Question 8 one step at a time, as you did in Question 7?

10. Exercise analysis chart

Analyze each exercise in the chart. Use one row for each joint involved that actively moves during the exercise. Do not include joints for which there is no active movement.

| Exercise | Phase | Joint, movement occurring | Force causing movement (muscle or gravity) | Force resisting movement (muscle or gravity) | Functional muscle group, type of contraction |
|---|---|---|---|---|---|
| Barbell press (overhead or military press) | Lifting phase |  |  |  |  |
|  |  |  |  |  |  |
|  |  |  |  |  |  |
|  | Lowering phase |  |  |  |  |
|  |  |  |  |  |  |
|  |  |  |  |  |  |

Exercise analysis chart (continued)

| Exercise | Phase | Joint, movement occurring | Force causing movement (muscle or gravity) | Force resisting movement (muscle or gravity) | Functional muscle group, type of contraction |
|---|---|---|---|---|---|
| Chest press (bench press) | Lifting phase | | | | |
| | | | | | |
| | | | | | |
| | Lowering phase | | | | |
| | | | | | |
| | | | | | |
| Chin-up (pull-up) | Pulling-up phase | | | | |
| | | | | | |
| | | | | | |
| | Lowering phase | | | | |
| | | | | | |
| | | | | | |
| Latissimus pull (lat pull) | Pull-down phase | | | | |
| | | | | | |
| | | | | | |
| | Return phase | | | | |
| | | | | | |
| | | | | | |

Chapter
**12**

Exercise analysis chart (continued)

| Exercise | Phase | Joint, movement occurring | Force causing movement (muscle or gravity) | Force resisting movement (muscle or gravity) | Functional muscle group, type of contraction |
|---|---|---|---|---|---|
| Push-up | Pushing phase | | | | |
| | | | | | |
| | | | | | |
| | Lowering phase | | | | |
| | | | | | |
| | | | | | |
| Dumbbell bent-over row (prone row) | Pull-up phase | | | | |
| | | | | | |
| | | | | | |
| | Lowering phase | | | | |
| | | | | | |
| | | | | | |
| Abdominal curl-up | Trunk flexion phase to curl-up position | | | | |
| | | | | | |
| | | | | | |
| | Rotating to right phase | | | | |
| | | | | | |
| | | | | | |
| | Return phase to curl-up position | | | | |
| | | | | | |
| | | | | | |
| | Return phase to starting position | | | | |
| | | | | | |
| | | | | | |

Exercise analysis chart (continued)

| Exercise | Phase | Joint, movement occurring | Force causing movement (muscle or gravity) | Force resisting movement (muscle or gravity) | Functional muscle group, type of contraction |
|---|---|---|---|---|---|
| Alternating prone extensions | Lifting phase | | | | |
| | | | | | |
| | | | | | |
| | | | | | |
| | Lowering phase | | | | |
| | | | | | |
| | | | | | |
| | | | | | |
| Squat | Lowering phase | | | | |
| | | | | | |
| | | | | | |
| | Lifting phase | | | | |
| | | | | | |
| | | | | | |
| Dead lift | Lifting phase | | | | |
| | | | | | |
| | | | | | |
| | Lowering phase | | | | |
| | | | | | |
| | | | | | |

Exercise analysis chart (continued)

| Exercise | Phase | Joint, movement occurring | Force causing movement (muscle or gravity) | Force resisting movement (muscle or gravity) | Functional muscle group, type of contraction |
|---|---|---|---|---|---|
| Rowing exercise | Arm pull/leg push phase | | | | |
| | | | | | |
| | | | | | |
| | | | | | |
| | | | | | |
| | | | | | |
| | Return phase to starting position | | | | |
| | | | | | |
| | | | | | |
| | | | | | |
| | | | | | |
| | | | | | |
| | | | | | |

# References

Adrian M: Isokinetic exercise, *Training and Conditioning* 1:1, June 1991.

Altug Z, Hoffman JL, Martin JL: *Manual of clinical exercise testing, prescription and rehabilitation*, Norwalk, CT, 1993, Appleton & Lange.

Andrews JR, Zarins B, Wilk KE: *Injuries in baseball*, Philadelphia, 1998, Lippincott-Raven.

Andrews JR, Wilk KE, Harrelson GL: *Physical rehabilitation of the injured athlete*, ed 3, Philadelphia, 2004, Saunders.

Booher JM, Thibodeau GA: *Athletic injury assessment*, ed 4, Dubuque, IA, 2000, McGraw-Hill.

Ellenbecker TS, Davies GJ: *Closed kinetic chain exercise: a comprehensive guide to multiple-joint exercise*, Champaign, IL, 2001, Human Kinetics.

Fahey TD: *Athletic training: principles and practices*, Mountain View, CA, 1986, Mayfield.

Fleck SJ: Periodized strength training: a critical review, *Journal of Strength and Conditioning Research*, 13(1):82–89, 1999.

Geisler P: Kinesiology of the full golf swing—implications for intervention and rehabilitation, *Sports Medicine Update* 11(2):9, 1996.

Hamilton N, Weimer W, Luttgens K: *Kinesiology: scientific basis of human motion*, ed 12, New York, 2012, McGraw-Hill.

Houglum PA, Bertoti DB: *Brunstom's clinical kinesiology*, ed 6, Philadelphia, 2012, FA Davis

Logan GA, McKinney WC: *Anatomic kinesiology*, ed 3, New York, 1982, McGraw-Hill.

Matheson O, et al: Stress fractures in athletes, *American Journal of Sports Medicine* 15:46, January–February 1987.

National Strength and Conditioning Association; Baechle TR, Earle RW: *Essentials of strength training and conditioning*, ed 3, Champaign, IL, 2008, Human Kinetics.

Northrip JW, Logan GA, McKinney WC: *Analysis of sport motion: anatomic and biomechanic perspectives*, ed 3, New York, 1983, McGraw-Hill.

Powers SK, Howley ET: Exercise physiology: *theory and application of fitness and performance*, ed 9, New York, 2015, McGraw-Hill.

Prentice WE: *Rehabilitation techniques in sports medicine*, ed 5, New York, 2011, McGraw-Hill.

Steindler A: *Kinesiology of the human body*, Springfield, IL, 1970, Charles C Thomas.

Torg JS, Vegso JJ, Torg E: *Rehabilitation of athletic injuries: an atlas of therapeutic exercise*, Chicago, 1987, Year Book.

Wilk KE, Reinold MM, Andrews JR, eds: *The athlete's shoulder*, ed 2, Philadelphia, 2009, Churchill Livingstone Elsevier.

Wirhed R: *Athletic ability and the anatomy of motion*, ed 3, St. Louis, 2006, Mosby Elsevier.

Chapter
**12**

# Worksheet Exercises

For in- or out-of-class assignments, or for testing, utilize this tear-out worksheet.

## Upright row exercise worksheet

List the movements that occur in each joint as the subject lifts the weight in performing upright rows and then lowers the weight. For each joint movement, list the muscles primarily responsible and indicate whether they are contracting concentrically, eccentrically, or isometrically with "C", "E," or "I", respectively.

| Lifting phase | | |
|---|---|---|
| Joint | Movement | Muscles |
| Hands | | |
| Wrists | | |
| Elbows | | |
| Shoulder joints | | |
| Shoulder girdles | | |
| Lowering phase | | |
| Wrists | | |
| Elbows | | |
| Shoulder joints | | |
| Shoulder girdles | | |

## Dip exercise worksheet

List the movements that occur in each joint as the subject moves the body up and down in performing dips. For each joint movement, list the muscles primarily responsible and indicate whether they are contracting concentrically, eccentrically, or isometrically with "C", "E," or "I", respectively.

| Lifting body up phase | | |
|---|---|---|
| Joint | Movement | Muscles |
| Hands | | |
| Wrists | | |
| Elbows | | |
| Shoulder joints | | |
| Shoulder girdles | | |
| Lowering body down phase | | |
| Wrists | | |
| Elbows | | |
| Shoulder joints | | |
| Shoulder girdles | | |

# Worksheet Exercises

For in- or out-of-class assignments, or for testing, utilize this tear-out worksheet.

## Selected sport skill analysis

Choose one skill to analyze in the left column below and circle it. In your analysis list the movements of each joint in each phase of the skill. You may prefer to list the initial positions that the joints are in for the stance phase. After each movement, list the muscle(s) primarily responsible for causing/controlling those movements or maintaining the position. Beside each muscle in each movement indicate the type of contraction as follows: I-isometric; C-concentric; E-eccentric.

| Skill | | Joint | Stance phase | Preparatory phase | Movement phase | Follow-through phase |
|---|---|---|---|---|---|---|
| Baseball pitch | | Cervical spine | | | | |
| Volleyball serve | | Lumbar spine | | | | |
| | | Shoulder girdle | | | | |
| Tennis serve | | Shoulder joint | | | | |
| Softball pitch | | Elbow | | | | |
| Tennis backhand | | Radio ulnar | | | | |
| | (R) | Wrist | | | | |
| Batting | | Fingers | | | | |
| Bowling | | Hip | | | | |
| Basketball free throw | | Knee | | | | |
| | | Ankle | | | | |
| Soccer kick | | Transverse tarsal/Subtalar | | | | |
| Sprint start | | Toes | | | | |

# Worksheet Exercises

For in- or out-of-class assignments, or for testing, utilize this tear-out worksheet.

## Selected sport skill analysis (*continued*)

Choose one skill to analyze in the left column below and circle it. In your analysis, list the movements of each joint in each phase of the skill. You may prefer to list the initial positions that the joints are in for the stance phase. After each movement, list the muscle(s) primarily responsible for causing/controlling those movements or maintaining the position. Beside each muscle in each movement indicate the type of contraction as follows: I-isometric; C-concentric; E-eccentric.

| Skill | | Joint | Stance phase | Preparatory phase | Movement phase | Follow-through phase |
|---|---|---|---|---|---|---|
| Baseball pitch | | Cervical spine | | | | |
| Volleyball serve | | Lumbar spine | | | | |
| | | Shoulder girdle | | | | |
| Tennis serve | | Shoulder joint | | | | |
| Softball pitch | | Elbow | | | | |
| Tennis backhand | | Radio ulnar | | | | |
| | (L) | Wrist | | | | |
| Batting | | Fingers | | | | |
| Bowling | | Hip | | | | |
| Basketball free throw | | Knee | | | | |
| | | Ankle | | | | |
| Soccer kick | | Transverse tarsal/Subtalar | | | | |
| Sprint start | | Toes | | | | |

# Worksheet Exercise

For in- or out-of-class assignments, or for testing, utilize this tear-out worksheet.

## Skill analysis worksheet

Using the techniques taught in this chapter and in Chapter 8, analyze the joint movements and muscles used in each phase of movement for selected skills in a given sport or physical activity. For each phase, list the initial and subsequent joint position and the joint motion that occurs with the approximate degrees of movement. List the muscles utilized in the appropriate cell as to whether contracting concentrically to cause movement or eccentrically to *control* movement or isometrically to *prevent* movement or maintain the position. For passive movements that may occur, place a dash across all three contraction types.

| Kinesiology skill analysis table (*Cells for six phases are provided, although some analyses may need only two or three. Delete any unneeded cells.*) | | | | |
|---|---|---|---|---|
| Introductory Paragraph(s) and Skill Overview | | | | |
| **I.** Phase name, general description | Describe position (angle) and activity of joints in this column. Describe changes in positions and joints angle from previous phases. | | List the muscle(s) actively involved (if any) and note whether contracting concentrically (C), eccentrically (E), or isometrically (I). For passive movement, use (P). | |
| | Right | Left | Right | Left |
| Toes (MCP, IP, PIP, & DIP) | | | | |
| Transverse tarsal/ Subtalar | | | | |
| Ankles | | | | |
| Knees | | | | |
| Hips | | | | |
| Lumbar spine | | | | |
| Cervical spine | | | | |
| Shoulder girdles | | | | |
| Glenohumeral | | | | |
| Elbows | | | | |
| Radioulnar | | | | |
| Wrists | | | | |
| Fingers (MCP, IP, PIP, & DIP) | | | | |

Chapter
**12**

| II. Phase name, general description | Describe position (angle) and activity of joints in this column. Describe changes in positions and joints angle from previous phases | | List the muscle(s) actively involved (if any) and note whether contracting concentrically (C), eccentrically (E), or isometrically (I). For passive movement, use (P). | |
|---|---|---|---|---|
| | Right | Left | Right | Left |
| Toes (MCP, IP, PIP, & DIP) | | | | |
| Transverse tarsal/ Subtalar | | | | |
| Ankles | | | | |
| Knees | | | | |
| Hips | | | | |
| Lumbar spine | | | | |
| Cervical spine | | | | |
| Shoulder girdles | | | | |
| Glenohumeral | | | | |
| Elbows | | | | |
| Radioulnar | | | | |
| Wrists | | | | |
| Fingers (MCP, IP, PIP, & DIP) | | | | |
| III. Phase name, general description | Describe position (angle) and activity of joints in this column. Describe changes in positions and joints angle from previous phases. | | List the muscle(s) actively involved (if any) and note whether contracting concentrically (C), eccentrically (E), or isometrically (I). For passive movement, use (P). | |
| | Right | Left | Right | Left |
| Toes (MCP, IP, PIP, & DIP) | | | | |
| Transverse tarsal/ Subtalar | | | | |
| Ankles | | | | |
| Knees | | | | |
| Hips | | | | |
| Lumbar spine | | | | |
| Cervical spine | | | | |
| Shoulder girdles | | | | |
| Glenohumeral | | | | |
| Elbows | | | | |
| Radioulnar | | | | |
| Wrists | | | | |
| Fingers (MCP, IP, PIP, & DIP) | | | | |

| IV. Phase name, general description | Describe position (angle) and activity of joints in this column. Describe changes in positions and joints angle from previous phases. | | List the muscle(s) actively involved (if any) and note whether contracting concentrically (C), eccentrically (E), or isometrically (I). For passive movement, use (P). | |
|---|---|---|---|---|
| | Right | Left | Right | Left |
| Toes (MCP, IP, PIP, & DIP) | | | | |
| Transverse tarsal/Subtalar | | | | |
| Ankles | | | | |
| Knees | | | | |
| Hips | | | | |
| Lumbar spine | | | | |
| Cervical spine | | | | |
| Shoulder girdles | | | | |
| Glenohumeral | | | | |
| Elbows | | | | |
| Radioulnar | | | | |
| Wrists | | | | |
| Fingers (MCP, IP, PIP, & DIP) | | | | |
| V. Phase name, general description | Describe position (angle) and activity of joints in this column. Describe changes in positions and joints angle from previous phases. | | List the muscle(s) actively involved (if any) and note whether contracting concentrically (C), eccentrically (E), or isometrically (I). For passive movement, use (P). | |
| | Right | Left | Right | Left |
| Toes (MCP, IP, PIP, & DIP) | | | | |
| Transverse tarsal/Subtalar | | | | |
| Ankles | | | | |
| Knees | | | | |
| Hips | | | | |
| Lumbar spine | | | | |
| Cervical spine | | | | |
| Shoulder girdles | | | | |
| Glenohumeral | | | | |
| Elbows | | | | |
| Radioulnar | | | | |
| Wrists | | | | |
| Fingers (MCP, IP, PIP, & DIP) | | | | |

| VI. Phase name, general description | Describe position (angle) and activity of joints in this column.  Describe changes in positions and joints angle from previous phases. | | List the muscle(s) actively involved (if any) and note whether contracting concentrically (C), eccentrically (E), or isometrically (I). For passive movement, use (P). | |
|---|---|---|---|---|
| | Right | Left | Right | Left |
| Toes (MCP, IP, PIP, & DIP) | | | | |
| Transverse tarsal/ Subtalar | | | | |
| Ankles | | | | |
| Knees | | | | |
| Hips | | | | |
| Lumbar spine | | | | |
| Cervical spine | | | | |
| Shoulder girdles | | | | |
| Glenohumeral | | | | |
| Elbows | | | | |
| Radioulnar | | | | |
| Wrists | | | | |
| Fingers (MCP, IP, PIP, & DIP) | | | | |

Chapter
12

# Appendix

## Appendix 1
### Range of motion for diarthrodial joints of the body

| Joint | Type | Motion | Range |
|-------|------|--------|-------|
| Sternoclavicular | Arthrodial | Protraction | Moves anteriorly 15° |
| | | Retraction | Moves posteriorly 15° |
| | | Elevation | Moves superiorly 45° |
| | | Depression | Moves inferiorly 5° |
| | | Upward rotation | 45° |
| | | Downward rotation | 5° |
| Acromioclavicular | Arthrodial | Protraction–retraction | 20°–30° rotational and gliding motion |
| | | Elevation–depression | 20°–30° rotational and gliding motion |
| | | Upward rotation–downward rotation | 20°–30° rotational and gliding motion |
| Scapulothoracic | Not a true synovial joint, all movement totally dependent upon AC and SC joints | Abduction–adduction | 25° total range |
| | | Upward rotation–downward rotation | 60° total range |
| | | Elevation–depression | 55° total range |
| Glenohumeral | Enarthrodial | Flexion | 90°–100° |
| | | Extension | 40°–60° |
| | | Abduction | 90°–95° |
| | | Adduction | 0° prevented by trunk, 75° anterior to trunk |
| | | Internal rotation | 70°–90° |
| | | External rotation | 70°–90° |
| | | Horizontal abduction | 45° |
| | | Horizontal adduction | 135° |
| Elbow (ulnohumeral and radiohumeral) | Ginglymus | Extension | 0° |
| | | Flexion | 145°–150° |
| Radioulnar | Trochoid | Supination | 80°–90° |
| | | Pronation | 70°–90° |

## Range of motion for diarthrodial joints of the body (continued)

| Joint | Type | Motion | Range |
|---|---|---|---|
| Wrist (primarily radiocarpal) | Condyloid | Flexion | 70°–90° |
| | | Extension | 65°–85° |
| | | Abduction | 15°–25° |
| | | Adduction | 25°–40° |
| Thumb carpometacarpal | Sellar | Flexion | 15°–45° |
| | | Extension | 0°–20° |
| | | Adduction | 0° |
| | | Abduction | 50°–70° |
| Thumb metacarpophalangeal | Ginglymus | Extension | 0° |
| | | Flexion | 40°–90° |
| Thumb interphalangeal | Ginglymus | Flexion | 80°–90° |
| | | Extension | 0° |
| 2nd, 3rd, 4th, and 5th metacarpophalangeal joints | Condyloid | Extension | 0°–40° |
| | | Flexion | 85°–100° |
| | | Abduction | Variable 10°–40° |
| | | Adduction | Variable 10°–40° |
| 2nd, 3rd, 4th, and 5th proximal interphalangeal joints | Ginglymus | Flexion | 90°–120° |
| | | Extension | 0° |
| 2nd, 3rd, 4th, and 5th distal interphalangeal joints | Ginglymus | Flexion | 80°–90° |
| | | Extension | 0° |
| Atlantooccipital | Condyloid | Flexion | 20° |
| | | Extension | 10° |
| | | Lateral flexion | 5° |
| Atlantoaxial | Trochoid | Flexion | 5° |
| | | Extension | 10° |
| | | Rotation unilaterally | 40° |
| Cervical C2-C7 | Arthrodial | Flexion | 45° |
| | | Extension | 45° |
| | | Lateral flexion | 45° |
| | | Rotation unilaterally | 60° |
| Thoracic | Arthrodial | Flexion | 30°–35° |
| | | Extension | 20°–25° |
| | | Lateral flexion | 25°–30° |
| | | Rotation unilaterally | 30°–35° |

## Range of motion for diarthrodial joints of the body *(continued)*

| Joint | Type | Motion | Range |
|---|---|---|---|
| Lumbar | Arthrodial | Flexion | 80° |
| | | Extension | 20°–30° |
| | | Lateral flexion | 35° |
| | | Rotation unilaterally | 45° |
| Hip (acetabulofemoral) | Enarthrodial | Flexion | 130° |
| | | Extension | 30° |
| | | Abduction | 35° |
| | | Adduction | 0°–30° |
| | | External rotation | 50° |
| | | Internal rotation | 45° |
| Knee (tibiofemoral) *For internal and external rotation to occur, the knee must be flexed 20° or more* | Ginglymus (trochoginglymus) | Extension | 0° |
| | | Flexion | 140° |
| | | Internal rotation | 30° |
| | | External rotation | 45° |
| Ankle (talocrural) | Ginglymus | Plantar flexion | 50° |
| | | Dorsal flexion | 15°–20° |
| Transverse tarsal (talonavicular and calcaneocuboid) and subtalar (talocalcaneal) | Arthrodial | Inversion | 20°–30° |
| | | Eversion | 5°–15° |
| Great toe metatarsophalangeal | Condyloid | Flexion | 45° |
| | | Extension | 70° |
| | | Abduction | Variable 5°–25° |
| | | Adduction | Variable 5°–25° |
| Great toe interphalangeal | Ginglymus | Flexion | 90° |
| | | Extension | 0° |
| 2nd, 3rd, 4th, and 5th metatarsophalangeal joints | Condyloid | Flexion | 40° |
| | | Extension | 40° |
| | | Abduction | Variable 5°–25° |
| | | Adduction | Variable 5°–25° |
| 2nd, 3rd, 4th, and 5th proximal interphalangeal joints | Ginglymus | Flexion | 35° |
| | | Extension | 0° |
| 2nd, 3rd, 4th, and 5th distal interphalangeal joints | Ginglymus | Flexion | 60° |
| | | Extension | 30° |

# Appendix 2

## Commonly used exercises for strengthening selected muscles

Some exercises may be more or less specific for certain muscles. In some cases, certain exercises are designed to emphasize specific portions of a particular muscle more than other portions. Some exercises may be modified slightly to further emphasize or deemphasize certain muscles or portions of muscles. In addition to the muscles listed, numerous other muscles in surrounding joints or other parts of the body may be involved by contracting isometrically to maintain appropriate position of the body for the muscles listed to carry out the exercise movement. Appropriate strength and endurance of these stabilizing muscles is essential for correct position and execution of the listed exercises. Finally, these exercises may be documented by different names by different authorities. Illustration and documentation of the proper techniques and indications for these exercises is beyond the scope of this text. Please consult several strength and conditioning texts for further details.

| Upper extremity | | |
|---|---|---|
| Muscle groups | Muscles | Exercise |
| Scapula abductors | Serratus anterior<br>Pectoralis minor | Front press<br>Dumbbell flys<br>Dumbbell press<br>Bench press<br>Close-grip bench press<br>Incline press<br>Push-ups<br>Incline dumbbell press<br>Pec deck flys<br>Cable crossover flys |
| Scapula adductors | Rhomboid<br>Trapezius, lower fibers<br>Trapezius, middle fibers | Bent-over lateral raises<br>Pec deck rear delt laterals<br>Seated rows<br>Bent rows<br>T-bar rows<br>Dead lifts<br>Sumo dead lifts |
| Scapula upward rotators | Serratus anterior<br>Trapezius, lower fibers<br>Trapezius, middle fibers | Dumbbell press<br>Dumbbell flys<br>One-arm dumbbell press<br>Lateral raises<br>Upright rows |
| Scapula downward rotators | Pectoralis minor<br>Rhomboid | Parallel bar dips<br>Dumbbell pullovers<br>Barbell pullovers<br>Chin-ups<br>Reverse chin-ups<br>Lat pulldowns<br>Back lat pulldowns<br>Close-grip lat pulldowns<br>Straight-arm lat pulldowns<br>One-arm dumbbell rows |
| Scapula elevators | Rhomboid<br>Levator scapulae<br>Trapezius, upper fibers<br>Trapezius, middle fibers | Upright rows<br>Barbell shrugs<br>Dumbbell shrugs<br>Machine shrugs<br>Dead lifts |
| Scapula depressors | Pectoralis minor<br>Trapezius, lower fibers | Parallel bar dips<br>Body dips |

| Upper extremity | | |
|---|---|---|
| Muscle groups | Muscles | Exercise |
| Shoulder flexors | Deltoid, anterior fibers<br>Deltoid, middle fibers<br>Pectoralis major, clavicular fibers<br>Coracobrachialis | Arm curls<br>Triceps dips<br>One-arm dumbbell press<br>Front raises<br>Low pulley front raises<br>One-dumbbell front raises<br>Barbell front raises |
| Shoulder extensors | Latissimus dorsi<br>Teres major<br>Triceps brachii, long head<br>Pectoralis major, sternal fibers<br>Deltoid, posterior fibers<br>Infraspinatus<br>Teres minor | Dumbbell pullovers<br>Barbell pullovers<br>Reverse chin-ups<br>Close-grip lat pulldowns<br>Straight-arm lat pulldowns<br>One-arm dumbbell rows |
| Shoulder abductors | Deltoid, anterior fibers<br>Deltoid, posterior fibers<br>Deltoid, middle fibers<br>Pectoralis major, clavicular fibers<br>Supraspinatus | Back press<br>Front press<br>Dumbbell press<br>One-arm dumbbell press<br>Lateral raises<br>Side-lying lateral raises<br>Low pulley lateral raises<br>Upright rows<br>Nautilus lateral raises<br>Upright rows |
| Shoulder adductors | Pectoralis major<br>Latissimus dorsi<br>Teres major<br>Subscapularis<br>Coracobrachialis | Triceps dips<br>Parallel bar dips<br>Chin-ups<br>Lat pulldowns<br>Back lat pulldowns |
| Shoulder internal rotators | Pectoralis major<br>Latissimus dorsi<br>Teres major<br>Subscapularis | Triceps dips<br>Side-lying internal rotations<br>Standing internal rotations at 90 degrees abduction |
| Shoulder external rotators | Infraspinatus<br>Teres minor | Side-lying external rotations<br>Standing external rotations at 90 degrees abduction |

| Upper extremity | | |
|---|---|---|
| Muscle groups | Muscles | Exercise |
| Shoulder horizontal abductors | Latissimus dorsi<br>Infraspinatus<br>Teres minor<br>Deltoid, middle fibers<br>Deltoid, posterior fibers | Bent-over lateral raises<br>Low pulley bent-over lateral raises<br>Pec deck rear delt laterals<br>Bent rows<br>T-bar rows |
| Shoulder horizontal adductors | Pectoralis major<br>Coracobrachialis | Dumbbell flys<br>Triceps dips<br>Dumbbell press<br>Bench press<br>Close-grip bench press<br>Incline press<br>Decline press<br>Push-ups<br>Incline dumbbell press<br>Incline dumbbell flys<br>Pec deck flys<br>Cable crossover flys<br>Seated rows |
| Elbow flexors | Biceps brachii<br>Brachialis<br>Brachioradialis | Arm curls concentration curls<br>Hammer curls<br>Low pulley curls<br>High pulley curls<br>Barbell curls<br>Machine curls<br>Preacher curls<br>Reverse barbell curls<br>Chin-ups<br>Reverse chin-ups<br>Lat pulldowns<br>Back lat pulldowns<br>Close-grip lat pulldowns<br>Seated rows<br>One-arm dumbbell rows<br>Bent rows<br>T-bar rows<br>Upright rows |
| Elbow extensors | Triceps brachii<br>Triceps brachii, lateral head<br>Triceps brachii, long head<br>Triceps brachii, medial head<br>Anconeus | Barbell pullovers<br>Bench press<br>Close-grip bench press<br>Decline press<br>Dumbbell press<br>Dumbbell pullovers<br>Dumbbell triceps extensions<br>Front press<br>Incline dumbbell press<br>Incline press<br>One-arm dumbbell triceps extensions<br>One-arm reverse pushdowns<br>Parallel bar dips<br>Pushdowns<br>Push-ups<br>Reverse pushdowns<br>Seated dumbbell triceps extensions<br>Seated ez-bar triceps extensions<br>Triceps dips<br>Triceps extensions<br>Triceps kickbacks |

| Upper extremity | | |
|---|---|---|
| Muscle groups | Muscles | Exercise |
| Wrist flexors | Flexor carpi radialis<br>Palmaris longus<br>Flexor carpi ulnaris<br>Flexor digitorum superficialis<br>Flexor digitorum profundus | Wrist curls |
| Wrist extensors | Extensor carpi radialis longus<br>Extensor carpi radialis brevis<br>Extensor carpi ulnaris | Reverse barbell curls<br>Reverse wrist curls<br>Reverse pushdowns |
| Finger flexors | Hand intrinsics<br>Flexor digitorum profundus<br>Flexor digitorum superficialis<br>Flexor pollicis longus | Ball squeezes<br>Putty squeezes<br>Rice bucket grips<br>Wrist curls<br>Dead lifts |
| Finger extensors | Extensor digitorum<br>Extensor digiti minimi<br>Extensor indicis | Reverse barbell curls<br>Reverse wrist curls<br>Rubber band stretches |

| Lower extremity | | |
|---|---|---|
| Muscle groups | Muscles | Exercise |
| Hip flexors | Rectus femoris<br>Iliopsoas<br>Pectineus<br>Tensor fasciae latae | Crunches<br>Sit-ups<br>Gym ladder sit-ups<br>Calves over bench sit-ups<br>Incline bench sit-ups<br>Specific bench sit-ups<br>Machine crunches<br>Incline leg raises<br>Leg raises<br>Hanging leg raises |
| Hip extensors | Gluteus maximus<br>Biceps femoris, long head<br>Semitendinosus<br>Semimembranosus | Stiff-legged dead lifts<br>Dead lifts<br>Back extensions<br>Dumbbell squats<br>Squats<br>Front squats<br>Power squats<br>Angled leg press<br>Good mornings<br>Lunges<br>Cable back kicks<br>Machine hip extensions<br>Floor hip extensions<br>Bridging<br>Prone arches |
| Hip abductors | Gluteus medius<br>Gluteus maximus<br>Tensor fascia latae | Cable hip abductions<br>Standing machine hip abductions<br>Floor hip abductions<br>Seated machine hip abductions |

| Lower extremity | | |
|---|---|---|
| Muscle groups | Muscles | Exercise |
| Hip adductors | Adductor magnus<br>Adductor longus<br>Adductor brevis<br>Gracilis | Sumo dead lifts<br>Power squats<br>Cable adductions<br>Machine adductions |
| Hip external rotators | Gluteus maximus<br>Piriformis<br>Gemellus superior<br>Gemellus inferior<br>Obturator externus<br>Obturator internus<br>Quadratus femoris | Hip turn-outs<br>Body turn-aways |
| Knee extensors | Vastus medialis<br>Vastus intermedius<br>Rectus femoris<br>Vastus lateralis | Leg extensions<br>Dead lifts<br>Sumo dead lifts<br>Dumbbell squats<br>Squats<br>Front squats<br>Angled leg press<br>Power squats<br>Hack squats<br>Lunges |
| Knee flexors | Semitendinosus<br>Biceps femoris, long head<br>Biceps femoris, short head<br>Semimembranosus<br>Gastrocnemius, lateral head<br>Gastrocnemius, medial head | Standing leg curls<br>Seated leg curls<br>Lying leg curls |
| Ankle dorsiflexors | Tibialis anterior<br>Extensor hallucis longus<br>Extensor digitorum longus<br>Peroneous tertius | Towel pulls<br>Elastic band pulls |
| Ankle plantar flexors | Gastrocnemius, lateral head<br>Soleus<br>Gastrocnemius, medial head | Standing calf raises<br>One-leg toe raises<br>Donkey calf raises<br>Seated calf raises<br>Seated barbell calf raises |
| Transverse tarsal/ subtalar inversion | Tibialis anterior<br>Tibialis posterior<br>Flexor digitorum longus<br>Flexor hallucis longus | Towel drags<br>Elastic band turn-ins |
| Transverse tarsal/ subtalar eversion | Extensor digitorum longus<br>Peroneus longus<br>Peroneus brevis<br>Peroneus tertius | Towel drags<br>Elastic band turn-outs |
| Toe extensors | Extensor hallucis longus<br>Extensor digitorum longus | Towel pulls<br>Elastic band pulls |
| Toe flexors | Flexor digitorum longus<br>Flexor hallucis longus<br>Foot intrinsics | Towel curls<br>Marble pickups<br>Pencil pickups |

| Cervical spine and trunk | | |
|---|---|---|
| Muscle groups | Muscles | Exercise |
| Cervical extensors | Splenius cervicis<br>Splenius capitus<br>Trapezius, upper fibers | Dead lifts<br>Neck extensions |
| Cervical flexors | Sternocleidomastoid | Chin tucks<br>Sit-ups |
| Cervical rotators | Sternocleidomastoid<br>Splenius cervicis<br>Splenius capitus | Machine neck rotations |
| Trunk extensors | Erector spinae | Back extensions<br>Alternating prone extensions<br>Prone arches |
| Trunk flexors | Rectus abdominis<br>External oblique abdominal<br>Internal oblique abdominal | Dead lifts<br>Crunches<br>Crunch twists<br>Sit-ups<br>V Sit-ups<br>Gym ladder sit-ups<br>Calves over bench sit-ups<br>Incline bench sit-ups<br>Specific bench sit-ups<br>High pulley crunches<br>Machine crunches<br>Incline leg raises<br>Leg raises<br>Hanging leg raises |
| Trunk rotators | External oblique abdominal<br>Internal oblique abdominal | Crunches<br>Crunch twists<br>Sit-up twists<br>V Sit-ups<br>Gym ladder sit-ups<br>Calves over bench sit-ups<br>Incline bench sit-ups<br>Specific bench sit-ups<br>High pulley crunches<br>Machine crunches<br>Incline leg raises<br>Leg raises<br>Hanging leg raises<br>Broomstick twists<br>Machine trunk rotations |
| Trunk lateral flexors | External oblique abdominal<br>Internal oblique abdominal<br>Quadratus lumborum<br>Rectus abdominis | Crunches<br>Crunch twists<br>Sit-ups<br>V Sit-ups<br>Gym ladder sit-ups<br>Calves over bench sit-ups<br>Incline bench sit-ups<br>Specific bench sit-ups<br>High pulley crunches<br>Machine crunches<br>Incline leg raises<br>Leg raises<br>Hanging leg raises<br>Broomstick twists<br>Dumbbell side bends<br>Roman chair side bends |

# Appendix 3

## Etymology of commonly used terms in kinesiology

Below are some of the most commonly used terms in naming the muscles, bones, and joints as well as some additional terms utilized in explaining their function. This etymology is provided in order to better understand the origin and historical development of these terms and to provide a more meaningful background as to how they came to be used in the study of the body and its movement today.

**abdominis**   Latin: belly

**abductor**   Latin: abducere

**acromion**   Greek: akron, extremity + omus, shoulder

**adductor**   Latin: adducere, adduct-, to bring to, contract

**amphiarthrodial**   Greek: ampho, both + arthron, joint + eidos, form, shape

**anconeus**   Greek: agkon, elbow

**antebrachial**   Latin: ante, before + brachium, arm

**antecubital**   Latin: ante, before + cubitum, elbow

**anterior**   Latin: comparative of ante, before

**appendicular**   Latin: appendere, to hang to

**arthrodial**   Greek: arthron, joint + eidos, form, shape

**axillary**   Latin: axilla

**axon**   Greek: axon, axis

**biceps**   Latin: two-headed, bi-, two; caput, head

**brachialis**   Latin: brachialis, brachial, arm

**brachii**   Latin: bracchium, arm

**brachioradialis**   Latin: bracchium, arm + radialis, radius

**brevis**   Latin: (adj.) short, low, little, shallow

**buccal**   Latin: cheek

**bursa**   Greek: a leather sack

**calcaneus**   Latin: calcaneus, heel, from Latin calcaneum, from calx, calc

**cancellous**   Latin: cancellus, lattice

**capitate, capitis**   Latin: caput, head

**carpal**   Latin: carpalis, from carpus, wrist

**carpus, carpi**   Latin: from Greek karpos, wrist

**caudal**   Latin: caudalis, tail

**celiac**   Greek: koilia, belly

**cephalic**   Greek: kephale, head

**cerebellum**   Latin: little brain

**cerebrum**   Latin: cerebrum, brain

**cervicis**   Latin: cervix, neck

**clavicle**   French: clavicule, collarbone; Latin: clavicula, little key

**coccyx**   Greek: kokkyz, cuckoo

**colli**   Latin: collare, necklace, band or chain for the neck, from collum, the neck

**concentric**   Latin: con, together with + centrum, center

**condyle**   Greek: kondylos, knuckle

**condyloidal**   Greek: kondylos, knuckle + eidos, form, shape

**coracobrachialis**   Greek: from coracoid korax, crow; eidos, form; Greek: brachion + Latin, radial

**coracoid**   Greek: korax, raven + eidos, form, shape

**coronal**   Greek: korone, crown

**coronoid**   Greek: korone, something curved, kind of crown + eidos, form, shape

**cortical**   Latin: rind

**costal**   Latin: costa, rib

**coxal**   Latin: coxa, hip

**cranial**   Latin: cranialis, cranium; Greek: kranion

**crest**   Latin: crista, crest

**crural**   Latin: cruralis, pertaining to leg or thigh

**cubital**   Latin: cubitum, elbow

**cuboid**   Greek: kybos, cube

**cuneiforms**   Latin: cuneus, wedge + forma, form

**deltoid, deltoidius**   Latin: deltoides: Greek deltoeides, triangular: delta, delta + -oeides, -oid

**dendrite**   Greek: dendrites, pertaining to a tree

**derma**   Greek: derma, skin

**dermatome**   Greek: derma, skin + tome, incision

**diaphysis**   Greek: diaphysis, a growing through

**diarthrodial**   Greek: dis, two + arthron, joint + eidos, form, shape

**digital, digitorum**   Latin: digitus, finger or toe

**distal**   Latin: distare, to be distant

**dorsal**   Latin: dorsalis, dorsualis, of the back, from dorsum, back

**dorsi**   Latin: dorsi, genitive of dorsum, back

**eccentric**   Greek: ek, out + kentron, center

**enarthrodial**   Greek: en, in + arthron, joint + eidos, form, shape

**endosteum**   Greek: endon, within + osteon, bone

**epiphyseal**   Greek: epi, above + phyein, to grow

**epiphysis**   Greek: a growing upon

**erector**   Latin: erigere, to erect

**extensor**   Latin: extendere, to stretch out

**external, externus**   Latin: externus, outside, outward

**fabella**   Latin: faba, little bean

**facet**   French: facette, small face

**fasciae**   Latin: fascia, band

**femoris**   Latin: femur, thigh, genitive of femur, thigh

**femur**   Latin: thigh

**fibers**   Latin: fibra, a fiber, filament, of uncertain origin, perhaps related to Latin: filum, thread

**fibrous**   Latin: fibra, fiber

**fibula**   Latin: clasp, brooch

**flexor**   Latin: bender

**foramen**   Latin: hole

**fossa**   Latin: ditch

**fovea**   Latin: pit

**frontal**   Latin: frontem (nom, frons), forehead, literally that which projects

**fusiform**   Latin: fusus, spindle + forma, shape

**gaster**   Greek: gaster, belly

**gastrocnemius**   Greek: gaster, belly + kneme, leg

**gemellus**   Latin: twin

**ginglymus**   Greek: ginglymos, hinge

**gluteus**   Greek: gloutos, buttock

**gomphosis**   Greek: bolting together

**goniometer**   Greek: gonia, angle + metron, measure

**gracilis**   Latin: graceful

**greater**   Middle English: grete; Old English: great, thick, coarse; French: grand, which is from Latin: magnus

**hallucis**   Latin: hallex, large toe

**hamate**   Latin: hamatus, hooked

**head**   Latin: from caput

**humerus**   Latin: a misspelling borrowing umerus, shoulder

**hyaline**   Greek: hyalos, glass

**iliacus**   Latin: ilium, flank

**iliocostalis**   Latin: ilium, flank + costa, rib

**ilium**   Latin: ilium, groin, flank, variant of Latin: ilia

**indicis**   Latin: forefinger, pointer, sign, list

**inferior**   Latin: inferior, lower

**infraspinatus, infraspinous**   Latin: infra, below 1 spina, spine

**inguinal**   Latin: inguinalis, groin

**insertion**   Latin: in, into + serere, to join

**intermediate**   Latin: intermediates, lying between; Latin: intermedius, that which is between; from inter, between + medius, in the middle

**intermedius**   Latin: inter, between, mediare, to divide that which is between

**internal, internus**   Latin: internus, within

**interossei**   Latin: inter, between + os, bone

**interspinalis**   Latin: inter, between + spina, spine

**intertransversarii**   Latin: inter, between + transverses, cross-direction

**ischium**   Greek: ischion, hip joint

**isokinetic**   Greek: isos, equal + kinesis, motion

**isometric**   Greek: isos, equal + metron, measure

**isotonic**   Greek: isos, equal + tonus, tone

**kinematic**   Greek: kinematos, movement

**kinesiology**   Greek: kinematos, movement + logos, word, reason

**kinesthesia**   Greek: kinematos, movement + aesthesis, sensation

**kyphosis**   Greek: humpback

**latae**   Latin: latus, side

**lateral, lateralis**   Latin: lateralis, belonging to the side

**latissimus**   Latin: latissimus, superlative of latus, wide

**lesser**   Middle English: lesse; Latin: minor

**levator**   Latin: levator, lifter

**lever**   Latin: levare, to raise

**linea**   Latin: linea, line

**longissimus**   Latin: longest, very long

**longus**   Latin: long

**lordosis**   Greek: lordosis, bending

**lumbar, lumborum**   Latin: lumbus, loin

**lumbricals**   Latin: lumbus, loin; referred to vermiform; Latin: vermis, worm + forma, form

**lunate**   Latin: lunatus, past participle of lunare, to bend like a crescent, from luna, moon

**magnus**   Latin: great

**major**   Middle English: majour; Latin: major

**malleolus**   Latin: malleolus, little hammer

**mammary**   Latin: mamma, breast

**mandible**   Latin: mandibula, jaw; Latin: mandere, to chew

**manubrium**   Latin: handle, from manus, hand

**margin**   Latin: marginalis, border

**maxilla**   Latin: upper jaw, of mala, jaw, cheekbone

**maximus**   Latin: greatest

**meatus**   Latin: meatus, passage

**medial, medialis**   Latin: medialis, of the middle; Latin: medius, middle

**medius**   Latin: middle

**mental**   Latin: mentum, chin

**metacarpal**   Greek: meta, after, beyond, over + Latin: carpalis, from carpus, wrist

**metatarsals**   Greek: meta, after, beyond, over + tarsos, flat surface

**middle**   Old English: middle; Latin: medium

**minimus, minimi**   Latin: minimum, smallest

**minor**   Latin: lesser, smaller, junior

**multifidus**   Latin: multus, many + clefts or segments

**muscle**   Latin: musculus

**myo**   Greek: mys, muscle

**myotome**   Greek: mys, muscle + tome, incision

**nasal**   Latin: nasus, nose

**navicular**   Latin: navicula, boat, diminutive of navis, ship

**neural**   Latin: neuralis, nerve

**neuron**   Greek: neuron, nerve, sinew

**notch**   French: noche, indention, depression

**nuchal**   Latin: nape (back) of the neck

**oblique, obliquus**   Latin: obliquus, slanted

**obturator**   Latin: obturare, to close

**occiput**   Latin: occiput (gen. occipitis), back of the skull, from ob, against, behind + caput, head

**olecranon**   Greek: elbow

**omos**   Greek: omos, shoulder

**opponens**   Latin: opponentem (nom. opponens), prp. of opponere, oppose, object to, set against

**oral**   Latin: oralis, mouth

**orbital**   Latin: orbita, track

**origin**   Latin: origo, beginning

**osseous**   Latin: osseus, bony

**otic**   Greek: otikos, ear

**palmar**   Latin: palma, palm of the hand

**palmaris**   Latin: palma, palm

**patella**   Latin: pan, kneecap

**pectineus**   Latin: pectin, comb

**pectoralis**   Latin: pectoralis, from pectus, pector-, breast; Middle English, French, Latin: pectorale, breastplate, from neuter of pectoralis

**pedal**   Latin: pedalis, foot

**pennate**   Latin: penna, feather

**perineal**   Greek: perinaion, perineum

**periosteum**   Greek: peri, around + osteon, bone

**peroneus**   Greek: perone, brooch

**phalanges**   A plural of phalanx Greek: phalangos, finger or toe bone

**phalanx**   Greek: phalangos, finger or toe bone

**piriformis**   Latin: piriformis, pear shaped

**pisiform**   Latin: pisa, pea + forma, form

**plantae**   Latin: planta, sole of the foot

**pollicis**   Latin: from pollex, thumb, big toe

**popliteus**   Latin: poples, ham of the knee

**posterior**   Latin: comparative of posterus, coming after, from post, afterward

**process**   Latin: processus, going before

**profundus**   Latin: deep, bottomless, vast

**pronator**   Latin: pronare, to bend forward

**proximal**   Latin: proximitatem (nom. proximitas), nearness, vicinity, from proximus, nearest

**psoas**   Greek: psoa, muscle of the loin

**pubis**   Latin: (os) pubis, bone of the groin

**quadratus**   Latin: quadratus, square

**quartus**   Latin: fourth

**quinti**   Latin: quintus, fifth

**radialis**   Latin: radialis, radius, beam of light

**radiate**   Latin: radiatre, to emit rays

**radius**   Latin: beam of light

**ramus**   Latin: branch

**rectus**   Latin: rectus, straight

**rhomboids, rhomboidus**   From the word rhombus; Latin: flatfish, magician's circle; from Greek: rhombos, rhombus

**rotatores**   Latin: rotare, to rotate

**sacrum**   Latin: os sacrum, sacred bone

**sartorius**   Latin: sartor, tailor

**scaphoid**   Latin: skiff, boat shaped + eidos, form

**scapulae, scapula**   Latin: shoulder; Latin: scapulae, the shoulder blades

**scoliosis**   Greek: scoliosis, crookedness

**sellar**   Latin: Turkish saddle

**semimembranosus**   Latin: semi, half + membrane, membrane

**semispinalis**   Latin: semi, half + spina, spine

**semitendinosus**   Latin: semi, half + tendere, to stretch

**serratus**   Latin: serratus, saw-shaped, from serra, saw

**sesamoid**   Latin: sesamoides, resembling a grain of sesame in size or shape

**sinus**   Latin: curve, hollow

**soleus**   Latin: solea, sandal

**somatic**   Greek: soma, body

**sphincter**   Greek: sphincter, hand

**spinae**   Latin: thorn

**spinal**   Latin: spinalis, spine

**splenius**   Greek: splenion, bandage

**sternocleidomastoid**   Greek: sternon, chest + kleis, key + mastos, breast + eidos, form

**sternum**   Greek: sternon, chest, breast, breastbone

**styloid**   Anglo-Saxon: stigan, to rise + eidos, form

**subscapularis**   Latin: sub, beneath + scapulae, shoulder blades

**sulcus**   Latin: groove

**superficialis**   Latin: superficies, of or pertaining to the surface

**superior**   Latin: superiorem, higher

**supinator**   Latin: reflectere, to bend back

**supraspinatus, supraspinous**   Latin: supra, above + spina, spine

**sural**   Latin: sura, calf

**suture**   Latin: sutura, a seam

**symphysis**   Greek: symphysis, growing together

**synarthrodial**   Greek: syn, together + arthron, joint + eidos, form, shape

**synchondrosis**   Greek: syn, together + chondros, cartilage + osis, condition

**syndesmosis**   Greek: syndesmos, ligament + osis, condition

**synovial**   Latin: synovia, joint fluid

**talus**   Latin: ankle, anklebone, knucklebone

**tarsal**   Greek: tarsalis, ankle

**temporal**   Latin: temporalis, of time, temporary, from tempus (temporis), time, season, proper time or season

**tendon**   Latin: tendo, tendon

**tensor**   Latin: tendere, to stretch

**teres**   Latin: rounded

**tertius**   Latin: third

**thoracic**   Greek: thorax, chest

**tibia**   Latin: shinbone

**tibialis**   Latin: tibia, pipe, shinbone

**transverse**   Latin: transverses, oblique

**trapezium**   Greek: trapezion, a little table

**trapezius**   Latin: trapezium, trapezium, from the shape of the muslces paired

**trapezoid**   Greek: trapezoeides, table shaped

**triceps**   Latin: three-headed; tri-, tri- + caput, head

**triquetrum**   Latin: neuter of triquetrus, three-cornered

**trochanter**   Greek: trokhanter, to run

**trochlear**   Greek: trokhileia, system of pulleys

**tubercle**   Latin: turberculum, a little swelling

**tuberosity**   Latin: tuberositas, tuberosity

**ulna**   Latin: elbow

**ulnaris**   Latin: ulna, elbow

**umbilical**   Latin: umbilicus, naval

**vastus**   Latin: immense, extensive, huge

**vertical**   Latin: verticalis, overhead, vertex, highest point

**visceral**   Latin: viscera, body organs

**volar**   Latin: vola, sole, palm

**xiphoid**   Greek: xiphos, sword + eidos, form, shape

**zygoma**   Latin: zygoma, zygomat-, from Greek zugoma, bolt, from zugoun, to join

# Appendix 4

## Determining if a muscle (or muscle group) is contracting and, if so, the contraction type (isometric, concentric, or eccentric)

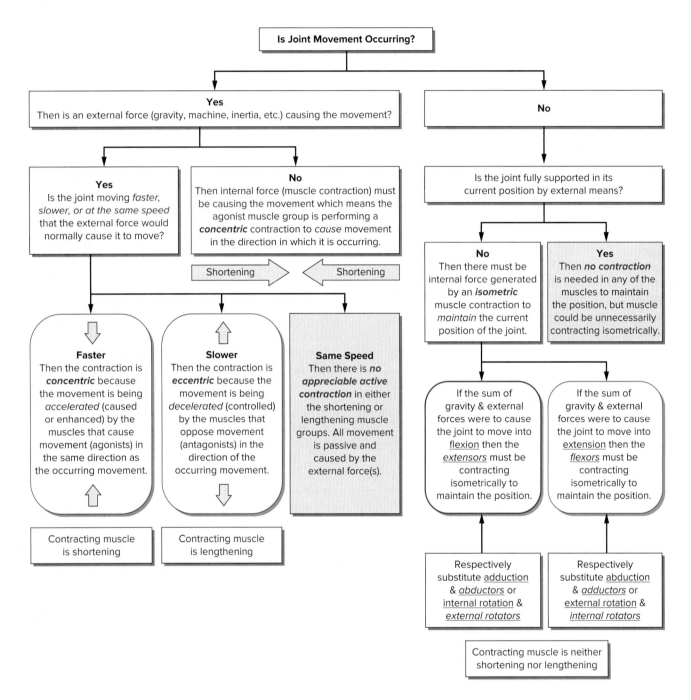

# Glossary

**abduction**  Lateral movement in the frontal plane away from the midline of the trunk, as in raising the arms or thighs to the side away from the anatomical position.

**acceleration**  The rate of change in velocity.

**accessory motion**  The actual change in relationship between the articular surface of one bone relative to another, characterized as roll, spin, and glide.

**action potential**  Electrical signal transmitted from the brain and spinal cord through axons to the muscle fibers in a particular motor unit providing the stimulus to contract.

**active insufficiency**  Point reached when a muscle becomes shortened to the point that it cannot generate or maintain active tension.

**active tension**  Tension in muscles that is generated via an active contraction of the respective muscle fibers in that muscle.

**adduction**  Movement medially toward and/or across the midline of the trunk, as in lowering the arm to the side or thigh back to the anatomical position in the frontal plane.

**afferent nerves**  Nerves that bring impulses from receptors in the skin, joints, muscles, and other peripheral aspects of the body to the central nervous system.

**aggregate muscle action**  Muscles working together in groups rather than independently to achieve given joint motions.

**agonist**  A muscle or muscle group that is described as being primarily responsible for a specific joint movement when contracting.

**all or none principle**  States that regardless of the number involved, the individual muscle fibers within a given motor unit will fire and contract either maximally or not at all.

**amphiarthrodial (amphiarthrosis) joints**  Joints that functionally allow only a very slight amount of movement such as synchondrosis (e.g., costochondral joint of the ribs with sternum), syndesmosis (e.g., distal tibiofibular), and symphysis (e.g., symphysis pubis) joints.

**amplitude**  Range of muscle fiber length between maximal and minimal lengthening.

**anatomical position**  The position of reference in which the subject is in the standing position, with feet together and palms of hands facing forward.

**angle**  Bend or protruding angular projection of a bone such as superior and inferior angle of scapula.

**angle of pull**  The angle between the muscle insertion and the bone on which it inserts.

**angular displacement**  The change in location of a rotating body.

**angular motion**  Motion involving rotation around an axis.

**antagonist**  A muscle or muscle group that counteracts or opposes the contraction of another muscle or muscle group.

**anterior axillary line**  A line parallel to the mid-axillary line which passes through the anterior axillary skinfold.

**anteroposterior axis**  The axis that has the same directional orientation as the sagittal plane of motion and runs from front to back at a right angle to the frontal plane of motion. Also known as the sagittal or AP axis.

**anteversion**  Abnormal or excessive rotation forward of a structure, such as femoral anteversion.

**aponeurosis**  A tendinous expansion of dense fibrous connective tissue, sheet- or ribbonlike in appearance and resembling a flattened tendon, which serves as a fascia to bind muscles together or as a means of connecting muscle to bone.

**appendicular skeleton**  The appendages, or the upper and lower extremities, and the shoulder and pelvic girdles.

**arthrodial joints**  Joints in which bones glide on each other in limited movement, as in the bones of the wrist (carpal) or the bones of the foot (tarsal).

**arthrokinematics**  Motion between the actual articular surfaces of the bones at a joint.

**arthrosis**  Joint or articulation between two or more bones.

**axial skeleton**  The skull, vertebral column, ribs, and sternum.

**axis of rotation**  The point in a joint about which a bone moves or turns to accomplish joint motion.

**axon**  An elongated projection that transmits impulses away from the neuron cell body.

**balance**  The ability to control equilibrium, either static or dynamic.

**biarticular muscles**  Those muscles that, from origin to insertion, cross two different joints, allowing them to perform actions at each joint.

**bilateral**  Relating to the right and left sides of the body or of a body structure such as the right and left extremities.

**biomechanics**  The study of mechanics as it relates to the functional and anatomical analysis of biological systems, especially humans.

**bipennate**  A type of pennate muscle with fibers running obliquely on both sides from a central tendon, such as the rectus femoris and flexor hallucis longus.

**border or margin**  Edge or boundary line of a bone such as lateral and medial border of scapula.

**brachial plexus**  Group of spinal nerves composed of cervical nerves 5 through 8, along with thoracic nerve 1; supplies motor and sensory function to the upper extremity and most of the scapula.

**cancellous bone**   Spongy, porous bone that lies under cortical bone.

**cardinal plane**   Specific planes that divide the body exactly into two halves.

**carpal tunnel**   A three-sided arch, concave on the palmar side and formed by the trapezium, trapezoid, capitate, and hamate. It is spanned by the transverse carpal and volar carpal ligaments creating a tunnel.

**carpal tunnel syndrome**   A condition characterized by swelling and inflammation with resultant increased pressure in the carpal tunnel, which interferes with normal function of the median nerve, leading to reduced motor and sensory function of its distribution; particularly common with repetitive use of the hand and wrist in manual labor and clerical work such as typing and keyboarding.

**carrying angle**   In the anatomical position, the angle formed by the forearm deviating laterally from the arm, typically 5 to 15 degrees.

**cartilaginous joints**   Joints joined together by hyaline cartilage or fibrocartilage, allowing very slight movement, such as synchondrosis and symphysis.

**caudocephalad**   Directionally from tail to head in the long axis of the body.

**cephalocaudal**   Directionally from head to tail in the long axis of the body.

**center of gravity**   The point at which all of the body's mass and weight are equally balanced or equally distributed in all directions.

**center of rotation**   The point or line around which all other points in the body move.

**central nervous system (CNS)**   The cerebral cortex, basal ganglia, cerebellum, brain stem, and spinal cord.

**cervical plexus**   Group of spinal nerves composed of cervical nerves 1 through 4; generally responsible for sensory and motor function from the upper part of the shoulders to the back of the head and front of the neck.

**circumduction**   Circular movement of a bone at the joint, as in movement of the hip, shoulder, or trunk around a fixed point. Combination of flexion, extension, abduction, and adduction.

**closed kinetic chain**   When the distal end of an extremity is fixed, preventing movement of any one joint unless predictable movements of the other joints in the extremity occur.

**coefficient of friction**   The ratio between the force needed to overcome friction over the force holding the surfaces together.

**collagen**   A protein in the body that forms fibrous connective tissues such as ligaments, tendons, cartilage, bone, and skin. Its elongated fibrils provide strength and flexibility to these tissues.

**concentric contraction**   A contraction in which there is a shortening of the muscle that causes motion to occur at the joints it crosses.

**concurrent**   Movement pattern allowing the involved biarticular muscle to maintain a relatively consistent length because of the same action at both of its joints.

**condyle**   Large, rounded projection that usually articulates with another bone, such as the medial or lateral condyle of the femur.

**condyloid joint**   Type of joint in which the bones permit movement in two planes without rotation, as in the wrist between the radius and the proximal row of the carpal bones or the second, third, fourth, and fifth metacarpophalangeal joints.

**contractility**   The ability of muscle to contract and develop tension or internal force against resistance when stimulated.

**contraction phase**   In a single muscle fiber contraction, it is the phase following the latent perion in which the muscle fiber actually begins shortening; lasts about 40 milliseconds.

**core training**   Strengthening and conditioning that focuses on the diaphragm, transversus abdominis, lumbar multifidus, and the muscles of the pelvic floor as well as the rectus abdominis, external obliques, internal obliques, and erector spinae.

**coronal axis**   Runs from side to side through the body and is at a right angle to the sagittal plane of motion. Also known as the frontal or lateral axis.

**cortex**   Diaphyseal wall of long bones, formed from hard, dense compact bone.

**cortical bone**   Harder, more compact bone that forms the outer bony surface of the diaphysis.

**countercurrent**   Movement pattern resulting from opposite actions occurring simultaneously at both joints of a biarticular muscle resulting in substantial shortening of the biarticular muscle.

**cranial nerves**   The group of 12 pairs of nerves originating from the undersurface of the brain and exiting from the cranial cavity through skull openings; they supply specific motor and sensory function to the head and face.

**crest**   Prominent, narrow, ridgelike projection of bone, such as the iliac crest of the pelvis.

**curvilinear motion**   Motion along a curved line.

**Davis's law**   States that ligaments, muscle, and other soft tissue when placed under appropriate tension will adapt over time by lengthening and conversely when maintained in a loose or shortened state over a period of time will gradually shorten.

**dendrite**   One or more branching projections from the neuron cell body that transmit impulses to the neuron and cell body.

**depression**   Inferior movement of the shoulder girdle, as in returning to the normal position from a shoulder shrug.

**dermatome**   A defined area of skin supplied by a specific spinal nerve.

**dexter**   Relating to, or situated to the right or on the right side of something.

**diagonal abduction**   Movement by a limb through a diagonal plane away from the midline of the body such as in the hip or glenohumeral joint.

**diagonal adduction**   Movement by a limb through a diagonal plane toward and across the midline of the body such as in the hip or glenohumeral joint.

**diagonal or oblique axis**   Axis that runs at a right angle to the diagonal plane. As the glenohumeral joint moves from diagonal abduction to diagonal adduction in overhand throwing, its axis runs perpendicular to the plane through the humeral head.

**diagonal plane**   A combination of more than one plane. Less than parallel or perpendicular to the sagittal, frontal, or transverse plane. Also known as oblique plane.

**diaphysis**   The long cylindrical portion or shaft of long bones.

**diarthrodial (diarthrosis) joints**   Freely movable synovial joints containing a joint capsule and hyaline cartilage and lubricated by synovial fluid.

**displacement**   A change in position or location of an object from its original point of reference.

**distal**   Farthest from the midline or point or reference; the fingertips are the most distal part of the upper extremity.

**distance**   The path of movement; refers to the actual sum length of units of measurement traveled.

**dislocating component**   When the angle of pull is greater than 90 degrees, the force pulls the bone away from its joint axis, thereby increasing joint distraction forces.

**dorsal (dorsum)**   Relating to the back, being or located near, on, or toward the back, posterior part, or upper surface of; also relating to the top of the foot.

**dorsiflexion (dorsal flexion)**   Flexion movement of the ankle resulting in the top of foot moving toward the anterior tibia.

**duration**   An exercise variable usually referring to the number of minutes per exercise bout.

**dynamic equilibrium**   Occurs when all of the applied and inertial forces acting on the moving body are in balance, resulting in movement with unchanging speed or direction.

**dynamics**   The study of mechanics involving systems in motion with acceleration.

**eccentric contraction**   A contraction in which the muscle lengthens in an attempt to control the motion occurring at the joints that it crosses, characterized by the force of gravity or applied resistance being greater than the contractile force.

**eccentric force**   Force that is applied in a direction not in line with the center of rotation of an object with a fixed axis. In objects without a fixed axis, it is an applied force that is not in line with the object's center of gravity.

**efferent nerves**   Nerves that carry impulses to the outlying regions of the body from the central nervous system.

**elasticity**   The ability of muscle to return to its original length following stretching.

**elastin**   A protein in the body that forms connective tissue. It has a highly elastic quality and will return to its original state after stress, whether compressed or stretched.

**electromyography (EMG)**   A method utilizing either surface electrodes or fine wire/needle electrodes to detect the action potentials of muscles and provide an electronic readout of the contraction intensity and duration.

**elevation**   Superior movement of the shoulder girdle, as in shrugging the shoulders.

**enarthrodial joint**   Type of joint that permits movement in all planes, as in the shoulder (glenohumeral) and hip joints.

**endochondral bones**   Long bones that develop from hyaline cartilage masses after the embryonic stage.

**endosteum**   Dense, fibrous membrane covering the inside of the cortex of long bones.

**epicondyle**   Projection located above a condyle, such as the medial or lateral epicondyle of the humerus.

**epiphyseal plate**   Thin cartilage plate separating the diaphysis and epiphysis during bony growth; commonly referred to as growth plate.

**epiphysis**   The end of a long bone, usually enlarged and shaped to join the epiphysis of an adjacent bone, formed from cancellous or trabecular bone.

**equilibrium**   State of zero acceleration in which there is no change in the speed or direction of the body.

**eversion**   Turning of the subtalar and transverse tarsal joints of the foot outward or laterally, as in standing with the weight on the inner edge of the foot.

**extensibility**   The ability of muscle to be stretched back to its original length following contraction.

**extension**   Straightening movement resulting in an increase of the angle in a joint by moving bones apart, as when the hand moves away from shoulder during extension of the elbow joint.

**external rotation**   Rotary movement around the longitudinal axis of a bone away from the midline of the body. Also known as rotation laterally, outward rotation, and lateral rotation.

**extrinsic muscles**   Muscles that arise or originate outside of (proximal to) the body part on which they act.

**facet**   Small flat or shallow bony articular surface such as the articular facet of a vertebra.

**fascia**   Sheet or band of fibrous connective tissue that envelops, separates, or binds together parts of the body such as muscles, organs, and other soft tissue structures of the body.

**fibrous joints**   Joints joined together by connective tissue fibers and generally immovable, such as gomphosis, sutures, and syndesmosis.

**fibular**   Relating to the fibular (lateral) side of the knee, leg, ankle or foot.

**first-class lever**   A lever in which the axis (fulcrum) is between the force and the resistance, as in the extension of the elbow joint.

**flat muscles**   A type of parallel muscle that is usually thin and broad, with fibers originating from broad, fibrous, sheet-like aponeuroses such as the rectus abdominus and external oblique.

**flexion**   Movement of the bones toward each other at a joint by decreasing the angle, as in moving the hand toward the shoulder during elbow flexion.

**follow-through phase**   Phase that begins immediately after the climax of the movement phase, in order to bring about negative acceleration of the involved limb or body segment; often referred to as the deceleration phase. The velocity of the body segment progressively decreases, usually over a wide range of motion.

**foramen**   Rounded hole or opening in bone, such as the foramen magnum in the base of the skull.

**force**   The product of mass times acceleration.

**force arm**   The perpendicular distance between the location of force application and the axis. The shortest distance from the axis of rotation to the line of action of the force. Also known as the moment arm or torque arm.

**force couple**   Occurs when two or more forces are pulling in different directions on an object, causing the object to rotate about its axis.

**force magnitude**   Amount of force usually expressed in newtons.

**fossa**   Hollow, depressed, or flattened surface of bone, such as the supraspinous fossa or iliac fossa.

**fovea**   Very small pit or depression in bone, such as the fovea capitis of the femur.

**frequency**   An exercise variable usually referring to the number of times exercise is conducted per week.

**friction**   Force that results from the resistance between the surfaces of two objects moving upon one another.

**frontal plane**   Plane that bisects the body laterally from side to side, dividing it into front and back halves. Also known as the lateral or coronal plane.

**fundamental position**   Reference position essentially the same as the anatomical position, except that the arms are at the sides and the palms are facing the body.

**fusiform muscles**   A type of parallel muscle with fibers shaped together like a spindle with a central belly that tapers to tendons on each end, such as the brachialis and the biceps brachii.

**gaster**   The central, fleshy, contractile portion of the muscle that generally increases in diameter as the muscle contracts.

**ginglymus joint**   Type of joint that permits a wide range of movement in only one plane, as in the elbow, ankle, and knee joints.

**glide (slide) (translation)**   A type of accessory motion characterized by a specific point on one articulating surface coming in contact with a series of points on another surface.

**Golgi tendon organ (GTO)**   A proprioceptor, sensitive to both muscle tension and active contraction, found in the tendon close to the muscle tendon junction.

**gomphosis**   A type of immovable articulation, as of a tooth inserted into its bony socket.

**goniometer**   Instrument used to measure joint angles or compare the changes in joint angles.

**goniometry**   Measuring the available range of motion in a joint or the angles created by the bones of a joint.

**ground reaction force**   The force of the surface reacting to the force placed on it, as in the reaction force between the body and the ground when running across a surface.

**hamstrings**   A common name given to the group of posterior thigh muscles: biceps femoris, semitendinosus, and semimembranosus.

**head**   Prominent, rounded projection of the proximal end of a bone, usually articulating, such as the humeral or femoral head.

**heel-strike**   First portion of the walking or running stance phase characterized by landing on the heel with the foot in supination and the leg in external rotation.

**horizontal abduction**   Movement of the humerus in the horizontal plane away from the midline of the body.

**horizontal adduction**   Movement of the humerus in the horizontal plane toward the midline of the body.

**hyaline cartilage**   Articular cartilage; covers the end of bones at diarthrodial joints to provide a cushioning effect and reduce friction during movement.

**hyperextension**   Extension beyond the normal range of extension.

**impingement syndrome**   Occurs when the tendons of the rotator cuff muscles, particularly the supraspinatus and infraspinatus, become irritated and inflamed as they pass through the subacromial space between the acromion process of the scapula and the head of the humerus, typically resulting in pain, weakness, and loss of movement.

**impulse**   The product of force and time.

**inertia**   Resistance to action or change; resistance to acceleration or deceleration. Inertia is the tendency for the current state of motion to be maintained, regardless of whether the body segment is moving at a particular velocity or is motionless.

**innervation**   The supplying of a muscle, organ, or body part with nerves.

**insertion**   The distal attachment or point of attachment of a muscle farthest from the midline or center of the body, generally considered the most movable part.

**instantaneous center of rotation**   The center of rotation at a specific instant in time during movement.

**intensity**   An exercise variable usually referring to a certain percentage of the absolute maximum that a person can sustain.

**internal rotation**   Rotary movement around the longitudinal axis of a bone toward the midline of the body. Also known as rotation medially, inward rotation, and medial rotation.

**interneurons**   Central or connecting neurons that conduct impulses from sensory neurons to motor neurons.

**intrinsic muscles**   Muscles that are entirely contained within a specified body part; usually refers to the small, deep muscles found in the foot and hand.

**inversion**   Turning of the subtalar and transverse tarsal joints medially into adduction in the frontal plane; characterized by the soles of the feet being turned inward, as in standing with the weight on the outer edge of the foot.

**irritability**   The property of muscle being sensitive or responsive to chemical, electrical, or mechanical stimuli.

**isokinetic**   Type of dynamic exercise usually using concentric and/or eccentric muscle contractions in which the speed (or velocity) of movement is constant and muscular contraction (usually maximal contraction) occurs throughout the movement.

**isometric contraction**   A type of contraction with little or no shortening of the muscle resulting in no appreciable change in the joint angle.

**isotonic**   Contraction occurring in which there is either shortening or lengthening in the muscle under tension; also known as a dynamic contraction, and classified as being either concentric or eccentric.

**joint capsule**   Sleevelike covering of ligamentous tissue surrounding diarthrodial joints.

**joint cavity**   The area inside the joint capsule of diarthrodial or synovial joints.

**kinematics**   The description of motion, including consideration of time, displacement, velocity, acceleration, and space factors of a system's motion.

**kinesiology**   The science of movement, which includes anatomical (structural) and biomechanical (mechanical) aspects of movement.

**kinesthesis**   The awareness of the position and movement of the body in space; sense that provides awareness of bodily position, weight, or movement of the muscles, tendons, and joints.

**kinetic friction**   The amount of friction occurring between two objects that are sliding upon one another.

**kinetics**   The study of forces associated with the motion of a body.

**Krause's end bulbs**   A proprioceptor sensitive to touch and thermal changes found in the skin, subcutaneous tissue, lip and eyelid mucosa, and external genitals.

**kyphosis**   Increased curving of the spine outward or backward in the sagittal plane.

**latent period**   In a single muscle fiber contraction, it is the brief period of a few milliseconds following the stimulus before the contraction phase begins.

**lateral axis**   Axis that has the same directional orientation as the frontal plane of motion and runs from side to side at a right angle to the sagittal plane of motion. Also known as the frontal or coronal axis.

**lateral epicondylitis**   A common problem quite frequently associated with gripping and lifting activities that usually involves the extensor digitorum muscle near its origin on the lateral epicondyle; commonly known as tennis elbow.

**lateral flexion**   Movement of the head and/or trunk laterally away from the midline; abduction of spine.

**law of acceleration**   A change in the acceleration of a body occurs in the same direction as the force that caused it and is directly proportional to the force causing it and inversely proportional to the mass of the body.

**law of reaction**   For every action there is an opposite and equal reaction.

**lever**   A rigid bar (bone) that moves about an axis.

**ligament**   A type of tough connective tissue that attaches bone to bone to provide static stability to joints.

**line**   Ridge of bone less prominent than a crest, such as the linea aspera of the femur.

**linea alba**   Tendinous division and medial border of the rectus abdominis running vertically from the xiphoid process through the umbilicus to the pubis.

**linear displacement**   The distance that a system moves in a straight line.

**linear motion**   Motion along a line; also referred to as translatory motion.

**linea semilunaris**   Lateral to the rectus abdominis, a crescent, or moon-shaped, line running vertically that represents the aponeurosis connecting the lateral border of the rectus abdominis and medial border of the external and internal abdominal obliques.

**lordosis**   Increased curving of the spine inward or forward in the sagittal plane.

**lumbar kyphosis**   A reduction of its normal lordotic curve, resulting in a flat-back appearance.

**lumbar plexus**   Group of spinal nerves composed of L1 through L4 and some fibers from T12, generally responsible for motor and sensory function of the lower abdomen and the anterior and medial portions of the lower extremity.

**mass**   The amount of matter in a body.

**maximal stimulus**   A stimulus strong enough to produce action potentials in all of the motor units of a particular muscle.

**meatus**   Tubelike passage within a bone, such as the external auditory meatus of the temporal bone.

**mechanical advantage**   The advantage gained through the use of machines to increase or multiply the applied force in performing a task; enables a relatively small force to be applied to move a much greater resistance; determined by dividing the load by the effort.

**mechanics**   The study of physical actions of forces; can be subdivided into statics and dynamics.

**medial epicondylitis**   An elbow problem associated with the medial wrist flexor and pronator group near their origin on the medial epicondyle; frequently referred to as golfer's elbow.

**median**   Relating to, located in, or extending toward the middle, situated in the middle, mesial.

**medullary cavity**   Marrow cavity between the walls of the diaphysis, containing yellow or fatty marrow.

**Meissner's corpuscles**   A proprioceptor sensitive to fine touch and vibration found in the skin.

**mid-axillary line**   A line running vertically down the surface of the body passing through the apex of the axilla (armpit).

**mid-clavicular line**   A line running vertically down the surface of the body passing through the midpoint of the clavicle.

**mid-inguinal point**   A point midway between the anterior superior iliac spine and the public symphysis.

**midsagittal (median) plane**   Cardinal plane that bisects the body from front to back, dividing it into right and left symmetrical halves.

**midstance**   Middle portion of the walking or running stance phase characterized by pronation and internal rotation of the foot and leg; may be divided into loading response, midstance, and terminal stance.

**mid-sternal line**   A line running vertically down the surface of the body passing through the middle of the sternum.

**momentum**   The quality of motion, which is equal to mass times velocity.

**motor neurons**   Neurons that transmit impulses away from the brain and spinal cord to muscle and glandular tissue.

**motor unit**   Consists of a single motor neuron and all of the muscle fibers it innervates.

**movement phase**   The action part of a skill, sometimes known as the acceleration, action, motion, or contact phase. Phase in which the summation of force is generated directly to the ball, sport object, or opponent, and is usually characterized by near-maximal concentric activity in the involved muscles.

**multiarticular muscles**   Those muscles that, from origin to insertion, cross three or more different joints, allowing them to perform actions at each joint.

**multipennate muscle**   A type of pennate muscle that has several tendons with fibers running diagonally between them, such as the deltoid.

**muscle spindle**   A proprioceptor sensitive to stretch and the rate of stretch that is concentrated primarily in the muscle belly between the fibers.

**myotatic or stretch reflex**   The reflexive contraction that occurs as a result of the motor neurons of a muscle being activated from the CNS secondarily to a rapid stretch occurring in the same muscle; the knee jerk or patella tendon reflex is an example.

**myotome**   A muscle or group of muscles supplied by a specific spinal nerve.

**neuromuscular junction**   Connection between the nervous system and the muscular system via synapses between efferent nerve fibers and muscle fibers.

**neuron**   Nerve cell that is the basic functional unit of the nervous system responsible for generating and transmitting impulses.

**neuron cell body**   Portion of a neuron containing the nucleus but not including the axon and dendrites.

**neutralizers**   Muscles that counteract or neutralize the action of other muscles to prevent undesirable movements; referred to as neutralizing, they contract to resist specific actions of other muscles.

**nonrotary component (horizontal component)**   Component (either stabilizing or dislocating) of muscular force acting parallel to the long axis of the bone (lever).

**notch**   Depression in the margin of a bone such as the trochlear and radial notch of the ulna.

**open kinetic chain**   When the distal end of an extremity is not fixed to any surface, allowing any one joint in the extremity to move or function separately without necessitating movement of other joints in the extremity.

**opposition**   Diagonal movement of the thumb across the palmar surface of the hand to make contact with the hand and/or fingers.

**origin**   The proximal attachment or point of attachment of a muscle closest to the midline or center of the body, generally considered the least movable part.

**osteoblasts**   Specialized cells that form new bone.

**osteoclasts**   Specialized cells that resorb new bone.

**osteokinematic motion**   Motion of the bones relative to the three cardinal planes, resulting from physiological movements.

**Pacinian corpuscles**   A proprioceptor sensitive to pressure and vibration found in the subcutaneous, submucosa, subserous tissues around joints, external genitals, and mammary glands.

**palmar flexion**   Flexion movement of the wrist in the sagittal plane with the volar or anterior side of the hand moving toward the anterior side of the forearm.

**palpation**   Using the sense of touch to feel or examine a muscle or other tissue.

**parallel muscles**   Muscles that have their fibers arranged parallel to the length of the muscle, such as flat, fusiform, strap, radiate, or sphincter muscles.

**parasagittal plane**   Planes parallel to the midsagittal plane.

**passive insufficiency**   State reached when an opposing muscle becomes stretched to the point where it can no longer lengthen and allow movement.

**passive tension**   Tension in muscles that is due to externally applied forces and is developed as a muscle is stretched beyond its normal resting length.

**pennate muscles**   Muscles that have their fibers arranged obliquely to their tendons in a manner similar to a feather, such as unipennate, bipennate, and multipennate muscles.

**periodization**   The intentional variance of overload through a prescriptive reduction or increase in a training program to bring about optimal gains in physical performance.

**periosteum**   The dense, fibrous membrane covering the outer surface of the diaphysis.

**peripheral nervous system (PNS)**   Portion of the nervous system containing the sensory and motor divisions of all the nerves throughout the body except those found in the central nervous system.

**pes anserinus**   Distal tendinous expansion formed by the sartorius, gracilis, and semitendinosus and attaching to the anteromedial aspect of the proximal tibia below the level of the tibial tuberosity.

**physiological movement**   Normal movements of joints such as flexion, extension, abduction, adduction, and rotation, accomplished by bones moving through planes of motion about an axis of rotation at the joint.

**plane of motion**   An imaginary two-dimensional surface through which a limb or body segment is moved.

**plantar**   Relating to the sole or undersurface of the foot.

**plantar flexion**   Extension movement of the ankle, resulting in the foot and/or toes moving away from the body.

**plica**   An anatomical variant of synovial tissue folds that may be irritated or inflamed with injuries or overuse of the knee.

**posterior axillary line**   A line parallel to the mid-axillary line which passes through the posterior axillary skinfold.

**preparatory phase**   Skill analysis phase, often referred to as the cocking or wind-up phase, used to lengthen the appropriate muscles so that they will be in position to generate more force and momentum as they concentrically contract in the next phase.

**primary mover**   Muscles that contribute significantly to causing a specific joint movement when contracting concentrically.

**process**   Prominent projection of a bone, such as the acromion process of the scapula or the olecranon process of the humerus.

**pronation**   Internally rotating the radius so that it lies diagonally across the ulna, resulting in the palm-down position of the forearm; term also refers to a combination of ankle dorsiflexion, subtalar eversion, and forefoot abduction (toe-out).

**proprioception**   Feedback relative to the tension, length, and contraction state of muscle, the position of the body and limbs, and movements of the joints provided by internal receptors located in the skin, joints, muscles, and tendons.

**protraction**   Forward movement of the shoulder girdle away from the spine; abduction of the scapula.

**proximal**   Nearest to the midline or point of reference; the forearm is proximal to the hand.

**proximodistal**   From the center of the body out towards the distal ends of appendages.

**Q angle (quadriceps angle)**   The angle at the patella formed by the intersection of the line of pull of quadriceps with the line of pull of the patella tendon.

**quadriceps**   A common name given to the four muscles of the anterior aspect of the thigh: rectus femoris, vastus medialis, vastus intermedius, and vastus lateralis.

**radial**   Relating to the radial (lateral) side of the forearm or hand.

**radial deviation (radial flexion)**   Abduction movement at the wrist of the thumb side of the hand toward the forearm.

**radiate muscles**   A type of parallel muscle with a combined arrangement of flat and fusiform muscle in that they originate on broad aponeuroses and converge onto a tendon such as the pectoralis major or trapezius. Also described sometimes as being triangular, fan-shaped, or convergent.

**ramus**   Part of an irregularly shaped bone that is thicker than a process and forms an angle with the main body such as the superior and inferior ramus of pubis.

**range of motion (ROM)**   The specific amount of movement possible in a joint.

**reciprocal inhibition**   Activation of the motor units of the agonists, causing a reciprocal neural inhibition of the motor units of the antagonists, which allows them to subsequently lengthen under less tension. Also referred to as reciprocal innervation.

**recovery phase**   Skill analysis phase used after followthrough to regain balance and positioning to be ready for the next sport demand.

**rectilinear motion**   Motion along a straight line.

**recurvatum**   Bending backward, as in knee hyperextension.

**reduction**   Return of the spinal column to the anatomic position from lateral flexion; spine adduction.

**relaxation phase**   In a single muscle fiber contraction, it is the phase following the contraction phase in which the muscle fiber begins relaxing; lasts about 50 milliseconds.

**reposition**   Diagonal movement of the thumb as it returns to the anatomical position from opposition with the hand and/or fingers.

**resistance**   Component of the lever that is typically being attempted to be moved, usually referred to as load, weight, or mass.

**resistance arm**   The distance between the axis and the point of resistance application.

**retinaculum**   Fascial tissue that retains tendons close to the body in certain places such as around joints like the wrist and ankle.

**retraction**   Backward movement of the shoulder girdle toward the spine; adduction of the scapula.

**retroversion**   Abnormal or excessive rotation backward of a structure, such as femoral retroversion.

**roll (rock)**   A type of accessory motion characterized by a series of points on one articular surface contacting with a series of points on another articular surface.

**rolling friction**   The resistance to an object rolling across a surface, such as a ball rolling across a court or a tire rolling across the ground.

**rotation**   Movement around the axis of a bone, such as the turning inward, outward, downward, or upward of a bone.

**rotary component (vertical component)**   Component of muscular force acting perpendicular to the long axis of the bone (lever).

**rotator cuff**   Group of muscles intrinsic to the glenohumeral joint, consisting of the subscapularis, supraspinatus, infraspinatus, and teres minor, that is critical in maintaining dynamic stability of the joint.

**Ruffini's corpuscles**   A proprioceptor sensitive to touch and pressure found in the skin, subcutaneous tissue of fingers, and collagenous fibers of the joint capsule.

**sacral plexus**   Group of spinal nerves composed of L4, L5, and S1 through S4, generally responsible for motor and sensory function of the lower back, pelvis, perineum, posterior surface of the thigh and leg, and dorsal and plantar surfaces of the foot.

**sagittal plane**   Plane that bisects the body from front to back, dividing it into right and left symmetrical halves. Also known as the anteroposterior, or AP plane.

**scalar**   Mathematical quantities are described by a magnitude (or numerical value) alone such as speed, length, area, volume, mass, time, density, temperature, pressure, energy, work, and power.

**scaption**   Movement of the humerus away from the body in the scapula plane. Glenohumeral abduction in a plane halfway between the sagittal and frontal plane.

**scapula line**   A line running vertically down the posterior surface of the body passing through inferior angle of the scapula.

**scapular plane**   In line with the normal resting position of the scapula as it lies on the posterior rib cage, movements in the scapular plane are in line with the scapular which is at an angle of 30 to 45 degrees from the frontal plane.

**scoliosis**   Lateral curving of the spine.

**second-class lever**   A lever in which the resistance is between the axis (fulcrum) and the force (effort), as in plantarflexing the foot to raise up on the toes.

**sellar joints**   Type of reciprocal reception that is found only in the thumb at the carpometacarpal joint and permits ball-and-socket movement, with the exception of rotation.

**sensory neurons**   Neurons that transmit impulses to the spinal cord and brain from all parts of the body.

**sesamoid bones**   Small bones embedded within the tendon of a musculotendinous unit that provide protection as well as improve the mechanical advantage of musculotendinous units as in the patella.

**shin splints**   Slang term frequently used to describe an often chronic condition in which the tibialis posterior, tibialis anterior, and extensor digitorum longus muscles are inflamed, typically a tendinitis of one or more of these structures.

**sinister**   Relating to, or situated to the left or on the left side of something.

**sinus**   Cavity or hollow space within a bone, such as the frontal or maxillary sinus.

**somatic nerves (voluntary)**   Afferent nerves, which are under conscious control and carry impulses to skeletal muscles.

**speed**   How fast an object is moving, or the distance an object travels in a specific amount of time.

**sphincter muscle**   A type of parallel muscle that is a technically endless strap muscle with fibers arranged to surround and close openings upon contraction, such as the orbicularis oris. Also referred to as circular muscles.

**spin**   A type of accessory motion characterized by a single point on one articular surface rotating clockwise or counterclockwise about a single point on another articular surface.

**spinal cord**   The common pathway between the central nervous system and the peripheral nervous system.

**spinal nerves**   The group of 31 pairs of nerves that originate from the spinal cord and exit the spinal column on each side through openings between the vertebrae. They run directly to specific anatomical locations, form different plexuses, and eventually become peripheral nerve branches.

**spine (spinous process)** Sharp, slender projection of a bone, such as the spinous process of a vertebra or spine of the scapula.

**stability** The resistance to a change in the body's acceleration; the resistance to a disturbance of the body's equilibrium.

**stabilizers** Muscles that surround the joint or body part and contract to fixate or stabilize the area to enable another limb or body segment to exert force and move; known as fixators, they are essential in establishing a relatively firm base for the more distal joints to work from when carrying out movements.

**stabilizing component** When the angle of pull is less than 90 degrees, the force pulls the bone toward its joint axis, thereby increasing joint compression forces.

**stance phase** Skill analysis phase that allows the athlete to assume a comfortable and balanced body position from which to initiate the sport skill; emphasis is on setting the various joint angles in the correct positions with respect to one another and to the sport surface.

**static equilibrium** The body at complete rest or motionless.

**static friction** The amount of friction between two objects that have not yet begun to move.

**statics** The study of mechanics involving the study of systems that are in a constant state of motion, whether at rest with no motion or moving at a constant velocity without acceleration. Involves all forces acting on the body being in balance, resulting in the body being in equilibrium.

**strap muscles** A type of parallel muscle with fibers uniform in diameter and arranged with essentially all fibers in a long parallel manner, such as the sartorius.

**stretch-shortening cycle** An active stretch via an eccentric contraction of a muscle followed by an immediate concentric contraction of the same muscle.

**submaximal stimuli** Stimuli that are strong enough to produce action potentials in multiple motor units, but not all motor units of a particular muscle.

**subthreshold stimulus** Stimulus not strong enough to cause an action potential and therefore does not result in a contraction.

**sulcus (groove)** Furrow or groovelike depression on a bone, such as the intertubercular (bicipital) groove of the humerus.

**summation** When successive stimuli are provided before the relaxation phase of the first twitch is complete allowing the subsequent twitches to combine with the first to produce a sustained contraction generating greater tension than a single contraction would produce on its own.

**supination** Externally rotating the radius to where it lies parallel to the ulna, resulting in the palm-up position of the forearm; term is also used in referring to the combined movements of inversion, adduction, and internal rotation of the foot and ankle.

**suture** Line of union between bones, such as the sagittal suture between the parietal bones of the skull.

**swing** Phase of gait that occurs when the foot leaves the ground and the leg moves forward to another point of contact.

**syndesmosis joint** Type of joint held together by strong ligamentous structures that allow minimal movement between the bones, such as the coracoclavicular joint and the inferior tibiofibular joint.

**synergist** Muscles that assist in the action of the agonists but are not primarily responsible for the action; known as guiding muscles, they assist in refined movement and rule out undesired motions.

**synergists (helping)** Muscles that have an action common to each other, but also have actions antagonistic to each other; they help another muscle move the joint in the desired manner and simultaneously prevent undesired actions.

**synergists (true)** Muscles that contract to prevent an undesired joint action of the agonist and have no direct effect on the agonist action.

**synovial joints** Freely movable diarthrodial joints containing a joint capsule and hyaline cartilage and lubricated by synovial fluid.

**tendinous inscriptions** Horizontal indentations that transect the rectus abdominus at three or more locations, giving the muscle its segmented appearance.

**tendon** Fibrous connective tissue, often cordlike in appearance, that connects muscles to bones and other structures.

**tetanus** When stimuli are provided at a frequency high enough that no relaxation can occur between muscle contractions.

**third-class lever** A lever in which the force (effort) is between the axis (fulcrum) and the resistance, as in flexion of the elbow joint.

**threshold stimulus** When the stimulus is strong enough to produce an action potential in a single motor unit axon and all of the muscle fibers in the motor unit contract.

**tibial** Relating to the tibial (medial) side of the knee, leg, ankle or foot.

**toe-off** Last portion of the walking or running stance phase characterized by the foot returning to supination and the leg returning to external rotation.

**torque** Moment of force; the turning effect of an eccentric force.

**transverse plane** Plane that divides the body horizontally into superior and inferior halves; also known as the axial or horizontal plane.

**treppe** A staircase effect phenomenon of muscle contraction that occurs when rested muscle is stimulated repeatedly with a maximal stimulus at a frequency that allows complete relaxation between stimuli, the second contraction produces a slightly greater tension than the first, and the third contraction produces greater tension than the second.

**triceps surae** The gastrocnemius and soleus together; triceps referring to the heads of the medial and lateral gastrocnemius and the soleus; surae referring to the calf.

**trochanter** A very large bony projection, such as the greater or lesser trochanter of the femur.

**trochoidal joint** Type of joint with a rotational movement around a long axis, as in rotation of the radius at the radioulnar joint.

**tubercle** A small, rounded, bony projection, such as the greater and lesser tubercles of the humerus.

**tuberosity** A large, rounded, or roughened, bony projection, such as the radial tuberosity or tibial tuberosity.

**ulnar** Relating to the ulnar (medial) side of the forearm or hand.

**ulnar deviation (ulnar flexion)** Adduction movement at the wrist of the little finger side of the hand toward the forearm.

**uniarticular muscles** Those muscles that, from origin to insertion, cross only one joint, allowing them to perform actions only on the single joint that they cross.

**unipennate muscles** A type of pennate muscle with fibers that run obliquely from a tendon on one side only, such as the biceps femoris, extensor digitorum longus, and tibialis posterior.

**valgus** Outward angulation of the distal segment of a bone or joint, as in knock-knees.

**varus** Inward angulation of the distal segment of a bone or joint, as in bowlegs.

**vector** Mathematical quantity described by both a magnitude and a direction such as velocity, acceleration, direction, displacement, force, drag, momentum, lift, weight, and thrust.

**velocity** Includes the direction and describes the rate of displacement.

**ventral** Relating to the belly or abdomen, on or toward the front, anterior part of.

**vertebral line** A line running vertically down through the spinous processes of the spine.

**vertical axis** Axis that runs straight down through the top of the head and spinal column and is at a right angle to the transverse plane of motion. Also known as the longitudinal or long axis.

**visceral nerves (involuntary)** Nerves that carry impulses to the heart, smooth muscles, and glands; referred to as the autonomic nervous system.

**volar** Relating to the palm of the hand or the sole of the foot.

**Wolff's law** States that bone in a healthy individual will adapt to the loads it is placed under. When a particular bone is subjected to increased loading, the bone will remodel itself over time to become stronger to resist that particular type of loading.

# Index

Page numbers in *italics* refer to tables and illustrations.